"Agrawal is a rarity: a female structural engineer in an adamantly male profession. A self-proclaimed 'geek,' she shares her discoveries far above and below ground...She will inspire young women who are considering a career in engineering." —*The New York Times Book Review*

"A full-throated celebration of structural engineering...[*Built*] is globally inclusive, provides personal insight into the life and achievements of a broadly accomplished female structural engineer, and teaches key engineering concepts in an approachable and engaging way." —*Science*

"Agrawal has a gift for rendering complex phenomena in simple terms... A real treat." —*Financial Times*

"A global romp through the ages...[Agrawal] weaves accessible explanations of scientific principles together with engaging historical stories and personal anecdotes...Timely and impassioned." —*The Guardian*

"[Agrawal] delves into the history of her discipline through Brunel's tunnels and Bazalgette's sewers and lingers tactfully over the hidden flaws and engineering disasters of the past from which so much has been learned." —*The Times Literary Supplement*

"A fascinating celebration of the impact that structural engineering has on all our lives...[An] eye-opening book." —*The Sunday Times*

"Interweaving science, history, illustrations, and personal stories, *Built* offers a simply fascinating and impressively informative history of architecture, a subject that makes up the foundation of our everyday lives." —*Midwest Book Review*

"Makes engineering concepts easy to understand...[Agrawal] asks the reader to dream with her, look at the world before a problem is solved, then rejoice in the solution and all that it has meant to us as heirs of earlier genius. In short, she does the almost impossible; she makes engineering fun." —*Manhattan Book Review*

"A delightful introduction to the science of engineering and those key in its development." —*Kirkus Reviews*

"Structural engineer Agrawal introduces engineering to the masses in this enthusiastic, easy-to-read primer to her field of work." —*Publishers Weekly*

"A highly engaging debut book that will appeal to the engineer and the inquisitive layperson alike." —*E&T Magazine*

BUILT

BUILT

*The Hidden Stories
Behind our Structures*

Roma Agrawal

BLOOMSBURY PUBLISHING

NEW YORK · LONDON · OXFORD · NEW DELHI · SYDNEY

BLOOMSBURY PUBLISHING
Bloomsbury Publishing Inc.
1385 Broadway, New York, NY 10018, USA

BLOOMSBURY, BLOOMSBURY PUBLISHING, and the Diana logo are
trademarks of Bloomsbury Publishing Plc

First published in 2018 in Great Britain
First published in the United States 2018
This paperback edition published 2019

Bloomsbury Publishing Plc does not have any control over, or responsibility for,
any third-party websites referred to or in this book. All internet addresses
given in this book were correct at the time of going to press. The author and
publisher regret any inconvenience caused if addresses have changed or sites
have ceased to exist, but can accept no responsibility for any such changes.

ISBN: HB: 978-1-63557-022-9; PB: 978-1-63557-023-6; eBook: 978-1-63557-021-2

Library of Congress Cataloging-in-Publication Data

Names: Agrawal, Roma, 1983- author.
Title: Built : the hidden stories behind our structures / Roma Agrawal.
Description: First U.S. edition. | New York : Bloomsbury USA, 2018. |
Includes bibliographical references and index.
Identifiers: LCCN 2018000409| ISBN 9781635570229 (hardcover) |
ISBN 9781635570236 (pbk.) | ISBN 9781635570212 (ebook)
Subjects: LCSH: Structural engineering–Popular works. | Building—Popular works.
Classification: LCC TA633 .A36 2018 | DDC 624.1–dc23 LC record available
at https://lccn.loc.gov/2018000409

2 4 6 8 10 9 7 5 3 1

Typeset by Newgen Knowledge Works Pvt. Ltd., Chennai, India
Printed and bound in the U.S.A. by Berryville Graphics Inc., Berryville, Virginia

To find out more about our authors and books visit
www.bloomsbury.com and sign up for our newsletters.

Bloomsbury books may be purchased for business or promotional use.
For information on bulk purchases please contact Macmillan Corporate and
Premium Sales Department at specialmarkets@macmillan.com.

For Maa,
and for little Samuel.

CONTENTS

STOREY

With one hand, I clutched my precious stuffed-toy cat, afraid that I would lose it. With the other, I clung to my mother's skirt. Terrified and exhilarated by the new, strange, unknown world in constant motion around me, I held on to the only two things that felt familiar.

When I think of Manhattan now, I am always taken back to my first visit, as an impressionable toddler: the funny smell of the car exhausts, the shouts of the streetside lemonade vendors, the swarm of people rushing by, bumping into me unapologetically. It was an overwhelming experience for a child who lived far from the big city. Here, instead of open sky, I saw towers of glass and steel blocking out the sun. What were these monstrous things? How could I climb them? What did they look like from above? I turned my head left and right as my mother dragged me along the busy streets. Stumbling after her with my head raised, I was transfixed by these pillars that reached towards the clouds.

At home, with my miniature cranes, I stacked building blocks to recreate what I had seen. At school, I painted tall rectangles on big sheets of paper in bright, bold colours. New York became part of my mental landscape as I visited and revisited the place over the years, admiring new towers that appeared on the ever-changing skyline.

For a few years we lived in America, while my father worked as an electrical engineer. We didn't live in one of the soaring sky-scrapers that so impressed me on my visits to Manhattan, however, but in a creaking wooden house among the hills upstate. When I was six, my father gave up engineering to look after the family business in Mumbai, and I went to live in a seven-storey concrete tower that looked out towards the Arabian Sea. When my Barbie dolls finally arrived safe and sound at my new home, after a long sea journey in a storage container, it was of course essential that they were made comfortable. Pop helped me reassemble my cranes, laying out a large white sheet so I wouldn't lose any pieces. Making loud, whirring noises, I lifted long plastic tubes and manoeuvred pieces of card into position, building a house for my dolls. My first step, perhaps, towards a career in engineering.

Having an American accent and – as you'll soon discover if you haven't already – a tendency to be a bit geeky, I found my new school a challenge at first. I was teased by some for being a 'scholar'. But gradually I found friends and teachers that 'got' me. Through large gold-framed glasses, I eagerly read physics, maths and geography textbooks, and I loved art class, although I struggled with chemistry, history and languages. Mom, who had studied maths and science at university and had worked as a computer programmer, encouraged my growing interest in science and maths, assigning me extra homework and reading. Throughout my school years I loved these subjects best and I resolved to be an astronaut or an architect when I grew up.

Back then, I'd never even heard the term 'structural engineer', and never imagined that one day I would play a part in designing a magnificent skyscraper – The Shard.

Since I loved learning so much, my family decided I should finish my schooling in another country, as it would be a great opportunity to broaden my horizons. And so, aged fifteen, I moved to London to study maths, physics and design at A level. Another new school in a new country, but this time I quickly sought out kindred spirits – girls who found Faraday's law as fascinating as I did, and who experimented in the lab just for fun. Brilliant teachers paved my way to studying physics at university, and I moved to Oxford.

At school, physics made sense to me. At university, it didn't – at least to begin with. Light was both a wave and a collection of particles? Space-time could be curved?? Time travel was mathematically possible?! I was hooked, but it was tough stuff to get my head around. Academically, I always felt like I was a few steps behind my peers. It was a real reward when I finally figured out how something worked. I balanced hours in the library with ballroom and Latin dance lessons, learning to wash clothes and to cook (though perhaps not all that skilfully, as you'll see), and generally fending for myself. I was enjoying physics; my childhood dreams of going into space or becoming an architect became distant memories. At the same time, however, I had little idea of what I wanted to do with my life.

Then, one summer, I worked in the physics department at the University of Oxford, drawing up plans of all the fire-safety features in the various buildings. The task in itself was hardly world-changing, but the people who sat around me were working on projects that were. They were engineers, and their job was to design the equipment that physicists could use to seek out the particles that define how our world works. As you might imagine, I badgered them with questions and was astonished at what their jobs entailed. One was designing a metal holder for a glass lens – a simple task, you might think, except that the whole apparatus had to be cooled to -70° Celsius. Metal contracts more than glass, and unless the holder was cleverly and carefully designed, the cooling metal would crush the lens. It was just a tiny piece in a immense maze of machinery, but a complex and creative challenge. I spent hours of my free time trying to figure out how I might solve the problem.

Suddenly, it became very clear to me: I wanted to use physics and maths to solve practical problems and, in the process, help the world in some way. And it was at this point that my childhood love of skyscrapers re-emerged from the depths of my memory. I would be a structural engineer and design buildings. To make the transition from physicist to engineer, I studied at Imperial College London for a year, graduated, got a job – and began my life as an engineer.

As a structural engineer, I am responsible for making sure that the structures I design stand up. In the past decade I have worked on an amazing variety of constructions. I was part of the team that designed The Shard – the tallest tower in Western Europe – spending six years working out the sums for its open-air spire and foundations; I worked on a fancy footbridge in Newcastle, and the curving canopy at Crystal Palace station in London. I've designed hundreds of new apartments, brought a Georgian townhouse back to its former glory, and ensured an artist's sculpture was stable. Whilst my job involves using maths and physics to create things (which in itself is incredible fun), it is also so much more. For a start, a modern engineering project is an enormous piece of teamwork. In the past, engineers like Vitruvius (who wrote the first treatise on architecture) or Brunelleschi (who built the breathtaking dome that crowns Florence's cathedral) were known as master builders. They knew about every discipline necessary for construction. Nowadays structures are more complex and technically advanced, and no single person can design every aspect of a project. Each of us has an area of specialisation, and the challenge is to bring everybody together in an intricate and quietly frenetic dance that weaves together materials, physical effort and mathematical calculations. With the architects and other engineers, I brainstorm design problems. Our drawings assist site managers, and surveyors calculate costs and consider logistics. Workers on site receive materials and reshape them to create our vision. At times, it's hard to imagine that all this sometimes chaotic activity will resolve into a solid structure that will last for decades, or even centuries.

For me, each new structure I design becomes personal, as 'my' building grows and takes on its own individual character. At first we communicate through a few rough drawings, but gradually I discover what will prop it up, and how it will stand tall and be able to evolve with the changing times. The more time I spend with it, the more I come to respect, even love it. Once complete, I get to meet her in person, and walk around her. Even after that, as far as I'm concerned, we have an ongoing commitment to one another, and I watch from afar as other people take my place and develop their own relationships with my creation, making the building their home or workplace, protected from the outside world.

Of course, my feelings for the structures I have worked on are particularly personal, but in fact all of us are intimately connected to the engineering that surrounds us – the streets we walk on, the tunnels we rush through, the bridges we cross. We use them to make our lives easier, and we look after them. In return, they become a silent but crucial part of our existence. We feel charged and professional when we walk into a glass skyscraper with neat rows of desks. The speed of our journey is emphasised by steel rings flying past the windows of an underground train. Uneven brick walls and cobbled stone pathways remind us of the past, of the history that has gone before us. Structures shape and sustain our lives and provide the canvas of our existence. We often ignore or are unaware of them, but structures have stories. The tense cables stretching above a massive bridge across a river; the steel skeleton beneath the glass skin of a tall tower; the conduits and tunnels burrowing beneath our feet – these things make up our built world, and they reveal a lot about human ingenuity, as well as our interactions with each other, and with Nature. Our ever-changing, engineered universe is a narrative full of stories and secrets that, if you have the ears to listen, and the eyes to see, is fascinating to experience.

My hope is that, through this book, you too will discover these stories and learn these secrets. That a new understanding of our surroundings will change the way you look at the hundreds of structures you move over, below and through every day. That you will see your home, your city, town or village, and the countryside beyond with a new sense of wonder. That you will see your world through different eyes – the eyes of an engineer.

FORCE

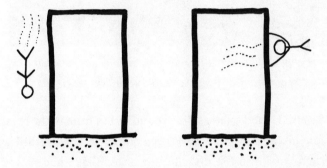

It's a peculiar feeling when you step onto or into something you've designed. My first project after leaving university was the Northumbria University Footbridge in Newcastle, England. For two years, I worked with the architects' plans, helping to make their vision a reality, covering hundreds of pages with calculations and creating countless computer models. Eventually it was constructed. Once the cranes and diggers had moved on, I finally had the opportunity to stand on the steel structure I had helped to create.

The Northumbria University Footbridge, built in 2007 to link the two main sites of the university in Newcastle upon Tyne, England.

Briefly, I stood on the solid ground just in front of the bridge, before taking a step forward. I remember that moment: I was excited but I also felt disbelief – amazed that I had played a role in making this beautiful bridge stand, so that hundreds of people could walk across it every day. I looked up at its tall steel mast and the cables radiating from it, supporting the slim deck safely above the motorway traffic – it held its own weight, and mine, effortlessly. Balustrades, carefully angled to make them difficult to climb, reflected the cold sunlight. Below me, cars and trucks whizzed past, oblivious to this young engineer standing proudly on 'her' bridge, marvelling at her first physical contribution to the world.

It was, of course, steadfast beneath my feet. After all, those numbers and models I had carefully executed to calculate the forces my bridge would be subjected to had been checked and

Standing upon the Northumbria University Footbridge, my first project as an engineer.

re-checked. Because, as engineers, we can't afford to make mistakes. I'm conscious that every day thousands of people will use structures that I have designed: they will cross them, work in them, live in them, oblivious to any concern that my creations could let them down. We put our faith and our feet (often quite literally) on engineering, and it is the engineer's responsibility to render things robust and reliable. For all that, history has shown us that things can go wrong. On the afternoon of 29 August 1907 residents of Quebec City thought they had just been shaken by an earthquake. In fact, 15 km away, something far more unthinkable was happening. On the banks of the Saint Lawrence River the sound of ripping metal tore through the air. The rivets that held together a bridge under construction had snapped, catapulting over the heads of terrified workers. The steel supports for the structure folded as if they were made from paper, and the bridge – with most of its builders – plunged into the river. One of the worst bridge collapses in history, it is a brutal example of how mismanagement and miscalculation can end in disaster.

*

Bridges expand cities, bring people together and promote commerce and communication. The idea of building a bridge across the Saint Lawrence had been debated in Parliament since the 1850s. The technical challenge, though, was huge: the river was three kilometres wide at its narrowest point, with deep, fast-moving water. In winter the water froze, creating piles of ice as high as 15 m in the river channel. Nonetheless, the Quebec Bridge Company was eventually set up to undertake the project, and work on the foundations began in 1900.

The company's chief engineer, Edward Hoare, had never before worked on a bridge longer than 90m (even the original plans for this project called for a 'clear span length' – i.e. a length of bridge without any supports – of just over 480m). So the fateful decision was made to enlist the services of Theodore Cooper as consultant. Cooper was widely regarded as one of the best bridge builders in America, and had written an award-winning paper on the use of steel in railway bridges. Theoretically he must have seemed like the ideal candidate. But there were problems from the start. Cooper lived far away in New York, and his ill health meant he rarely visited the site. Yet he insisted on being personally responsible for inspecting the steel fabrication and construction. He refused to have his design checked by anyone and relied on his relatively inexperienced inspector, Norman McLure, to keep him informed of progress on site. Construction on the steel structure began in 1905, but over the next two years McLure became increasingly worried about how the build was progressing. For a start, the pieces of steel arriving from the factory were heavier than he expected. Some of them were even bowed rather than straight because they were buckling under their own weight. Even more worryingly, many of the steel pieces installed by the workers had already deformed even before the bridge was complete, a sign that they were not strong enough to carry the forces flowing through them.

This deformation was the result of Cooper's decision to change the design of the bridge away from its original plans, increasing the length of the central span (the unsupported middle of the bridge) to nearly 549m. Ambition may have clouded Cooper's judgement: in making the decision he might have hoped to become the engineer responsible for the longest-spanning cantilever bridge in the world, an honour held at the time by the Forth Bridge in Scotland. The larger the span of a bridge, the more material you need to build it, and the heavier it becomes. Cooper's new design was about 18 per cent heavier than the original, yet without paying enough attention to the calculations, he decided that the structure was still strong enough to carry this extra weight. McLure disagreed, and the two men argued about it in an exchange of letters. But nothing was resolved.

Finally, McLure became so concerned that he suspended construction and set off by train to New York to confront Cooper. In his absence, an engineer on site overturned his instructions and the workforce went back to assembling the bridge, with tragic results. In just 15 seconds, the entire south half of the bridge – 19,000 tonnes of steel – collapsed into the river, killing 75 of the 86 people working on the structure.

The scene of devastation following the collapse during construction of the Quebec Bridge in 1907, spanning the Saint Lawrence River, Quebec City, Canada.

Many problems and mistakes contributed to the bridge's collapse. In particular, the disaster revealed the dangers of putting huge power in the hands of one engineer without supervision. In Canada and elsewhere, organisations of professional engineers were set up to regulate the profession and try and prevent a repeat of the Quebec Bridge mistakes. Ultimately, however, much of the responsibility lies with Theodore Cooper, who underestimated the weight of the bridge. In the end, the way it was engineered meant it was just too feeble to hold itself up.

*

The abrupt devastation of the Quebec Bridge demonstrates the catastrophic effect gravity can have on a faulty human construction. A major part of the engineer's job is figuring out how structures can withstand the manifold forces determined to push, pull, shake, twist, squash, bend, rend, split, snap or tear them apart. Grappling with gravity is therefore a key consideration on many projects. It is the omnipresent force that holds the solar system together, and which attracts everything on our planet towards its centre. This creates a force within every object, which we call its weight. This force *flows* through the object. Think about the weight of different parts of your body. The weight of your hand acts on your arm, pulling on your shoulder then pushing into your spine. Flowing down the spine, the force reaches your hips, and here, at the pelvic bone, the force splits into two, flowing into each of your legs and down into the ground. In much the same way, if you build a tower from straws and pour water on top of it, the water will stream through the different pathways it finds, dividing where more than one option is available.

When planning a structure, then, it is vital for an engineer to understand where the force is flowing, what kind of force it is, and then make sure that the structure transmitting the force is strong enough for the job.

There are two main types of forces that gravity (and also other phenomena such as wind and earthquakes) creates in structures: *compression* and *tension*. If you roll a piece of thick paper into a cylindrical tube, stand it vertically on a table, and then put a book on top of it, the book pushes down on the tube. The *force* with which it does this (which is its mass multiplied by gravity) flows through the tube down to the table – just as your weight flows through your leg. The tube (like your leg) is in *compression*.

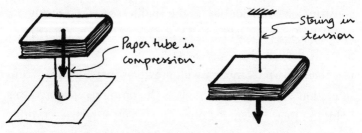

Supporting a book using compression (above left) and tension (above right).

Conversely, if you take a piece of string, tie the book to one end and hold the other, the suspended book – still experiencing the force of gravity – is now pulling on the string. The force of the book flows up into the string, which is said to be in *tension*. This is the same effect that the weight of your hand has on your arm.

In the first example, the book doesn't crash down onto the table because the paper tube is strong enough to resist the *compression* it feels. In the second example, it remains safely suspended because the piece of string is strong enough to resist the *tension* it feels.

To cause a collapse, use a heavier book. The new force exerted by this book on its support is larger because the weight of the book has increased. The tube is no longer strong enough, so it crushes and the book falls to the table. Similarly, if you try suspending the heavier book, the tension is too big for the string. The string snaps and the book plummets.

The forces in a bridge arise from its own weight, and from the weight of the people and vehicles that travel over it. When working on the Northumbria University Footbridge, I did calculations to find out where the forces were in the structure. As a result, I knew exactly how much compression or tension was at work in each part of it. I used a computer model to test every section of my bridge, then calculated how big the steel needed to be so it didn't bend excessively, crush or snap.

*

The type of force and the way it flows depends on how the structure is assembled. There are two main ways this can be done. The first is known as the *load-bearing* system and the second as the *frame* system.

Our early ancestors' mud huts – which they made by forming mud into thick walls arranged in a circle or square – were built using the first method. The walls of these single-storey dwellings were solid, forming a load-bearing system: the weight of the structure was free to flow as compression throughout the mud walls. This is similar to the book resting on the paper tube, in which all sides of the tube are uniformly in compression. If additional storeys were added to the hut, at some point the compression

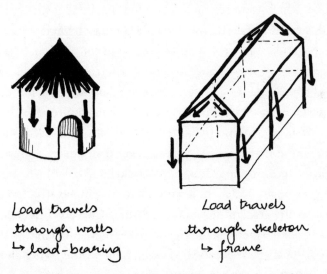

Load travels
through walls
↳ load-bearing

Load travels
through skeleton
↳ frame

Two ways to build a home, using load-bearing walls (above left) or a skeleton frame (above right).

would become too much for the load-bearing mud walls and they would crumble, just like the heavier book crushes the paper tube.

When our ancestors had access to trees, they built their homes using the frame system – by tying timber logs together to create a network or skeleton through which the forces are channelled. To protect the inside from the elements, animal skins or woven vegetation were suspended across the logs. Where mud huts had solid walls that bore the forces *and* protected the residents, the timber home had two distinct systems: the logs that carried the forces *plus* the 'walls' or the animal skins which carried no weight. The way in which forces are channelled is the fundamental difference between load-bearing and frame structures.

Over time, the materials we used to create load-bearing walls and frames for structures became more and more sophisticated. Load-bearing structures were made from brick and stone, which were stronger than mud. In the 1800s, after the Industrial Revolution, iron and steel could be manufactured at a large scale, and we started to use metals for building, rather than just for vessels and weaponry. Concrete was rediscovered (the Romans had known how to manufacture it, but that knowledge was subsequently lost when their empire fell). These moments of evolution changed our structures forever. Since steel and concrete are so much stronger than timber, and well-suited to creating large frames, we could build taller towers and longer bridges. Today, the largest and most complex structures – such as the graceful steel arch of Sydney Harbour Bridge, the triangular geometry of the Hearst Tower in Manhattan, and the iconic 'Bird's Nest' National Stadium built for the 2008 Beijing Olympics – are created using the frame system.

*

When I start designing a new building, I study the carefully crafted drawings from the architects which convey their vision of what the construction will look like once it's finished. Engineers soon develop a kind of X-ray vision, enabling them to see through the building in the picture to the skeleton it would need in order to resist gravity and the other forces that test it. I visualise where the building's spine will go, where the supporting bones need to be connected, and how big these need to be so the skeleton is stable. With a black marker pen, I sketch over the architects' drawings, adding bones to the flesh. The thick, black lines I add to the colourful drawings add a certain solidity. Inevitably, there is much discussion – sometimes quite spirited – between me and the architects: we need to compromise to find a solution. Often, I need a column where they have shown an open space; at other times they think I need structure where I don't – so I can give them more area. We have to understand each other's perspectives when technical problems arise: we must reach a balance between visual beauty and technical integrity. Eventually, we arrive at a design in which structure and aesthetic vision are (almost) in perfect harmony.

The frames in our structures are made up of a network of columns, beams and braces. Columns are the vertical sections of the skeleton; beams are the horizontal ones, and the pieces at other angles – the braces – are usually called 'struts'. If you look at a photograph of Sydney Harbour Bridge, for example, you'll see that it's formed of pieces of steel at all sorts of angles – a melee of columns, beams and struts. By understanding how columns and beams interact and support one another, how they attract forces and, most importantly, how they break, we can design them so they won't fail.

The Sydney Harbour Bridge, completed in 1930, built to carry rail, vehicular and pedestrian traffic between the North Shore and the central business district of Sydney, Australia.

Although columns have been used to resist gravity for millennia, the Greeks and Romans turned them into an art form. Much of the beauty and the solidity of the Parthenon in Athens comes from its outer row of fluted Doric marble columns. The remains of the Forum in Rome are dominated by monumental columns that support the fragile remnants of temples, or which simply strike upwards, stunted, towards the sky. Of course, the columns fulfilled a very important practical function – holding up structures – but this didn't stop ancient engineers from decorating them with carvings inspired by nature and mythology. The Corinthian column, with its capital decorated by intricately curled leaves, was supposedly invented by the Greek sculptor Callimachus after he noticed an acanthus plant growing through and around a basket left upon the grave of a maiden of Corinth. There are dozens of examples of it dotted around the Forum, and it has remained a classic of civic architecture for centuries, grandly gracing the façade of the United States Supreme Court Building for example and, more humbly, the entrance to the Victorian block of flats where I live.

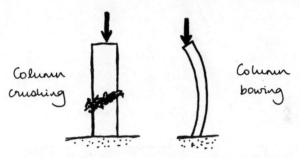

Two of the ways in which a column can fail, through crushing (above left) and bowing (above right).

Columns generally work by countering compression. One way they can fail is when they are squashed so much that the material they are made of simply gives up and crushes or crumbles – this is what happens to the paper tube when the heavier book is placed on top. The other way columns can fail is by bowing. Take a plastic ruler, stand it vertically on a table and then press down on it with the palm of your hand: you'll see it begin to bow. The more you push down, the further the ruler will bow – until finally it snaps.

There is a delicate balance to be struck when designing a column. You want it to be thin so that it doesn't take up too much space, but if it's too slender the load it carries can cause it to bow. At the same time, you want to use a material that's strong enough

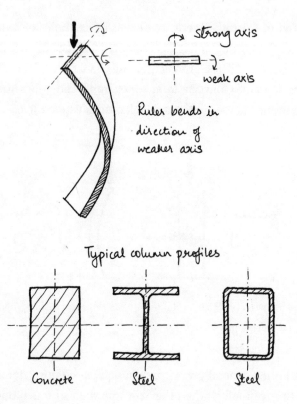

Flexing a ruler shows how a slender structure bends along its weaker axis (top), whereas a column, whether it's made from concrete or steel, is shaped to resist bending in both axes (bottom).

to prevent crushing. The columns used in ancient structures tended to be stocky, chunky things most often made from stone, and were unlikely to fail by bowing. By contrast, our modern steel or concrete columns tend to be far more slender, making them mostly susceptible to bowing.

A ruler is wide in one direction and flat in the other: as you'll have seen as you pressed down, it bows about its much weaker axis. To stop this effect, modern steel columns are usually made in an H shape, and concrete columns in squares or rectangles so that both axes are comparably stiff – so the columns can resist larger loads.

*

Beams work differently. They form the skeleton of the floors. When we stand on a beam, it flexes slightly, channelling our weight across to the columns that support it. The columns in turn compress and transmit our weight to the ground. If you stand on the centre of a beam, half of your weight, and half the weight of the beam is transmitted to each end. The column then transmits that load downward. We don't want beams to bend too much when we stand on them, partly because it feels uncomfortable when the floor is moving below our feet – but also because they can fail. We need to make beams appropriately stiff; using depth, geometry or specific materials to strengthen them.

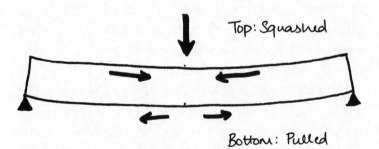

A beam flexes when it bears any weight, with the top of the beam being squashed and the bottom of the beam being pulled.

Typical beam profiles

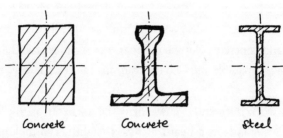

Concrete   Concrete   Steel

To resist this flexing, beams are made in specific shapes.

When a beam bends under a load, the load flows unevenly through it. The top portion of the beam is squashed, while the bottom portion is pulled: the top of the beam is in compression and the bottom is in tension. Try bending a carrot in your hands: as you curve it into a U-shape, the bottom eventually splits. This happens when the tension force in the bottom of the carrot is too big for the carrot to resist. If you repeat this with carrots of different diameters, you'll find that thinner carrots bend more easily. A carrot with a bigger diameter needs more force to bend it the same amount. Similarly, the deeper the beam, the stiffer it is, so the less it distorts under load.

Using clever geometry is another way to make a beam stiff. The highest compression force a beam experiences is right at the top, and the highest tension is right at the bottom. So the more material you put in the top or bottom of a beam, the more force it can take. By combining these two principles – depth and geometry – we arrive at the best shape for a beam: an I (i.e. in cross-section it resembles that letter), because the greatest amount of material is at the top and bottom, where the greatest forces flow. Most steel beams are I-shaped. (They are subtly different from H-shaped columns because they are deeper than they are wide, whereas H-shaped columns are closer to squares.) Concrete beams can also be made like this, but it is easier to pour concrete into a rectangular shape, so for reasons of cost and practicality most concrete beams are simple rectangles.

Large bridges like the Quebec Bridge are just too long to use a 'normal' I-shaped beam. To span the distance, such a beam would have to be so deep and heavy that it would be impossible to lift into place. Instead, we use another type of structure that harnesses the stability of triangles: the *truss*.

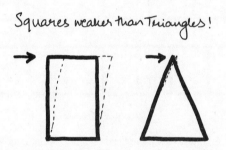

A square is an inherently weaker shape than a triangle.

Take four sticks and tape the corners together to make a square. Then push it sideways: the square becomes a diamond and collapses. Triangles, on the other hand, do not deform and collapse in the same way. A truss is a network of triangles made up of beams, columns and struts, which cleverly channels forces through its members. And in creating a truss we use smaller and lighter pieces with gaps in the middle, so we use less material than we would for an equivalent I-beam.

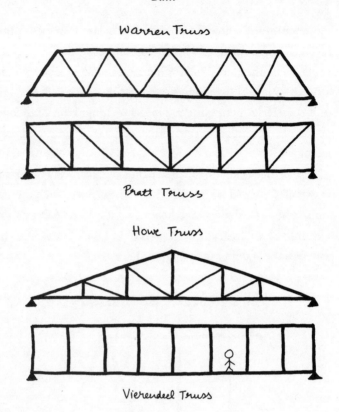

Most trusses are made up of smaller triangular shapes, although occasionally some do use squares.

Trusses are easier to build because smaller pieces of steel can be transported to the construction site and then joined together. Most large bridges have trusses somewhere. Take a look at the Golden Gate Bridge, for example: a pattern in the metal runs along the sides at road level for the length of the bridge. It looks like the letter N followed by a reversed N, one after another – a careful arrangement of triangles forming a truss.

*

Gravity exerts a predictable pull on objects on the surface of the Earth. An engineer knows what it is, and can design columns, beams and trusses to resist it. But other, equally destructive forces are not so easily reduced to equations. One of these is the wind. Random, fluctuating, unpredictable, wind has challenged engineers throughout history, and it remains a problem all engineers have to solve if their structures are going to remain stable.

When I visited Athens, one of the monuments that I was most excited to see was a large white marble octagonal tower in the Roman Agora, just north of the Acropolis. Built around 50 BC by Andronicus of Cyrrhus, a Macedonian astronomer, the Horologion of Andronikos Kyrrhestes or 'Tower of the Winds' was a timepiece with eight sundials, a water clock and a wind vane. Taking a slow walk round the tower I could see that each of its sides had a relief at the top depicting one of eight wind gods, winged figures flying forwards with a stern or benign expression, and sometimes an amphora or garland of flowers in their arms. Originally a bronze statue of Triton stood on top of the tower and acted as a weathervane, pointing towards whichever wind god was blowing.

The tower is a testament to the respect the Romans had for the wind and its potentially destructive force. The Roman master builder Marcus Vitruvius Pollio (born 80 BC), who is sometimes called 'the first architect', talks extensively about the importance of considering wind in *De Architectura*, his hugely influential ten-volume treatise on the design of structures. In Book 1 he tells us about the four main directions: *Solanus* (east), *Auster* (south), *Favonius* (west), *Septentrio* (north) – and the other four, which act in directions between the primary winds.

To me it seems amazing that Roman engineers already had such a deep understanding of how wind acts differently in different directions. Even though the way engineers calculate this is now much more sophisticated, the basis of our work was carved into the sculptures on that octagonal tower 2,000 years ago.

*

Wind acts on structures everywhere on our planet. When I am working on a construction that is less than 100m tall, I typically use a wind map. This is essentially a weather map with contours that tells me what the basic wind speed is at a particular location, created using data measured over decades. I take the basic wind speed and combine it with other numbers that define, say, how far the place is from the sea, how high up it is, and the variability of the surrounding terrain (how hilly it is or how many buildings there are). Formulae combine these factors to tell me how much wind a structure will feel in 12 different directions – every 30° around a circle – which is not far off the eight directions enumerated by Vitruvius and featured on the Horologion's reliefs.

But when I design a larger structure, such as a skyscraper, the numbers on the wind maps no longer hold. Wind is not linear: it doesn't change in a predictable way the higher you go into the atmosphere. Trying to extrapolate the data, or using mathematical trickery to adjust the numbers for 100m towers to fit 300m towers, will only produce unrealistic results. Instead, the structure has to be tested in a wind tunnel.

When I was working on the design of a 40-storey tower near the Regent's Canal in London, I visited one such facility. The miniaturised world of the wind-tunnel testers is a marvel in itself. In Milton Keynes, modelmakers had created a scaled replica of my

Horologion of Andronikos Kyrrhestes (Tower of Winds) built in the 2nd–1st centuries BC *in Athens, Greece.*

building that was 200 times smaller than the real thing would be. Not only that, they had also created tiny versions of all the other structures in the area, and the whole model sat on a turntable. The structures around my building were crucial to the data. If my tower was in the middle of a field, it would be hit directly by the force of the wind, unimpeded by any other object. In the middle of a metropolis, however, the densely textured cityscape with its mix of different structures affects the wind flow and turbulence, so the forces my building feels would differ.

I stood behind the model of my building and peered down the 'tunnel' – a long, square, smooth-walled passageway – towards the gigantic fan at the other end. It was set at the wind speed the building would feel from that particular direction. Once the cables connected to the apparatus were checked and the operatives ready, the fan was switched on. I braced myself as the blades whirred and a blast of chilly air shot through the miniature city in front of me, and hit me in the face. Inside the model of my building, thousands of sensors detected how much they were being pushed or pulled, and sent the numbers to a computer. The turntable was rotated by 15° and the process repeated until the system had logged data from 24 directions. Over the next few weeks, engineers at the facility organised the data and prepared a report. I entered their numbers in my computer model to test my building. It was imperative that my structure remained stable against all the different effects that the wind could have on it, in every direction.

There are three ways in which wind can adversely affect a structure. First, if the structure above ground is light, wind can make it topple over, like the scattered traffic cones you see after a storm. Second, if the ground is weak, wind can cause the building to move and sink. Think of a sailboat on a windy day. The strength of the wind pushes the boat across the water – which of course is the desired effect if you're out sailing. But you wouldn't want your building or bridge to move sideways in the soil as the wind hits it. Now, soil is not as fluid as water, so you wouldn't see a building floating past you in a storm (if you do see this, take my professional advice and run the other way). But soil can still be squashed and moved around, so engineers need to provide an anchor – foundations – to keep their buildings in place.

The third effect is similar to a boat rocking at sea. Like trees, all buildings sway back and forth in the wind, depending on how strongly it is blowing – this is normal and safe. Unlike trees, however, buildings don't move so much that you can easily see the displacement. Towers are generally designed to bend through a maximum distance of their height divided by 500 – so a 500m-tall tower won't move more than 1m; but if this sway happens too quickly it could make you feel seasick.

One way to prevent a structure from toppling over is to make it heavy enough. In the past, most buildings were relatively modest in height and, because they were made from stone or brick, contained enough weight to resist the threat of the wind. But the higher you build, the stronger the wind is that you encounter. In the twentieth century, as we began to build taller and lighter structures, the force of the wind became a force to be reckoned with.

And so, in the modern skyscraper, weight alone is not always enough to keep it upright. Instead, the engineer must find a way to make the structure stiff enough to resist the wind. If you've ever watched a tree bending in a high wind and seen how it's able to withstand such a force, then you already understand the principle engineers use to keep modern buildings upright, even if it's blowing a gale outside. Just as a tree's stability depends on a solid, well-rooted but pliable trunk, so a building's stability often depends on a *core*, made from steel or concrete.

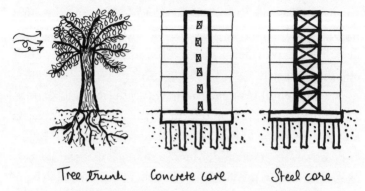

Tree trunk   Concrete core   Steel core

The core of a building, whether it's concrete or steel, is designed to provide the stable 'trunk' of the structure and so must be well-rooted in the ground.

The core – which, as the name suggests, tends to be in the centre of a tower – is an arrangement of walls in a square or rectangle that extends vertically throughout the height of a tower – like the spine in the human body. The floors of the building are joined to the core walls. The reason we don't generally notice cores is because they are well hidden, and usually themselves hide the essential services that are needed, like elevators, stairs, air ventilation ducts, electricity cables and water pipes.

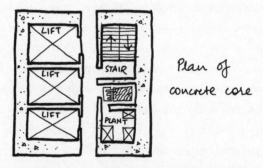

Plan of concrete core

Arranging the core of a building, usually hidden within the centre of the structure, which in turn provides a suitable place for essential services.

When wind hits the building, its force is channelled into and through the core. A building's core is a cantilever – a structure, like a diving board, that is clamped firmly at one end and free to move at the other. The core is designed to flex a little and allow the wind forces to flow down into the foundations, stabilising the core and the building – much as a tree's roots help it to withstand and disperse the wind's power.

The walls in a concrete core will be made of solid concrete (apart from holes in specific places for elevator or stair doors), making the core inherently stiff. Steel cores are different: simply replacing concrete walls with steel ones would be incredibly expensive and heavy; the sheer weight of the steel would make them impossible to build. So instead of solid walls, steel columns and beams are arranged in formations of triangles and rectangles to create a frame or vertical truss.

The force in each steel section or concrete wall depends on which direction the wind is blowing. My computer model has the wind force values for 24 different directions from the wind tunnel report. The forces create compression and tension in the beams, columns and struts that make up the frame in a steel core, or the walls in a concrete one. The computer then works out the compression and tension in every bit of the core, for every orientation. We then design each steel section or concrete wall using the largest compression and tension figures. We vary the size of the steel, or the thickness of the concrete, depending on the force in each. The core thus keeps the tower stable irrespective of wind direction. It's a complicated procedure to check the force in just one area for 24 different wind effects, let alone an entire core. Fortunately, computing power nowadays does the hard work, making it somewhat simpler for the engineer.

The building at 30 St Mary Axe in London, which is 41 storeys tall and shaped like a gherkin (hence its nickname), has
a different way of remaining stable in the face of wind. The
elegantly curved cylinder of shaded blue glass is surrounded
by large criss-crossing pieces of steel in the shape of big
diamonds.

*Completed in 2012, 30 St Mary Axe, London – otherwise known as 'the
Gherkin' – has a steel exoskeleton to protect it from external forces.*

A core is like a spine or a skeleton, giving a building integrity from the inside, but 30 St Mary Axe is surrounded by an exoskeleton. This exoskeleton – or, to use the technical term, *external braced frame* or *diagrid* – is like the shell of a turtle. Instead of an internal structure that resists the forces trying to push it over, it's the shell or frame around the building that does the protecting. As it is buffeted by wind, the network of steel that forms the diagrid transmits the wind force to the foundations to keep the building stable.

Another spectacular example of the external braced frame is the Centre Pompidou in Paris. Architects Renzo Piano, Richard Rogers and Gianfranco Franchini envisioned what is, in effect, an inside-out building. All its arteries – the stuff that's usually hidden away, like fresh-water and waste pipes, electricity cables, ventilation ducts, and even the stairs, elevators and

The Pompidou Centre, Paris, has an external braced frame composed of a network of steel rods.

escalators – are on the outside of the building. It's these details that catch the eye and which people remember: the snaking pipes painted white, blue or green; the translucent tube of the escalator zigzagging upwards. But take a second look and you'll notice that the whole structure is clad in a network of large X-shaped rods, which are there to keep it stable against the wind. An exoskeleton, among the air ducts and waste pipes.

As a structural engineer, I like that I can see how the building works, and understand where the loads are going. Instead of hiding or disguising all the seemingly unglamorous but essential systems that make a building run smoothly, exposed systems like the Centre Pompidou's are delightfully honest, and treat us to an insight into the character of a structure.

*

Diagrids and cores, however, are not incorporated into buildings just to stop them toppling over – they also control sway. It might seem strange that our seemingly solid structures, made from steel and concrete, move – but they do. The swaying in itself is not a problem: what's important is how fast the building sways, and for how long. Through years of experiments we've been able to determine the levels of acceleration (a measure of how quickly the speed of an object is changing) at which humans can feel this movement. Take travelling in an aeroplane, for example: even though it flies extremely fast, in calm air you hardly feel you're moving at all. When you hit turbulence, however, the speed starts to change suddenly and quickly, and you feel it. Buildings are similar: they can move by quite a large amount, and you won't feel anything so long as the acceleration is small. But if the acceleration is large, then even if the building is only moving a small amount you could feel queasy.

It's not just the acceleration that affects us. How long the building continues to sway – how long it oscillates or moves side-to-side – can also make us feel unsteady. To use a diving-board analogy once more: when you bounce on the board and take a dive, the board oscillates before it stops moving. A thick board that is strongly clamped at its end only oscillates a short distance and stops after just a few oscillations. A thinner, weaker board that isn't as strongly clamped will oscillate a greater distance and for a longer time.

When I design a tall tower, I have to make sure that the acceleration of the sway is outside the range of human perception, and that the oscillation stops quickly.

The same computer model that helps me design a structure that can resist gravity and wind also helps me with this challenge. I enter the materials, shape and size of the beams, columns and core into the programme. The software then analyses the wind force, the materials' stiffness and the geometry of the structure, and tells me what the acceleration is. If it is below the threshold that people can feel, then nothing more needs to be done. If, however, the acceleration is greater, then I need to make the structure stiffer. We can achieve this by increasing the thickness of the walls of a concrete core, or using bigger steel struts in a steel one. I then rerun the model, sometimes many times, until the target acceleration is reached.

The taller and more slender the tower, the more pronounced the sway. Sometimes it isn't possible to stiffen the structure enough to control the acceleration and how long it oscillates. So although the building is perfectly safe, it wouldn't feel safe. In that case, the sway of the tower is artificially controlled using a form of pendulum called a *tuned mass damper*, which moves in the opposite direction to the tower.

Every object, including buildings, has a natural frequency: the number of times it vibrates in one second when it is disturbed. An opera singer can shatter a wine glass because the glass has its own natural frequency. If the singer can hit a note with the same frequency as the glass, the energy of her voice causes the glass to vibrate dramatically until it rips itself apart. Similarly, wind (and earthquakes) can shake buildings at a particular frequency. If the natural frequency of the building is the same as that of the gusts of wind or the earthquake, the building will vibrate dramatically, and will be damaged. This phenomenon – an object vibrating dramatically at its natural frequency – is called *resonance*.

A pendulum – which is basically a weight suspended by cables or springs – oscillates back and forth. Depending on the length of the cable, or the stiffness of the springs, it swings a fixed number of times in a fixed period. When using a pendulum to cancel out a skyscraper's sway, the trick is to calculate the skyscraper's frequency (using a computer model), and

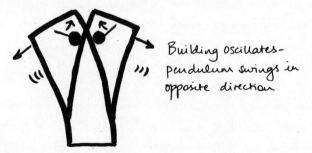

Building oscillates – pendulum swings in opposite direction

A pendulum cancels out the sway of a tall building by swinging in the opposite direction.

then to install a pendulum with a similar frequency at the top. When wind or an earthquake hits the skyscraper, it starts to move back and forth. This causes the pendulum to oscillate as well – but in the opposite direction to the tower.

You can stop the vibration of a tuning fork – and therefore its sound – just by touching one of its prongs. Your finger absorbs the energy of the vibration. The same process is at work in our

Standing at 509 metres tall, the Taipei 101 tower dominates the skyline of Taipei City, Taiwan.

swaying skyscraper. The building is like the tuning fork and the pendulum acts like your finger, absorbing the energy created by the movement of the skyscraper, which moves less and less. The movement of the structure is said to be 'damped' (hence the term 'tuned mass damper'), so the people inside can't feel it.

Taipei 101, the 509m tower in Taipei City in Taiwan, was the tallest building in the world when it was completed in 2004. It is deservedly famous for its distinct architectural aesthetic: inspired by pagodas and stalks of bamboo, the building is composed of eight trapezoidal sections that give it a ridged, organic feel, as though it has pushed its way out of the ground like the stem of a plant – an illusion reinforced by the tinted windows, which give it a green hue.

But the tower is also famous for the huge ball of steel that hangs between the 92nd and 87th floors. At 660 tonnes, this steel pendulum is the heaviest in any skyscraper in the world. It is a huge

The pendulum in Taipei 101 is how the building survives earthquakes.

tourist attraction (its sheer scale, geometrical elegance and bright yellow colour make it look like something from a sci-fi film), but its real purpose is to protect the tower from the typhoons and earthquakes that can hit the city. When the building is shaken by a storm, or by an earthquake vibrating the ground beneath it, the pendulum swings into action, oscillating to absorb the movement of the tower. In August 2015, Typhoon Soudelor swept across Taiwan, gusting to at least 170km/h, but Taipei 101 escaped undamaged. Its saviour, the pendulum, recorded movement of up to 1m – its largest-ever movement.

*

Engineers use a pendulum to defend against wind and earthquakes because both are random forces that act in a horizontal direction. But earthquakes can have far more devastating effects, so we often need other precautions too. The terrifying, annihilative power of the earthquake gave rise to all manner of explanations for its origins. Ancient Indian mythology says that the Earth shakes when the four elephants that carry it on their backs move or stretch. According to Norse myths, the Earth trembles when Loki (the God of Mischief, imprisoned in a cave for his misdeeds) wrestles with his restraints. The Japanese blame the giant catfish, Namazu, which lives underneath the Earth in mud, guarded by a god who holds it down with a huge stone. Sometimes, however, the god becomes distracted and allows Namazu to thrash about. Nowadays we have a less colourful but more accurate explanation for the periodic vibration of the Earth. Earthquakes happen when different layers of the Earth's crust move relative to one another. A wave of energy explodes from a single point: the epicentre. The energy spreads outward from this point, shaking everything on the surface, including our structures. The waves of energy from the tremors that affect our structures are unpredictable and irregular – they strike without warning.

Engineers study the frequencies of earthquakes in historical records, then they use a computer model to compare these to the natural frequency of the building to be constructed. Just like we did for wind, we must ensure that the two frequencies aren't too similar, otherwise the building will resonate and could be damaged, or even collapse. If they are, the natural frequency of the building can be changed by adding more weight to it, or by making the core or frame of the structure stiffer.

Another way to mitigate the effects of an earthquake's energy waves is to use special rubber 'feet' or 'bearings'. If you sit in your living room with powerful speakers busting out some bass, you feel vibrations transmit from the speakers, into the floors, through the sofa and finally into your body. Put some rubber feet on the underside of the speakers and the effect lessens, because the feet absorb most of the vibrations. Similarly, we can install big rubber bearings at the bottom of the columns of a building, which then absorb an earthquake's vibrations.

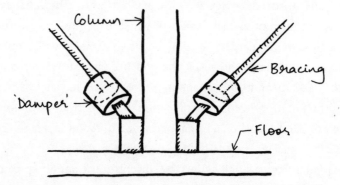

Dampers protecting the Torre Mayor skyscraper, Mexico City, Mexico.

Earthquake energy can also be absorbed in the connections between beams, columns and diagonal braces. The Torre Mayor skyscraper in Mexico City employs a very clever system to do this. In this 55-storey structure, 96 hydraulic dampers or shock absorbers – like pistons in a car – are arranged in X shapes all around the building and across its full height (creating a diagrid) to act as extra bracing against earthquakes. When an earthquake occurs, the whole building sways and the movement is absorbed into these dampers so the structure itself doesn't move too much. In fact, very soon after the Torre Mayor was completed, an earthquake recorded at a magnitude of 7.6 shook Mexico City, causing widespread damage. The Torre Mayor building survived unscathed; it's said that the occupants did not even realise there had been an earthquake.

And this, in a way, is the engineer's ideal – a building so well-designed, and so secure, that its occupants carry on comfortably with their business, completely unaware of the amount of complicated technology tackling all the forces the structure has to withstand each day.

FIRE

On the morning of 12 March 1993, I went to school in the Juhu district of Mumbai as usual, with my hair tied neatly back, wearing a crisp white blouse and grey pinafore. My teeth were hidden by braces, which were interwoven with my choice of green bands; definitely not cool (yes, even at nine I was already the class nerd). At 2.00pm Mum picked up my sister and me in our lime-green Fiat and took us home. While she was parking the car, we raced up four flights of stairs in our daily competition to see who could make it to our front door first. But something felt different. We stopped at the last step; we couldn't get to the door because our neighbour was standing there, nervously fiddling with her *dupatta*, looking distressed.

43

We soon discovered why. While Mum was collecting us from school there had been a bomb attack on the Bombay Stock Exchange – the building where my father and uncle worked.

Panicking, we ran into the flat and switched on the television. Every news channel was covering the mayhem. Bombs continued to explode around the city. Hundreds had been killed and injured. This was before the advent of mobile phones, so we had no way of knowing if my father and uncle were alive and safe.

The Bombay Stock Exchange is a 29-storey concrete tower in the heart of Mumbai's financial district. A car carrying a bomb had made its way into the basement garage and then detonated. Many lives were lost; many more people were hurt. I stood in front of the television horrified, watching images of weeping people covered in blood and dust running from billowing smoke. Police cars, fire trucks and ambulances raced to the tower, sirens blaring. We could see that the offices on the ground and first floor nearest to the explosion had been destroyed. It was clear that no one in that part of the building could have survived. Dazed people from the higher floors clambered down stairs and out of the tower. At home, we looked at each other and didn't utter a word, but I knew the same thought was running through all our minds. My dad and uncle worked on the eighth floor. We quietly hoped for the best.

As I learned later, my dad had been sitting at his desk, shouting down a poor phone line to one of his clients when a huge bang rattled the building. At first he thought an electricity generator or a large cooling unit had exploded. He jumped out of his seat, telling his staff to stay calm and remain in the office. Seconds later, however, he heard terrified people running down the stairs. Many screamed that there had been a bomb and that everybody should get out as quickly as possible. My father, uncle and their colleagues left their office, to scenes of horror.

Hundreds of people were filing down the stairs. There was barely any space to move. Head down, he focused on taking one step at a time, trying not to look at the dismembered bodies – the arms, the legs, the blood – that lay just beyond the staircase. Finally, he arrived at the ground floor. Emergency vehicles, trying to deal with the injured, blocked the street. My father and uncle fled the area and got on a bus to my grandmother's house. About two hours after we'd come home from school – the longest two hours of my life – Pop called us to tell us they were both safe.

Years later, while studying for my master's in structural engineering, I attended a class in which we discussed how to protect towers against explosions. Suddenly, the events of that terrible day in March came rushing back. For the first time a thought occurred to me: given that it was rocked by a serious explosion right at the base of the structure, and fires broke out afterwards, why didn't the whole Bombay Stock Exchange tower collapse?

I know now that there are two main reasons for this. The first is that engineers design certain buildings to resist explosions, so even if it is hit and damaged, it doesn't collapse like a house of cards. There is a minimum standard of safety governing the design of *all* structures, but the more vulnerable ones – tall, iconic buildings, for example, or those with particularly large numbers of people inside – are designed specifically for a range of possible explosion scenarios. The second reason is that all structures should be designed to stop fires rapidly engulfing them, providing enough time for occupants to escape, and for the fire to be tackled or burn out – contained in a small area – before it causes significant structural failure.

But we didn't start out building this way; we have learned from disasters of the past.

*

After waking early on the morning of 16 May 1968, Ivy Hodge went to the kitchen to make a cup of tea. She turned on the gas hob, struck a match – and the next thing she knew she was flat on her back, looking at the sky. A wall of her kitchen and a wall of the living room had disappeared.

In Ivy's flat on the 18th floor of a 22-storey tower block in Canning Town, London, there had been an explosion. Occurring in peacetime in a quiet residential neighbourhood, it was an event without precedent in the city, and it profoundly influenced how we would build future structures.

The tower had been constructed quickly as part of the regeneration desperately needed in the aftermath of the Second World War. The neighbourhood had lost about a quarter of its homes to bombing, and the destruction, coupled with the large post-war population increase, meant there was a severe housing shortage. To build rapidly and efficiently, new forms of construction were being experimented with. This particular structure was the second of nine identical towers being built to create an estate called Ronan Point.

The tower had been thrown together hastily by 'prefabrication'. Instead of pouring wet concrete on a construction site and waiting for it to solidify to form walls and floors (like most other concrete construction required), room-sized panels of concrete were made in a factory. The panels were then driven to site and lifted in to place with a crane. It was like building a house of cards: put up the walls of the ground floor, carefully place the horizontal panels on top of them to create the first floor, and so on, up and up. The panels were joined together with a small amount of wet concrete on site. The weight of the building was being channelled through these large load-bearing panels; there was no skeleton or frame. This novel prefabricated system produced lower costs, quicker construction times and required less labour, all important economic factors to consider in recovering post-war Britain.

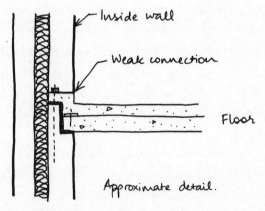

Poor detailing, such as that used at Ronan Point, where only a small amount of wet concrete was used to join together prefabricated panels during construction.

In Ivy Hodge's flat, gas had been leaking steadily from her recently installed but defective boiler system. The match flame had lit the escaped gas and BOOM!, the wall panels making up the corner of her flat blew out. With nothing now supporting them, the wall panels of the flat above fell, hitting the level below. One by one, each floor on that corner of the tower block collapsed, taking a great chunk out of the structure, from top to bottom. Four people, asleep in their flats, died.

Oddly, the explosion did not perforate Ivy's eardrums, which suggests that its force wasn't that large – since it doesn't require much pressure to damage them. In fact, subsequent investigations showed that even an explosion with just a third of the force of the actual event would have dislodged the wall panels. Since the panels were just sitting one on top of the other, without being tied together properly, there was little to stop them blowing out. The designers had relied on friction between the panels and the little bit of wet concrete 'glue' to hold them in place. It wasn't enough. When the explosion pushed out on the wall, the force of the push was bigger than the resistance of the friction and the concrete, and it flew out. Then, because the load from the walls above had nowhere to channel itself, the walls simply fell.

There was another unusual thing about this collapse. Normally I would expect an explosion at the base of a building to cause the most damage, because there are many storeys above it which can come crashing down. In this case, however, if the same explosion had happened at the base of the building, the collapse might not have happened at all.

Friction depends on weight. The heavier the load acting at the junction between two surfaces, the greater the friction. Close to the top of the tower (where Ivy was), there were only

The disproportionate collapse of floors following an explosion at Ronan Point, London, in 1968.

four storeys of weight at the junction between wall and floor, so the friction was low. The pressure of the explosion overcame the friction and sent concrete panels flying. But at the base of the tower, the weight of more than twenty storeys of panels created greater friction between wall panels (it's the reason why pulling a magazine out of the base of a stack is much harder than extracting one from higher up). So counter-intuitively, the explosion near the top was the event with disastrous results. This is not a very common occurrence now – especially because, as we'll see, buildings aren't built like this any more.

The debacle at Ronan Point had two important lessons for future construction. Firstly, it was vital to tie structures together, so that if a wall or floor panel were pushed with a force bigger than expected, the ties would stop the panels from sliding out. (At Ronan Point, steel rods, for example, tying together the prefabricated wall panels between floors could have helped the building withstand the blast; variations of this tie-system are used in modern prefabricated buildings.) Even for structures built in a more traditional way, with all the concrete poured, or steel being fixed on site, it is essential to make sure that the beams and columns have robust connections. In the case of steel frames, the bolts used to join pieces of steel together should be strong enough not only to resist normal loads exerted by wind and gravity, but also to keep the structure bound together.

Secondly, engineers had to prevent disproportionate effect. At Ronan Point, a single explosion on the 18th floor caused the corner of the tower to collapse at all levels. This domino effect was disproportionate to the cause, and a new term, *disproportionate collapse,* was born. If an event like an explosion happens, then of course damage will occur, but the effect of an explosion on one storey shouldn't propagate throughout the structure. The problem at the Canning Town tower block was that the loads didn't have anywhere to go. So the key is to ensure that the forces have somewhere to go, even if part of a structure disappears. It's like sitting on a stool: theoretically, only a quarter of your weight is transmitted through each of the four legs. But if, like many people, you're inclined to tilt the stool so all your weight is going down only two legs, you've just doubled the load the leg is designed for – the legs fail, you hit the ground, and you bruise your backside. But if structural engineers anticipate this sort of behaviour and design every leg for double the load, then you're safe.

Thus the idea of consciously creating new paths for loads to travel through was born. In my computer model, I will delete a column, record the larger forces in neighbouring columns, and design for this higher load. Then I know that even if that column is gone, its neighbours will do its job. Then I put that column back in and remove another one, trying different combinations to check my structure is stable in the face of explosions. Never challenge a structural engineer to a game of Jenga: we know which blocks to remove – how to take chunks out of a structure so that it doesn't crash.

*

Throughout history, engineers and civic authorities have been engaged in battle – against the fires that threaten to raze our towns and cities to the ground. Roman houses were often made with timber frames, floors and roofs, which caught alight easily, and fires were common. The Great Fire of Rome in AD 64 laid waste to two-thirds of the city. Originally, timber was not protected with anything to resist fire like it is now, and walls were made from wattle and daub. Wattle, a lattice woven from narrow wooden strips which looked a bit like a straw basket, was coated – daubed – with a mixture of wet soil, clay, sand and straw. Such a construction was highly flammable, enabling fire to spread quickly. The narrow streets aggravated the situation because flames could easily jump the small distance between one building and another.

In the first century BC, Marcus Licinius Crassus was born into the upper echelons of Roman society. He grew up to become a respected general (he helped quash the slave revolt of Spartacus) and a notorious businessman. Crassus was a man who spotted opportunities: observing the devastation caused by Rome's fires, he created the world's first fire brigade, made up of over 500 slaves who were trained to fight fires. He ran it as a private business, rushing his team to burning buildings, where they intimidated and drove away rival firefighters, then stood about until Crassus had negotiated a price to put out the fire with the building's distraught owners. If no deal could be reached, the firefighters simply allowed the structures to burn to the ground. Crassus would then offer the owners a derisory sum to purchase the smoking site. This meant that he quickly managed to buy up much of Rome, and amassed a fortune as a result. Fortunately, modern-day fire brigades work on a more honest basis.

After the Great Fire of Rome, Nero ordered several changes to the city. Streets were made wider, apartment buildings limited to six storeys, and bakers' or metal workers' shops separated from residential units, using double walls with air gaps. He proclaimed that balconies should be made fireproof to make escape easier, and invested in improving the water supply, so it could be used to extinguish fires. The Romans learned from tragedy, and we too have benefited from that hard-won wisdom. Thousands of years later, these simple principles – separating rooms, flats and buildings with fire-resistant materials and installing air gaps – are still used to prevent fires ravaging modern structures.

*

On 11 September 2001, the world watched in horror as two planes collided with the World Trade Center towers in New

York. I was in Los Angeles on holiday before starting at university, and was scheduled to fly to New York the next day. Paralysed, I sat watching the news, shocked as the towers collapsed an hour after being hit. A few days later, I went directly back to London, already feeling part of a changed world.

Looking at the events from an engineer's point of view, the events of that appalling day had a ripple effect on the design and construction of skyscrapers. Reading about the structural failures that led to the collapse of the towers, I was surprised to learn that it wasn't just the impact of the planes that caused the devastation, it was also the fires that followed.

New York is filled with spectacular skyscrapers, yet the World Trade Center's twin towers (opened in 1973) were among the city's most iconic symbols. Visually, each of the towers was very simple – a perfect square from a bird's-eye view, 110 storeys high. Each had a large central core made of steel columns. But this spine wasn't responsible for keeping the towers stable: they used the 'turtle-shell'-style exoskeleton instead.

Vertical columns, spaced just over a metre apart all around the perimeter of the square, were joined up at each storey with beams. The beams and columns together formed a robust frame, similar to the construction of the Gherkin we saw earlier, but with giant rectangles instead of triangles. The connections between the beams and columns were very stiff. This external frame kept the building strong against the force of the wind.

When the planes crashed into the towers, giant holes opened up in the exoskeleton. A number of columns and beams were destroyed. Engineers had in fact planned for the possibility of some form of impact by aeroplane. They had studied what might happen if a Boeing 707 (the largest commercial aircraft in operation at the time of construction) hit the building, and they had designed accordingly. The beams and columns had been constructed with extra-strong connections tying them together, so even though some of the structure was gone, the loads found somewhere else to go: they flowed *around* the hole (using the principle of preventing disproportionate collapse, which engineers had learned from Ronan Point).

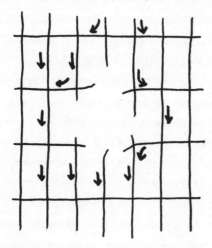

Loads within a building find new routes as the forces are channelled through alternative load paths.

The planes that hit the Twin Towers were not the Boeing 707s that engineers had planned for nearly 30 years earlier; they were larger 767s, carrying more aviation fuel. On impact, the fuel caught fire, and the conflagration of the fuel, aircraft parts, desks and other flammable material inside the building made the steel columns very hot. When steel gets hot, it behaves badly: the tiny crystals which make up the material become excited, vibrate and begin to move around, and the normally strong bonds between them are loosened. Loose bonds mean soft metal. So hot steel is weaker than cold steel, and cannot bear the same load. On 9/11, the columns just next to the holes were supporting a larger load than usual, because they were channelling not just their own forces but also those their neighbours had once carried. The steel columns and floor beams had been sprayed with a special paint mixed with mineral fibres, designed to insulate the steel from the heat of a fire and prevent it from getting too hot. But the crash of the plane and the projectile debris had chipped away areas of the protective paint, leaving big patches of exposed steel. The temperature of the columns around the perimeter of the tower rose ever higher.

The steel columns which made up the core also became unnaturally hot. Two layers of gypsum board (a panel made of gypsum plaster pressed between two thick sheets of paper) separated the core from the rest of the building. The idea was that a fire in the office space couldn't infiltrate the core past these boards, so people could run into this safe zone and to the stairs to escape. But this board was damaged, leaving the core columns susceptible and the intended safe passage exposed.

The columns became weaker and weaker, and as temperatures reached about 1,000° Celsius, they gave up. They couldn't carry the forces any more and they bowed.

In the end, the columns failed completely and the structure above it was then left vulnerable to the effects of gravity. The floor above the failed columns came crashing down. But the level on which it landed wasn't strong enough to resist the falling load and it too failed. One after the other – in a domino effect reminiscent of the Canning Town disaster but on an even more shockingly huge scale – all the floors failed and the towers came down. The fire protection – paint and boards – was no match for the size and intensity of the fire.

The way we design skyscrapers has changed since that day. Now, we make sure that escape routes are protected more robustly. The easiest way to do this is to build the core in concrete instead of steel, so that instead of weak gypsum boards standing between the fire and safety, you have a solid wall of concrete.

Concrete is not a good conductor: it doesn't transmit heat well, which means it takes longer to heat up. To strengthen concrete, however, we insert steel reinforcement bars into it; these are excellent conductors of heat, which creates a problem for the engineer. In a fire the steel bars heat up, and the heat energy spreads quickly through their length, while the surrounding concrete heats up slowly. The hot steel expands more quickly than the colder concrete, causing the outer layers of concrete to crack and burst off. This is similar to how thick glass tumblers crack if you pour hot water into them: the inner layers of the glass get very hot and expand, but the outer layers remain cold because glass, like concrete, is not a good conductor of heat. As the inner layers expand against the colder outer layers, the outermost cracks.

Through testing and experimentation, we know how long it takes for concrete to conduct heat to steel bars, and then for the steel bars to heat up and make the concrete burst. So we bury the steel deep enough in the concrete to ensure that the fires can be put out before the outer layer of concrete is damaged. This buys enough time for people to leave the building through the concrete core, or for firefighters to get the flames under control, without the structure collapsing. The taller or larger the building, the longer it takes to escape, so the deeper the steel is embedded in the concrete. Just a few centimetres make a tremendous difference.

So concrete cores perform a dual function: keeping the building stable against wind loads, and forming a protected escape route for the occupants. Today, even if we use an exoskeleton to resist wind (which means we don't *need* an internal core), we still often install concrete walls to safeguard escape routes. And the protection for steel columns and beams against fire has also been improved dramatically: fire-resistant boards and intumescent paint (which expands when heated and insulates the metal) are much more robust now than ever before. They stop steel getting too hot too quickly, so it remains strong.

Learning from disasters is fundamental to engineering: part of the engineer's job is a constant process of improvement, endeavouring to build structures that are better, stronger and safer than they were before. Thanks to such lessons we now anticipate the removal of columns, and check in advance that a building will not collapse. The Bombay Stock Exchange tower had been built in such a way that even though the structure in the immediate vicinity of the car bomb was severely impacted, the loads it was carrying found somewhere else to go. The damaged part of the building remained stable enough because it was tied into the rest of the structure, so – unlike Ronan Point – the floors above didn't come crashing down. The steel bars buried in the concrete walls and columns held their strength in the face of the fires that blazed after the explosion.

It was the lessons engineers learned from history, and the new way of designing for the unanticipated, which saved my dad's life that day.

CLAY

I love baking, which is perhaps not surprising, given that it has a lot in common with engineering. I like the way you have to follow an ordered series of processes to construct a cake. I like that you work in a very patient and precise fashion, otherwise you won't get the right shape and texture. I like the hopeful wait, that quiet period when my work is done and it slowly takes shape in the oven. Usually, I find all this incredibly satisfying. But there are moments of perplexed frustration – like the time I opened the oven door ready to slide out a delicious pineapple upside-down cake and was confronted instead with chunks of uncooked fruit swimming listlessly in a greasy sea of butter. Forget soggy bottom, this was a soggy disaster. Cursing the oven and recipe (after all, it could hardly have been my fault), I slung it straight in the bin: useless – except as a valuable reminder that in baking, as in engineering, the right choice of materials, combined in the correct way, is crucial to the outcome.

When designing a building or bridge, materials are one of my foremost concerns. In fact, different materials can entirely change the way the frame of a structure is arranged, how intrusive it feels, and how physically heavy and expensive it is. They must serve the purpose of the building or bridge correctly: I need to weave in the skeleton of the structure without it becoming obtrusive to the people using it. The materials must also resist the stresses and strains of loads that assail a building, and perform well in the face of movement and temperature fluctuations. Ultimately, my choice of material has to ensure that the structure survives as long as possible in its environment. Luckily, my engineering creations are more successful than my baking endeavours.

The science of materials has long obsessed humans, and since ancient times we have theorised about what makes up 'stuff'. The Greek philosopher Thales (c. 600 BC) contended that water was the primordial substance of all things. Heraclitus of Ephesus (c. 535 BC) said it was fire. Democritus (c. 460 BC) and his follower Epicurus suggested it was the 'indivisibles': the precursors to what we now call atoms. In Hinduism, the four elements – earth, fire, water and air – described matter, and a fifth – *akasha* – encompassed that beyond the material world. Roman engineer Vitruvius writes in *De Architectura* agreeing that matter is made up of the same four elements, adding that the behaviour and character of a material depends on the proportions of these elements within it.

This idea – that there were a limited number of fundamental ingredients which in different proportions could explain every colour, texture, strength and other property of any material – was revolutionary. The Romans surmised that materials which were soft must have a larger proportion of air, and that tougher materials had more earth. Water in large proportions made a material resistant to it, and brittle materials were ruled by fire.

Ever curious and inventive, the Romans manipulated these materials to better their properties, which is how they made their renowned concrete. They may not have had the periodic table (it would be a while before Dmitri Mendeleev published the original version of the table in 1869), but they knew that the properties of a material depended on the proportions of its elements, and they could be changed by exposing it to other elements.

For a long time, however, humans simply built from the materials that Nature provided, without changing their fundamental properties. Our ancient ancestors' dwellings were made from whatever they could find in their immediate surroundings: materials that were readily available and could be easily assembled into different shapes. With a few simple tools, trees could be felled and logs joined to create walls, and animal skins could be tied together and suspended to form tents.

If there were no trees, humans created homes from mud. As we developed our tools and became more innovative and daring, we took this one step further – we tried to make the mud better by shaping it into rectangular cuboids of various sizes using wooden moulds. We discovered that by allowing the mud to dry in the sun (according to Roman philosophy, letting the water escape and the earth take over, using fire), the result was a much tougher unit. Humans had created the brick.

Bricks were already in use around 9000 BC in an expanse of desert in the Middle East. In the deep valley of the River Jordan, hundreds of metres below sea-level, Neolithic man created the city of Jericho. The residents of this ancient city baked hand-moulded flat pieces of clay in the sun and built homes with them in the shape of beehives. As early as 2900 BC the Indus Valley Civilisation was building structures using bricks baked in kilns. It was a process that required skill and precision: if it wasn't heated for long enough, the shaped mud wouldn't dry out properly. Heated too much and too quickly, it would crack. But if baked at the right temperature for just the right length of time, the mud became strong and weather-resistant.

Archaeological remains from the Indus Valley Civilisation have been found in the ruins of Mohenjo-daro and Harappa, in modern-day Pakistan. Every brick they used, no matter what its size, was in the perfect ratio of 4 : 2 : 1 (length : width : height) – a ratio that engineers still (more or less) use, because it allows the brick to dry uniformly, it's a handy size to work with, and it has a good proportion of surface area that can be bound to other bricks with whatever form of glue or mortar is used. At about the same time as the Indus Valley Civilisation, the Chinese were also manufacturing bricks on a large scale. But for the humble brick to become one of Western civilisation's most used materials, we had to wait for the rise of one of its greatest empires.

*

The energy and inventiveness of Roman engineering is, for me, a source of wonder and inspiration. So it was with not a little excitement that I took a train south from Naples, along the coast, to one of the most famous archaeological sites in the world. Wearing matching sandals, my husband and I alighted at our destination and put on matching safari hats to keep the scorching summer sun at bay. In great anticipation, we strode towards the ancient ruins of Pompeii.

Along the cobbled streets were shopfronts with counters studded with holes in which conical pots or *amphorae* were once stored. On the ground was a dramatic floor mosaic of writhing fish and sea creatures. Another showed a ferocious canine and was inscribed with the legend 'Cave canem' – 'Beware of the dog'. Alongside these were well-laid-out homes, like Menander's (a Greek writer), with its spacious atrium, baths and garden surrounded by a beautifully proportioned colonnaded walkway or *peristyle*. All these gave a powerful impression of what a glorious, bustling town it must have been in its heyday.

Among the things that most caught my eye, though, were the blood-red bricks. They were everywhere. They peeked surreptitiously from columns on which the decorations that originally hid them from view had crumbled away. They looked proudly on from the walls, where they were arranged in thin layers of three, alternating with sharply contrasting layers of white stone. But my favourite brick-built features were without doubt the arches.

Arches are important building components. They are curved – they are a part of a circle or an ellipse, or even a parabola. They are strong shapes. Take, for example, an egg: if you squeeze an egg in your hand with a uniform grip, you'll find it nearly impossible to break because the curved shell channels the uniform force of your hand around itself in compression, and the shell is strong in resisting it. To crack the shell, you normally have to use a sharp edge, such as the blade of a knife, on one side, creating a non-uniform load. When you load an arch, the force is channelled around its curved shape, putting all portions of the arch in compression. In ancient times, stone or brick were commonly used building materials – these are great under these squashing loads but not tension loads. The Romans understood both the properties of such materials and the virtues of the arch, and they realised they could bring the two things together in perfect union. Until then, flat beams were used to span distances, whether in bridges or buildings. As we saw earlier, when loaded, beams experience compression in the top and tension in the bottom – and since stone and brick aren't very strong in tension, the beams the ancients used tended to be large and often unwieldy. This limited the length of the beams' spans. But by using the high compression resistance of stone in an arch, the Romans could create stronger and larger structures.

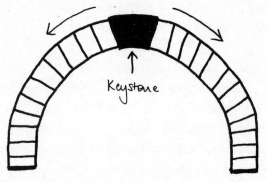

Keystone

Forces channel around the curve of the arch; it is all in compression all of the time.

The brick arches surrounding me had survived millennia, and made me think of the beautiful ancient Arabic saying 'Arches never sleep.' They never sleep because their components are continuously in compression, resisting the weight they bear with endless patience. Even when Mount Vesuvius spewed lava over Pompeii, smothering its people and buildings, the arches remained the watchers of the city. They may have been buried, but they never stopped doing their job.

The ruins of Pompeii show us that the Romans used brick in almost every form of construction in the lands they conquered. In Italy and elsewhere, legions operated mobile kilns, spreading this practice as far as what are now the British Isles and Syria. You won't be surprised to learn that Vitruvius had an opinion on the material needed to make a perfect brick, the description for which he outlined in *De Architectura*. Creating a brick is much like creating a cake, so here's my take on a recipe for The Ancient Brick, courtesy of a range of ancient engineers – one that even I would be able to follow.

RECIPE FOR THE ANCIENT BRICK

Ingredients

Clay

'They should not be made of sandy or pebbly clay, or of fine gravel, because when made of these kinds they are in the first place heavy; and secondly, when washed by the rain as they stand in walls, they go to pieces and break up, and the straw in them does not hold together on account of the roughness of the material.

They should rather be made of a white and chalky or red clay, or even of a coarse-grained gravelly clay. These materials are smooth and therefore durable; they are not heavy to work with, and are readily laid.'

The waters of fruit

Warmth, in the form of the sun or a kiln

Method

1. Throw a lump of clay into knee-deep water and then stir and knead forty times with your feet.

2. Wet the clay with the waters of pine, mango and tree bark, and the water of three fruits, and continue kneading it for a month.

3. Form the clay, mixed with a little water, into large, flat rectangles using a wooden mould. (The Greek Lydian brick – typically used by the Romans, as per Vitruvius – is a foot and a half long and one foot wide.) Once formed, remove the bricks from the moulds.

4. Heat the clay gently and gradually. If made in the summer the bricks will be defective because the heat of the sun will cause their outer layers to harden quickly, while leaving the insides soft and vulnerable. The outer, drier layers will shrink more than the moist inner layers, causing the bricks to crack. On the other hand, if you make the bricks during the spring or the autumn they will dry out uniformly, due to the milder temperature.

5. After an interval of between two and four months, throw the bricks into water, take them out and allow them to dry completely.

Patience is key, as it takes up to two years for bricks to dry completely. Younger bricks will not have dried out completely, so may shrink over time. A wall made from such bricks and then plastered over will be seen to crack. Vitruvius alerts us to this: 'This is so true that at Utica in constructing walls they use brick only if it is dry and made five years previously, and approved as such by the authority of a magistrate.'

Roman bricks were, in general, larger and flatter than those we use today. They looked more like tiles: the Romans favoured that shape because they realised that, with the tools and methods they used, flatter bricks would dry out more evenly – an essential feature of the ideal brick recipe. From the temples in the Forum in Rome to the Colosseum and the monumental walls and arches of the Red Basilica in their ancient city of Pergamon, bricks formed the basis of their most impressive structures.

The ruins of the Red Basilica made from Roman bricks in modern-day Bergama, Turkey.

When the Roman empire fell in AD 476, the art of brick-making was lost to the West for several hundred years, only to be revived in the Early Middle Ages (between the sixth and tenth centuries), when they were used to build castles. During the Renaissance and Baroque periods (from the fourteenth to the early eighteenth centuries), exposing bricks in buildings went out of fashion, and instead they were hidden behind intricate plaster and paintings. Personally, I like seeing bricks on display, much as I like seeing the air ducts and escalators on the outside of the Centre Pompidou. I prefer my structures direct and honest: like my cakes, I enjoy being able to view the materials from which they are created (this has nothing to do with my complete lack of icing skills).

During the Victorian period in Great Britain (1837–1901), and between the World Wars, the use of brick peaked to its highest in recent history. One of my favourite buildings in London, George Gilbert Scott's grand Gothic fantasy the St Pancras Renaissance Hotel, is a spectacular example of an exposed brick structure. Up to 10 billion bricks were made annually in Britain. It seemed that all structures, from factories

The brickwork of a Roman arch at Pompei, southern Italy.

to houses, from sewers to bridges, were made from bricks, left exposed for all to see.

*

Such a timescale, stretching back millennia, is already hard to get your head around. But that's nothing compared to the dates involved in the creation of the raw material that makes up a brick. During the filming of *Britain Beneath Your Feet*, a two-part documentary about the ground and what's under it, I visited a clay mine in north-east London. There I was confronted by a vast clay cliff, sculpted by diggers from the ground on which London sits. The mine owner pointed to the top of it, which was the colour of rust. 'That clay is new, it's only twenty million years old.' My flabbergasted expression prompted him to continue. He explained that the 'newer' layers of clay had a much higher iron content, giving them a reddish tint. The stuff at the foot of the cliff was purer, so it had a blue-grey hue – a sure sign that it was older.

By 'older', he meant more than 50 million years old. Long ago, igneous (volcanic) rocks were weathered and transported by water, wind and ice. While the rocks and stones were being carried along, they picked up particles of other minerals such as quartz, mica, lime or iron oxide. This mixture of rock and minerals was deposited far from its original home in layers of sediment at the bottom of rivers, valleys and seas. In these environments, plants and animals thrived then died, adding a layer of organic matter that would then be covered by more rock, and so on. Gradually, over millions of years, under the right conditions of temperature and high pressure, these layers turned into sedimentary rock. And that's what the miners were busy digging out of the cliff face. The owner told me that, because of its incredible age, the clay is full of the fossils of tropical plants such as mangrove palms (which once flourished in British climes), and the ancestors of birds, turtles and crocodiles that no longer exist on Earth.

The mined clay is used for many things: crafting pots, art projects in schools and, of course, making bricks. For this, it is transported from the mine to factories where it is transformed into neat, solid cuboids. The principle of heating clay to create a brick hasn't changed from ancient times, but the method has. First, we treat the clay by adding extra sand or water to make it the right consistency: stiff but malleable. Then it is put into a machine that extrudes it through a mould or die (a bit like the hand-press in a Play-Doh Fun Factory, but on a slightly larger scale). The clay emerges in a long, rectangular column, which is chopped into brick-length pieces and conveyed to a dryer to gently remove as much moisture as possible – otherwise you end up with the cracked bricks that Vitruvius cautioned us about. The dryer is set at the relatively low temperature of 80–120° Celsius and is humid enough to stop the bricks from drying too quickly on the outside while the insides are still damp. And as they dry, they shrink.

If the process is stopped here, bricks similar to the ancient kiln-dried examples would be created. The next step is where the real difference between ancient and modern lies. The bricks are fired at temperatures of between 800° and 1,200° Celsius, fusing the particles of clay together so that they undergo a fundamental change. Clay turns into ceramic: more similar to glass than dried mud. This *fired* brick is far more durable than a *dried* brick, and that's what we use to build structures today. A fired brick is pretty strong: if we took the four elephants that support the Earth (and cause earthquakes when they stretch) from Indian mythology, added one more for luck, persuaded them to stand on top of each other and then tip toe onto a single brick, the brick would remain intact.

To turn a single brick into a usable structure, we need a special glue or mortar that can bind the units together to form a whole. The ancient Egyptians used the mineral gypsum to make a plaster (also known as plaster of Paris, since it was commonly found and mined in the Montmartre district of the city). Unfortunately, however, gypsum isn't stable in the presence of water, so gypsum-sealed structures will eventually suffer damage and degradation. Fortunately, the Egyptians also used a different mixture that had lime mortars. This hardened and strengthened as it dried (and absorbed carbon dioxide from the atmosphere) and is more resilient than the gypsum recipe. When made correctly, mortars give strength to the structures they form and can last a very long time: parts of the Tower of London were built largely with lime mortar, and are still standing strong more than 900 years later.

Other materials are often mixed into the mortar to give it different properties. In China, the mortar used to build the Great Wall had a small amount of sticky rice added to it. Rice is mainly composed of starch – this made the mortar bond well with the stone, but also allowed some flexibility, so it wouldn't crack easily if the wall moved slightly as it heated and cooled with the seasons. The Romans added the blood of animals to their mortars, believing it helped the mortar stay strong when it was hit by frost. The dome of the Taj Mahal is held together with *chuna*, a mixture of burnt lime, ground shells, marble dust, gum, sugar, fruit juice and egg white.

Bricks are used in most UK houses today because they are cheap. But they have their disadvantages. You need specialist labour to lay the units one at a time, and it's a relatively slow process. And because of the standard size of the unit, you have less flexibility in the shapes of the structures you can create. Brick structures are also very weak in tension: the mortar glue between bricks, and the bricks themselves, can crack if pulled apart. Bricks can only be used in structures in which they are being compressed most of the time. They aren't strong enough to carry the weight of taller structures (steel and concrete can take far more compression than brick, as we'll see) so are impractical for, say, high-rise buildings or the larger bridges. However, their popularity remains where cost is the driver. Approximately 1.4 trillion bricks are made each year around the world; China alone manufactures about 800 billion, and India about 140 billion. LEGO, by comparison, makes a mere 45 billion or so bricks per year.

This ancient building block, born from the earth and baptised by fire, is so versatile that it was used in the construction of pyramids, the Great Wall of China, the Colosseum, the medieval Castle of the Teutonic Order in Malbork, the famous dome of the Catedrale di Santa Maria del Fiore in Florence, and even my own house. I love that in our modern, fast-paced world, with all the technology we've developed, we continue to rely heavily on a building tool that has been in use for over 10,000 years, created from a material that was 50 million years in the making.

METAL

In Delhi in India, there is a pillar of iron that doesn't rust. This column stands discreetly within the Qutb complex, a historic compound filled with extraordinary examples of Islamic architecture. The cavernous tomb of Iltutmish, in which every inch of the arched walls is decorated with loops and whorls, and the imposing Qutb Minar, a gracefully ridged, tapering tower – and at 72.5m the tallest brick minaret in the world – are simply breathtaking. At first glance, the dark grey column – about as thick as a tree trunk and barely seven metres tall – seems insignificant and out of place: a stray cat in a zoo of exotic animals. But it made a big impression on me.

The pillar predates the architecture around it. It was made in around AD 400 by one of the kings of the Gupta dynasty, as an offering to Lord Vishnu, the Hindu god worshipped as the Preserver of the Universe. Originally it was topped with a statue of *Garuda* (Vishnu's part-human, part-eagle steed, believed to be large enough to block out the sun). People consider it lucky if you can stand with your back to the pillar and wrap your arms around it so your fingers touch, but a fence now protects the monument from tourist limbs. I wasn't interested in luck, though, I was fascinated by another peculiar property of the pillar: in defiance of its natural propensities, this iron hasn't rusted in over 1,500 years.

The iron pillar that never rusts at the Qutb complex, Delhi, India.

The Iron Age followed the Bronze Age, which came to an end as copper and tin, the raw materials for making the metal, became difficult to obtain. The Iron Age is believed to have started around 1200 BC in India, and in Anatolia (modern-day

Turkey). Archaeologists studying the ruins of Kodumanal, a small village in the middle of Tamil Nadu state in southern India, found a trench dating back to around 300 BC on the southern edge of the village. In this was a furnace that still contained some iron slag (a by-product left over from the smelting of metals). Indian iron – mentioned in the writings of Aristotle and in Pliny the Elder's *Historia Naturalis* – was famous for its excellent quality. It was exported as far as Egypt for use by the Romans, but its secret recipe was carefully guarded.

To build the Iron Pillar, the ancient Indians made discs of iron, which they then forged (heated up and hammered together), before striking and filing the outer surface to make it smooth. The iron used to forge the column was extraordinarily pure, except for the higher than usual amounts of phosphorus it contained; a result of the extraction process used by the ironmongers. It is the presence of phosphorus that prevents the pillar from rusting. Rust forms on iron when it is exposed to oxygen and moisture; at first, the metal would have corroded but, in the dry local climate of Delhi, the phosphorus was drawn to the interface between the rust and the metal surface, creating a very thin film. This film prevents air and moisture from reacting with the iron. And so the pillar hasn't rusted any further. Modern steel is not made with those relatively high levels of phosphorus because the steel would become susceptible to cracking when it is 'hot-worked', which is a typical part of the manufacturing process where the metal is deformed at high temperatures. Take a look at structures made from iron or steel that are exposed to the atmosphere and you'll see they are painted to prevent the formation of rust, which would weaken them. But the steel beams and columns in our air-controlled buildings are left unpainted – unless painting is necessary for fire protection – because the lack of humidity means they won't rust much.

While the ancients recognised the wonders of iron, it was mostly used to make household vessels, jewellery and weapons, because the iron they extracted was too soft to build with, and they didn't know how to strengthen it enough to create an entire building or bridge. There are nonetheless rare examples of structures that use it: in *A Record of Buddhistic Kingdoms*, the Chinese monk Fa Hsien wrote about suspension bridges held up by iron-link chains in India around the time the pillar in Delhi was made. And the monumental marble gateway to the Acropolis in Athens, the Propylaea (built in around 432 BC), has iron bars to strengthen the ceiling beams. That's how the ancient engineers used metal: in little snippets to help strengthen their stone and brick structures. Before iron (or its cousin steel) could be used in large-scale structures, scientists and engineers had to learn more about its character.

*

Bricks and mortar crack easily when pulled apart, but metals don't. They are fundamentally different because of their molecular structure. Like diamonds, metals are made from crystals – but not large shiny ones like we see shimmering on the dresses of glamorous Bollywood actresses. Metal crystals are tiny – so small, in fact, that you can't see them with the naked eye – and they are opaque.

These crystals are attracted to each other, and this attraction bonds them together in a matrix or grid. However, when you heat up a metal, the crystals vibrate faster and faster until the bonds weaken. The metal then becomes malleable, and may even melt into a liquid if the temperature is high enough. Because of the flexibility of the bonds, metals are *ductile*, which means they can stretch and move to a limit without breaking; the process of hot-working mentioned above makes sure this characteristic is retained. A thick plate of steel, say 100mm thick, can be rolled into a very thin sheet of 0.1mm thickness without splitting (like my pastry normally does). The matrix of crystals and the bonds between them can be softened, reshaped and moved around.

Another property the bonds give metals is *elasticity*. If a metal is pulled or squashed by a force (within a certain range), it adjusts back to its original shape when the force is removed. It's similar to when a stretched rubber band is released and returns to its normal size and shape – unless it's overstretched, in which case it deforms. The same thing can happen to metals.

In combination, these characteristics – the bonds, ductility, elasticity and malleability – make metals resistant to cracking. This gives them a very special property that makes them ideal for construction: they are good in tension. It was this property of metals that revolutionised the way we build. Before, structures had been designed mainly for compression, but now for the first time, we could create structures that could stand up to significant compression *and* tension.

While pure iron is good in tension, it's too soft to resist the immense loads in larger structures because the bond between its crystals is quite fluid and flexes. So engineers of the past could make decorative pillars, but pure iron was not strong enough for large, complex structures. It needed to be strengthened somehow. The crystals that make up iron are arranged in a lattice, so scientists and engineers began devising ways to stiffen it.

One way to do this is to add atoms to the lattice. A simple (and tasty) illustration of this involves taking lots of Maltesers and rolling them under your hand on a table, during which you'll find that they move around very easily. But if you then add a few chocolate-covered raisins to the mix, you won't be able to roll them as easily as before. Okay, you can eat the experiment now, but the point is that the 'impurities' – the raisins – lodge themselves in awkward positions and stop the Maltesers from moving around as smoothly. Similarly, if carbon atoms are added to iron they jam the crystal lattice.

There is a balance. Too few carbon atoms and the iron is still too soft. Too many, and the lattice becomes so stiff that it loses its fluidity and the material ends up very brittle, cracking easily. As if this wasn't complicated enough, iron naturally contains some carbon (and other elements like silicon) as an impurity – usually too much – but the amount varies, so the quality of the iron varies. Scientists had great difficulty trying to determine precisely how much carbon to remove to create iron that was neither too soft nor too brittle. Results of their experiments include cast iron (which, being resistant to wear, is good for cooking pots, but is not used much in buildings because it's brittle, like an Italian *biscotti*); wrought iron (which is not used much commercially any more, and which has a texture more like the soft, luxurious chocolate-chip cookies I used to eat as a child in America); and steel. While wrought iron was a decent enough building material – the Eiffel Tower is made from it – steel turned out to be the ideal compromise between strength and ductility. Steel is simply iron with about 0.2 per cent carbon content. The process of removing all but 0.2 per cent of the carbon was originally very expensive, so until someone worked out how to manufacture steel cheaply and on a large scale, it didn't make a splash in the structural world. Engineer Henry Bessemer finally solved this long-standing problem and revolutionised the steel-making process, facilitating the development of railways across the world and allowing us to begin building skywards.

*

Henry Bessemer's father, Anthony, ran a factory that manufactured typefaces for the printing press that he kept under lock and key. The protection was designed to safeguard his secrets from his competitors, but the young Henry often broke in to try and figure them out. Realising that his disobedient son was adamant about learning a trade, Anthony relented, and began training him in the factory. In 1828, when he was fifteen, Henry left school to work with his father. He loved it: he excelled at metalwork, had a natural talent for drawing and eventually began making his own inventions.

During the Crimean War (1853–1856), Henry Bessemer turned his attention to the guns the French and British were using against the Russians. The principal drawback of these guns was that they could only fire one shot before they had to be reloaded. An elongated shell that could carry more explosive seemed like a valuable improvement, so Henry tested this in the garden of his home in Highgate, North London (much to the annoyance of his neighbours). The British War Office, however, wasn't interested in his design, so he showed it to the French emperor, Napoleon Bonaparte, and his officers.

Although impressed by the shells, the officers pointed out that the extra firepower would make their brittle cast-iron guns explode. As far as they were concerned, the shells were too big. Bessemer disagreed: the problem was the guns, not the shells – so he took on the challenge of finding a better way to make them.

He decided to improve the quality of the iron being used to make the guns by developing another way of casting it. He set about formally experimenting in his homemade furnace, but the invention that made his name happened almost by mistake.

One day, in his workshop, Bessemer was heating pieces of iron in a furnace. Even though he turned up the heat, a few pieces on the top shelf refused to melt. Frustrated, he tried

MANUFACTURE OF STEEL: THE BESSEMER PROCESS.

The Bessemer process, developed for producing steel on an economic scale, led to radical developments in the construction industry.

blowing hot air into the top of the furnace, and then prodded the pieces with a bar to see if they had finally melted. To his surprise, they were not brittle like cast iron but instead were ductile and flexible. Noticing that they were the ones closest to the hot air, Bessemer realised that the oxygen in the air must have reacted with the carbon and other impurities in the iron – and removed most of them.

Until now, everyone had tried to purify iron by heating it with coal or other fuels in an open furnace. Bessemer decided to use a closed furnace with a current of warm air running through it – and without using any fuel. This is like blowing hot air into a pan which has a lid covering it, rather than heating up an open pan on a gas hob. You would normally expect burning gas to create more heat than hot air, but this is not what happened.

Bessemer must have watched cautiously as sparks emerged from the top of the furnace when the chemical reaction began. Then, a raging inferno started up – there were mild explosions and molten metal splashed around, erupting from the furnace. He couldn't even approach the machine to switch it off. Ten terrifying minutes later the explosions petered out. Bessemer discovered that what was left in the furnace was purified iron.

The furnace inferno was the result of an *exothermic* reaction: a chemical reaction that releases energy – usually in the form of heat – during the oxidation of impurities. After the silicon impurities had been quietly consumed, the oxygen in the air current reacted with the carbon in the iron, releasing a huge amount of heat. This heat raised the temperature of the iron far beyond what a coal-fired furnace was then capable of, so Bessemer didn't need to use external sources of heat. The hotter the iron became, the more impurities burned off, which made the iron hotter still, so it burned off even more impurities. This positive loop created pure, molten iron.

Now having pure iron to work with, Bessemer found it easy to add back precisely the right amount of carbon to create steel. Until his invention, steel's prohibitive manufacturing costs meant it was used to make cutlery, hand tools and springs, but nothing larger. Bessemer had just swept away that huge barrier.

He presented his work at the British Association meeting in Cheltenham in 1856. There was huge excitement about his process because his steel was almost six times cheaper than anything else available at the time. Bessemer received tens of thousands of pounds from factories all over the country to replicate his process. But his lack of understanding of chemistry was nearly the end of him.

When other manufacturers tried to reproduce Henry's methods, they failed. Furious at the amount they had spent on the licence to use the process, they sued Bessemer, and he returned all their money. He then spent the next two years trying to figure out why the process worked perfectly in his brick-lined furnace but not in others. Finally, he cracked it: the iron he was using contained only a small amount of phosphorus as an impurity. His peers, however, had been using high-phosphorus iron which, it appeared, didn't work in a brick kiln. So Bessemer experimented with changing the furnace lining, and realised that replacing brick with lime was the answer.

However, the perplexing and financially frustrating failure of his original process had bred a mistrust of Bessemer that meant no one believed him. Finally, he decided to open his own factory in Sheffield to mass-produce steel. Although it took a few years before suspicions faded, after that factories started manufacturing steel on a truly industrial scale. By 1870, fifteen companies were producing 200,000 tonnes of steel each year. When Bessemer died in 1898, 12 million tonnes of steel were being produced worldwide.

High-quality steel transformed the railway networks because it could be made into rails quickly and cheaply, and they lasted ten times longer than iron rails. As a result, trains could be bigger, heavier and faster, clearing up the clogged veins of transport. And because steel was cheaper, it could now be used in bridges and buildings – ultimately opening up the sky.

*

Without Bessemer's steel, I wouldn't have been able to design the Northumbria University Footbridge, which literally hangs on steel's ability to carry tension. The bridge was, in fact, the very first structure I worked on, fresh out of university. I can still vividly remember the first day of my brand new job, taking a packed Tube train to Chancery Lane in London, and being swept up and out of the station by the hurrying throngs of other professionals in suits. Feeling excited, nervous and a little awkwardly formal, I threaded my way along the pavement towards my destination – a five-storey office building clad in white stone.

My new boss was John, a slim man of average height, with straight, short dark hair, rimless glasses and a passionate love of cricket (something that, even though I grew up in India, I couldn't match). We went through some forms, a process made lively by his occasional ironic and funny observations; meanwhile I kept quiet about the fact that it was my 22nd birthday. Then he showed me his hand-drawn sketch of a new footbridge, made from steel, that was due to be built in Newcastle. The confident pencil marks showed that at the east end of the bridge a tall tower would support three pairs of cables. The cables in turn would hold up the main deck of the bridge. To counter-balance the weight of the bridge on this tower, a further set of cables would anchor it from behind. As I sat with John, looking at the drawings in front of me, I did a little dance inside. As far as I was concerned, this was as good a birthday present as a girl could get. I was thrilled that my first project was going to be this elegant and distinctive structure. Apart from its lovely aesthetic, however, this bridge had other nuances that made it, to my eyes, even more beautiful.

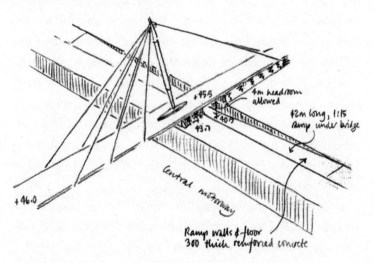

A working sketch of the Northumbria University Footbridge by John Parker.

The bridge is a 'cable-stayed' bridge, one famous example of which is the Millau Viaduct in France. Its gently curving deck is held in place by seven pillars, from which cables fan out in the shape of a sail, giving the impression that the bridge is floating 270m above the Tarn valley. Cable-stayed bridges have one or more tall towers to which cables are attached; the deck is pulled down by gravity, and is held in place by cables, which are always in tension. The tension forces are channelled through the cables directly into the tower. The tower in turn compresses and the forces flow down into the foundations on which the tower is supported; the foundations spread the forces out into the ground.

The *Millau Viaduct in France is an elegant example of a cable-stayed bridge.*

As a fresh-faced engineer, designing the cables for the Northumbria Footbridge (which were as thick as my fist) was a real challenge. If you take a metal ruler, representing the steel deck, and use three pairs of rubber bands to mimic the cables, you'll find that you have to pull on each band just the right amount before they're all taut and supporting the ruler evenly so it lies flat. If you pull too hard on the three bands on one side, the ruler tips over sideways. If you pull too tightly on the central pair, it bows upwards. Now imagine the same effect, but on a real, full-sized bridge.

I used software to create a three-dimensional computer model to recreate the bridge beams that run under the deck and the cables that run from the deck to the mast. Then I simulated gravity on the structure. I also had to consider the weight of all the people that would stand on the bridge, and that they might congregate on different parts of the bridge at different times. For example, during the Great North Run, in which athletes run along the motorway below the structure, cheering crowds might stand on one side as the runners approach them, then walk to the other side to watch them continue into the distance. I had to think about 'patterned loading' – I modelled people standing on the bridge in different configurations. No matter where people stood, the cables had to remain tight to support the deck. If the cables were not in tension they would become floppy, and the deck would lose its support. To stop this from happening, I added extra tension to the cables artificially.

Cables can be tightened up using a jack – which is a tube with clasps on each side. Each cable had at least one break in it where a jack could be installed. The clasps each held a bit of cable either side of the break. The jack can be adjusted to pull the ends closer together (to tighten the cable) or further apart (to loosen it), therefore altering the amount of force in the cable. If you look at the cables fanning out from the tower of my footbridge you'll see that they have connector pieces – where the cables look briefly thicker than the rest of their length: those are the points at which the jacks were temporarily connected. This is like replacing the rubber-band cables in our demonstration with shorter ones, but then stretching them out to the same length as before. This puts more stretch in the rubber bands – they contain a higher pulling or tension force.

The key to building a cable-stayed bridge is balance. If you use a thin piece of card as a deck and pull on the rubber bands, the card simply lifts up. If you replace the thin card with a book, then you can pull on the bands to make them taut without deforming the book. Once the stiffness and weight of the deck and the tension in the cables are reconciled and calibrated, you can then work out what the force is in the cables. When I did the drawings of the bridge, I added notes stating how much each cable needed to be tightened to stop it going slack.

The engineer's job is a lot like plate-spinning. You have to plan for, and control, a multitude of problems simultaneously. Take temperature: like all structures, my bridge is affected by it. Throughout the year, to varying degrees (depending on the season), it will be heated and cooled. Steel has a 'coefficient of thermal expansion' of 12×10^{-6}. This means that for every 1 degree of change in temperature on a piece of material 1mm long, the material will expand or contract by 0.000012mm. This may sound small, but my bridge was about 40m long and had to be designed for a temperature range of 40°. The savvy among you will argue that the British summer is not 40° warmer than the winter, and you would be correct, but the steel itself will get much hotter than the air as it absorbs heat from the sun. We're looking at the range of temperature experienced by the steel, not the air, in the most extreme (but reasonable) weather we can anticipate.

This adds up to an expansion of nearly 20mm. If I fixed the ends of the bridge to stop it from expanding or contracting, a large compression force would build up in the steel deck when it got warmer, and a large tension force would build up when it cooled down. The problem is that this expansion and contraction could happen thousands of times over the life of the bridge; this constant pulling and pushing can gradually damage not only the steel deck itself, but also the supports at either end.

To prevent this, I allowed one end to move. (In larger bridges, or bridges with many supports, you can create 'movement joints' in multiple places. You can sometimes feel your car 'boing' as you drive over them.) Because the movement on this bridge was relatively small, I used a 'rubber bearing' to absorb it. The steel beams which made up the deck were supported on these bearings, which were about 400mm wide, 300mm long and 60mm thick. When the steel expands or contracts, the bearings flex, letting the bridge move.

I also needed to think about vibration and resonance. I've already explored how an earthquake can cause a building to resonate, just as an opera singer can shatter a wine glass when she hits the right note. With the footbridge, I was concerned about whether resonance could make pedestrians feel uncomfortable. Heavy bridges, like those made from concrete, generally don't suffer from this problem because their weight stops them from vibrating easily. But the steel deck was light, and its natural frequency was close to the frequency of walking pedestrians, which meant it was in danger of resonating. So we connected tuned mass dampers with strong springs to the underside of the deck. These work in a similar way to the giant pendulum inside the Taipei tower, absorbing the sway and stopping the deck from vibrating too much. You can't see these tuned mass dampers unless you look carefully at the bottom of the deck from the road underneath the bridge (perhaps while stretching your legs on the Great North Run). If you do, you'll notice three steel box-like objects hidden between the bright-blue-painted beams.

A type of tuned-mass damper, similar to those used on the Northumbria University Footbridge.

Once I was sure that my bridge was stable in its final configuration, I had to work out exactly how it would be built. As it was too large to be transported to Newcastle fully constructed, I went to a steel fabricators' factory in Darlington. Amid showers of sparks cascading from a welder's arc, we discussed some options. We would have to bring the bridge to the site in pieces that fit on the back of lorries, so we looked at splitting it in various places, checking how those sections could be installed and supported safely until the cables had been tied in like a sculpture that would need to support itself even while each piece was being placed.

We also had to consider how to cause minimal disruption to the public. Since the structure was to span a motorway, we decided the best approach was to bring it to site in four pieces, connect them together, and then use a crane to lift the assembled bridge into place. A one-of-a-kind monster crane was booked to do the job.

Months of planning went into ensuring that the bridge was hitched up without a hitch. First, the crane itself arrived in pieces at the start of a bank-holiday weekend, and roads were closed off as it was assembled by swarms of steel fixers. Meanwhile, the four steel sections of the bridge were transported from Darlington to a nearby car park, where they were joined together, like a jigsaw puzzle, to make the deck.

The plan was to hoist the steel deck into place, and then to attach the cables. I had designed the deck such that it needed all three sets of cables to resist both its own weight and the weight of pedestrians crowded on top of it. This meant that, until the cables were in place, it needed extra support on site, so I had also calculated that the deck could stand up with a single support at its centre (it had less load on it in this configuration as the public wouldn't have access). We erected a temporary steel column in the central reservation of the motorway.

The motorway was closed. The crane swung into action. The deck was lifted up from the car park and lowered into place, its ends held up by their permanent concrete supports, and its centre by the temporary steel one. The deck was disengaged from the crane, and the motorway reopened. This complex operation took just three days.

Over the next few weeks the rest of the bridge was assembled. The mast was lifted into place using a crane and then anchored to its concrete base with bolts. The all-important cables could then be installed in pairs starting from one end of the bridge. Every time a new pair of cables was connected, the tension was adjusted using a jack. Once the cables were all in and adjusted one final time, the road was closed again, the temporary steel column removed, and the bridge was complete.

I'm not normally excited about getting up early, but my eyes were already wide open at 5am on the day I travelled to Newcastle to visit my completed bridge, which was now ready and open to the public. After taking a first small step, which felt to me like a giant leap, I walked back and forth across the bridge a number of times. I skipped and I ran. The solid steel beams, the taut cables, the rubber bearings, the tuned mass dampers – they all reminded me of the time, only a few months ago, when I had painstakingly designed them. Details that perhaps no one would notice except me – but they made me happy.

At one end of the bridge there was a bench. I sat there, grinning, for a while, watching bleary-eyed students walking across the deck from one lecture to another, all of them oblivious to the pleasure it gave me to experience my first physical contribution to the world.

ROCK

I've been known to stroke concrete. Others might feel the irresistible urge to pat a little kitten or handle an object in a museum, but for me it's concrete. It doesn't matter if it's a smooth, stark grey surface, or one with little stones visible, or even one left intentionally rough – I have to know what the texture feels like, how cold or warm it is. So you can imagine how I felt when I visited Rome and saw tonnes of ancient concrete above my head, but too far away to reach.

The Pantheon in the Piazza della Rotonda in Rome is one of my favourite structures. Built by the emperor Hadrian around AD 122 (at about the same time as he was building a wall to divide England from Scotland), it has stood strong ever since in a variety of guises – temple to the Roman gods, Christian church, tomb – though barbarians removed what they could and Pope Urban VIII even melted the ceiling panels to make cannons. A triangular pediment supported by a portico of 16 Corinthian columns greets you at its entrance. Inside, the rotunda is topped with a dome punctuated by a circular opening (*oculus* – Latin for eye) through which streams an almost otherworldly shaft of light. It's an atmospheric and beautifully proportioned building. I'm overwhelmed by its sheer scale when I wander around in it, bumping into people as I stare up at the beautiful roof. Even now, it's the largest unreinforced concrete dome in the world. The Romans really honed their craft, creating an engineering masterpiece from a revolutionary material they called *opus caementicium*.

The giant concrete dome and oculus at the Pantheon in Rome, Italy.

For me, what's special about concrete is that its form is indeterminate: it can be anything. It starts as rock, then becomes a lumpy grey liquid that can be poured into a mould of any shape and left while chemistry takes over, turning the liquid back into rock. The end product could be a circular column, a rectangular beam, a trapezoidal foundation, a thin curvy roof, a giant dome. Its amazing flexibility means it can be formed into any shape; because of its huge strength, and because it lasts an extremely long time, concrete is, after water, the most-used material on the planet.

If you crush most types of rock into a powder and add water, you end up with an uninteresting sludge, the two parts don't hold together. But something strange happens when you heat certain rocks up to really high temperatures. Take a mixture of limestone and clay, for example, and fire them in a kiln at about 1,450° Celsius, and they will fuse into small lumps without melting. Grind these lumps into a very fine powder and you've got the first ingredient of an incredible material.

The powder is called cement. It's a dull grey colour and might not look particularly impressive. But because it's been burned at very high temperatures, the parent materials are chemically changed. If you add water to this powder it doesn't turn into a sludge – instead, a reaction called *hydration* begins. The water reacts with the calcium and silicate molecules in the lime and clay to create crystal-like rods or fibres. These fibres give the material a jelly-like structure – a matrix – which is soft but stable. As the reaction continues, the fibres grow, and they bond to each other. The mixture becomes thicker and thicker until, ultimately, it solidifies.

So water + cement powder = cement paste. Cement paste hardens into a rock incredibly well, but it has its drawbacks. For a start, making it is expensive. The process also uses a large amount of energy. And importantly, hydration releases lots of heat. Once the chemical process finishes, the cement cools down, and as it cools it shrinks. And cracks.

Fortunately, engineers realised that cement paste binds solidly to other rocks, and began adding *aggregates* (small, irregular pieces of stone and sand of varying sizes) to the mixture. The aggregates help to reduce not only the amount of cement powder being used (and hence the amount of heat being released), but also the energy consumption and hence cost. The cement paste undergoes the same chemical reaction, creating fibres that in turn bind strongly to other fibres *and* the aggregates – and the whole mass solidifies to give us the concrete we are familiar with today. So water + cement powder + aggregate = concrete.

To make good concrete, the proportions of this mix need to be right: too much water, and not all of it will react with the cement powder – and the concrete will be weak. Too little water and all the powder doesn't react and, again, the concrete ends up weak. For the best result, *all* the water needs to react with *all* the cement powder. And the mixing itself needs to be right too: concrete can end up poorly if it isn't stirred properly. The larger, heavier stone aggregates settle to the bottom, leaving the fine sand and cement paste at the top, making the concrete inconsistent and weak. That's why concrete trucks have giant rotating drums – the mixture is continually sloshed around so that the aggregates are nicely distributed throughout.

Ancient engineers didn't have such trucks, but their formula for concrete was pretty similar to ours. They too burned limestone, and powdered it then added water to create a paste with which to bind stones, bricks and broken tiles. However, their mixture was much lumpier and thicker than ours is today. But then the Romans found something even better. In the land around Mount Vesuvius was an ash they called *pozzolana*. Instead of using burnt limestone as a cement, they tried this ready-made ash. When they mixed it with lime, rubble and water, their resulting concrete hardened as they expected. But this mixture also hardened underwater. That's because the *pozzolanic* chemical reaction did not need carbon dioxide from the air to help it along: the mixture could harden without it.

To begin with, the Romans didn't appreciate the amazing potential of the material they had made, and they only used it in small structures in a tentative way. They used it to strengthen the walls of their homes and monuments – sandwiching a layer of concrete between two layers of brick. After all, how did they know that it wouldn't crack and crumble in a few years like plaster did? As the years passed, of course, they realised that this incredibly resilient substance was nothing like plaster, and concrete became a commonly used material. And because it solidified underwater, they could build concrete foundations for bridges in rivers, solving the problem they'd had so far in trying to cross vast stretches of water.

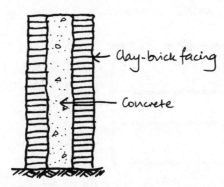

A Roman concrete sandwich. In Roman construction, the concrete wall was faced with a brick layer on both sides.

The Romans frequently used arches in their constructions, and concrete is a good material for arches. For one thing, it is incredibly strong. If a standard brick made from fired clay can carry the weight of five elephants, a similar brick made from relatively weak concrete can carry fifteen. In fact, a brick made from one of the stronger mixes of concrete can carry 80 elephants. And its strength can be changed, depending on the exact proportion of ingredients you add to the mix. Unlike bricks and mortar – where mortar is usually weaker than brick and more susceptible to crushing – concrete is cast monolithically (in large continuous chunks) and doesn't have weak links in the same way: its strength is maintained uniformly across its whole body. Ultimately, of course, if the compression load is large enough, concrete will crush and crumble, but it takes a lot of load (or a good number of elephants) to get to this point.

Concrete is, however, a fussy material. It loves compression, and for millennia it was used this way, being squashed in foundations or walls. But it dislikes being pulled apart. Its resistance to tension is minimal; in fact it cracks if tested at loads less than one-tenth of what it can resist in compression. This is another

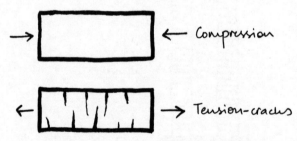

Fussy concrete prefers to be in compression. At even relatively low loads of tension, concrete will crack.

reason why the Pantheon impresses me so much. The Romans really understood how concrete works, and how domes work, and even though concrete wasn't the ideal material to use to build this immense structure, they still used it – and used it well.

To understand why making a dome from concrete is challenging, start by making an arch. If you bend a long, thin rectangular strip of card and place it on a table, you'll find it won't hold that curve on its own. It simply collapses. To make your arch stand up, position an eraser on the table against each of the outer edges of the curved card. The ends of the original, unsupported card arch pushed outwards, collapsing the structure; this time, however, although the arch still pushes outwards, the sideways friction between the eraser and the table reacts against the push from the base of the arch. This is Newton's third law of motion: every action has an equal and opposite reaction. The base of the arch exerts a pushing 'action' on its support – and its support keeps it stable by 'reacting' against this force.

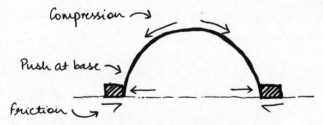

Forces flow around an arch, and then push out at the base.

Domes are similar to arches, but in three dimensions; the third dimension adds a layer of complexity. If instead of having one card you cut many long thin strips, then stack them one on top of the other and stuck a pin through the centre of the stack, you could still curve them downwards to create an arch. But you could also fan them out through 360° (so that they form lines a bit like the longitudes of the Earth), thereby creating the shape of a hemisphere or dome. This dome, though, will be no more stable than your original, unanchored arch: it won't retain that hemispherical shape on its own. To hold it in place you could arrange a ring of erasers on the table, one at the base of each strip. Or you could try something smarter, such as using rubber bands, arranged like the latitudes of the Earth, to tie the dome together. With rubber bands in place, you can remove the erasers and the dome still stands.

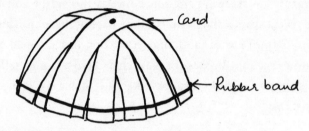

When 'tied' sufficiently, the forces that flow around a dome do not push out at the base.

What this means is that the supports for a dome do not feel any horizontal pushing force on them (unlike the arch). But you'll notice that the rubber bands are in tension: they are stretched and resist the push of the card strips. So yes, each of the strips is in compression individually along their 'longitudes', but you need tension to hold the strips together in the 'latitudes'.

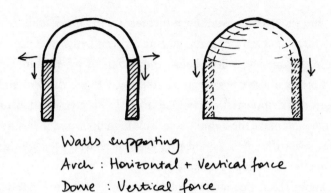

Walls supporting
Arch : Horizontal + Vertical force
Dome : Vertical force

The difference between where forces flow in an arch as opposed to a dome.

Viewed from the piazza, the Pantheon looks quite shallow, but in fact the inside is almost perfectly hemispherical. It appears shallow from the outside because the base is much thicker than the crown: the concrete at the top of the dome is only 1.2m thick, but by the time it reaches the base it has increased to more than 6m. Making it thicker towards the base meant the dome could resist higher tension forces – more material, more resistance.

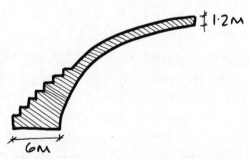

The widening stepped rings around its base act to strengthen the dome of the Pantheon.

But the Romans went even further, adding more stability in the form of seven stepped concentric rings (which you can see from the outside, a little below the oculus, if you're somehow airborne). These rings act in a similar way to the rubber bands from our demonstration, helping to resist some of the tension forces and make the dome stable. This ingenious design ensured that, even though concrete isn't great at resisting tension, the Romans succeeded in making it work.

While thicker concrete might solve some problems in resisting tension, it also creates problems of its own. The thicker the dome, the greater the cement content – which means it generates more heat, and the more it shrinks when it cools. As it shrinks, it pulls itself apart and, since concrete can't resist this tension, it cracks. The Romans were worried that the base of the Pantheon's dome would suffer extensive cracking. It's believed that the series of squares which are inset all around the inside of the dome, which are part of its unique visual aesthetic, are there to allow the concrete to cool down more quickly and evenly, minimising cracking. Even so, engineers studying the Pantheon have found cracks in the base of the dome (ancient ones that occurred while it was being built) – though they haven't undermined the integrity of this ancient building.

The first time I visited I was a teenager, and I loved this building for its beauty and sense of peace. The second time, as a trained engineer, I gazed – no less lovingly – at the recesses in its surface and searched for the fine cracks at its base. For a long time I watched the shaft of light coming through the oculus at the top of this amazing structure. I left astounded by the dome's scale and apparent simplicity of form, but conscious of how complex it must have been to construct it so many years ago.

I often wonder whether, like the Pantheon, the structures we design and build today will still be around, and in such good condition, in 2,000 years. It seems inconceivable.

*

After the collapse of the Roman Empire in the fifth century, the Dark Ages – or as I like to call them, the Crumbly Ages – began, as the Roman recipe for concrete was lost for almost 1,000 years. We reverted to a more primitive way of life and concrete only re-emerged in the 1300s. Even then, engineers continued to struggle with the fundamental problem of concrete cracking in tension. It was only centuries later that the true magic of concrete was discovered, by an unlikely hero, in the most unexpected of places.

In the 1860s, French gardener Joseph Monier became fed up with the fact that his clay pots would constantly crack. He tried making pots out of concrete instead but found that they fractured just as much. Randomly, he decided to reinforce the concrete by embedding a grillage of metal wires within it. This experiment could have failed for two key reasons: first, the concrete might not have actually bound to the metal reinforcement (there was no reason to think that it would), so the metal would only create more weak points in the pot. Second, during the change in seasons, the metal and concrete would expand and contract at different rates, creating yet more fissures. Unwittingly, Monier created a revolutionary pot that remained solid and barely cracked.

Like most metals, iron and steel (as we've seen) are elastic and ductile, and they're good in tension: they don't crack when pulled. Metals aren't brittle like brick or concrete. So by combining concrete (which breaks in tension) with iron (which can absorb tension loads), Monier had created a perfect marriage of materials. In fact, an ancient version of this principle can be found in

Morocco, where the walls of some Berber cities were made of mud with straw mixed in: a mixture known as *adobe*, also used by the Egyptians, Babylonians and Native Americans, among others. Straw fulfils a similar function to metal in concrete; it binds mud and plaster together and stops it from cracking too much because the straw resists tension forces. The plaster on the walls of my Victorian flat has horse hair mixed into it for the same reason.

Having exhibited his new material at the Paris Exposition in 1867, Monier then expanded its application to pipes and beams. Civil engineer Gustav Adolf Wayss from Germany saw the material and had visions of building entire structures with it. After buying the rights to use Monier's patent in 1879, he conducted research into concrete's use as a building material, and went on to build pioneering reinforced concrete buildings and bridges across Europe.

The marriage of steel (which replaced iron once the use of the Bessemer process spread) and concrete appears so obvious today that it seems almost inconceivable to me that the two weren't always used together in this way. In every concrete structure I design, I use steel *reinforcement bars* – long, textured rods between 8mm and 40mm in diameter that are bent

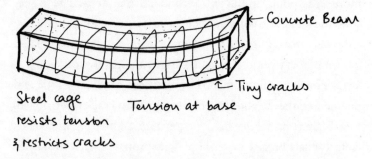

Steel cage resists tension & restricts cracks

Tension at base

← Concrete Beam

Tiny cracks

A perfect marriage of construction materials: a steel cage provides reinforcement for concrete, resisting tension and restricting cracking.

into different shapes and tied together to form a grid or mesh to bind the concrete. My calculations tell me where the concrete will be in tension and where it will be in compression, and I distribute steel bars within it accordingly.

Contractors take my drawings and set the dimensions and shapes of every single steel bar in the project, and calculate their weight. These schedules are sent to a factory, and a few weeks later real bars appear, which are fixed into shape before the concrete is poured around them.

As the chemical reaction in the concrete mixture progresses, steel and concrete form a strong bond. Just as cement paste binds strongly to aggregates in the mix, it also sticks to the steel. And once intertwined, steel and concrete are very difficult to separate. They have near-identical thermal coefficients – which is to say that they expand and contract by almost identical amounts under the same changes in temperature. When a concrete beam bends under gravity and is squashed at the top but pulled apart at the bottom, the concrete cracks at the bottom. These cracks are fractions of a millimetre wide and often not visible to the human eye – but they are there. Once this happens, the steel bars in the base of the beam are activated, and resist the tension loads keeping the beam stable.

Steel reinforcement is now part of the DNA of how we build modern structures. Many construction sites around London have small windows in the protective hoardings that surround them. As you can imagine, whenever I walk past one I can't resist taking a peek, curious to see what's going on inside. No matter what the site, I always see big piles of steel reinforcement bars ready to be tied together, or steel cages already made up inside wooden moulds. When the trucks with rotating drums appear, they pour a thick stream of concrete into the moulds, after which workers use short poles attached to a power supply to vibrate the concrete, mixing it to make sure that the different-sized aggregates are well-distributed throughout. Engineers like me have made sure that the gap between the steel bars is big enough to allow the concrete to flow easily around them. As a young engineer, my first boss John told me, 'If a canary can fly out of your steel cage, the bars are too far apart. If it suffocates, they're too close together.' It's a lesson I've never forgotten. (No canaries were harmed in this thought experiment.)

Once all the concrete has been poured and mixed thoroughly, the workers flatten the top of it with huge rakes and leave it to solidify. But this incredible material has one more secret in store. Over the next few weeks, the bulk of the chemical reaction will finish, it's tested, and results show that it has reached its target strength. In fact, its strength continues to grow – very slowly – over months, and even years, plateauing to a steady number far into the future.

*

Nowadays, we use concrete for many structures, creating skyscrapers, apartment blocks, tunnels, mines, roads, dams and countless others. In ancient times, different civilisations employed different materials and techniques that were suited to their indigenous skills, climate and surroundings. Today, concrete is universal.

Scientists and engineers are constantly innovating, trying to make concrete even stronger and longer-lasting than it already is. One recent invention has been 'self-healing' concrete, which contains tiny capsules with calcium lactate. These are mixed with the liquid concrete, but the capsules have a fascinating secret. Inside is a type of bacteria (normally found in highly alkaline lakes near volcanoes) that can survive without oxygen or food for 50 years. The concrete, mixed with these bacteria-filled capsules, hardens. If cracks form in the material and water seeps in, the water activates the capsules, releasing the bacteria. Habituated as they are to alkaline environments, these escapees don't die when they encounter the highly alkaline concrete. Instead, they feed on the capsules, combining the calcium with oxygen and carbon dioxide to form calcite, essentially pure limestone. With calcite filling the cracks in the concrete, the structure repairs itself.

There are other challenges. Five per cent of human-created carbon dioxide comes from making concrete. Using concrete in small amounts is not particularly unfriendly to the environment, but we use so much of it that the emissions quickly add up. Some of the CO_2 comes from the firing of limestone to create the cement, but the rest comes from the hydration reaction. The amount of cement being used in the mix can be reduced by replacing a proportion of the cement with suitable waste materials from other industrial processes, such as 'ground granulated blast furnace slag' (GGBS), which is created during the manufacture of steel. Using these waste materials doesn't affect concrete's strength too much but can save tonnes of carbon. You can't use them for all types of construction, because these ingredients have other effects on the mix. They can make the concrete take longer to solidify, or make it stickier, and hence harder to pump up many storeys, which is definitely a challenge when constructing skyscrapers.

'My' skyscraper, The Shard, uses concrete and steel in a really clever way that neatly reconciles the different requirements of office and residential areas. In typical office buildings, the aim is to create large, open spaces with few columns. Steel is often the material of choice because it behaves well in both tension and compression meaning that steel beams can span further than concrete ones of the same depth. Moreover, compared to apartment buildings, offices need a lot of air-conditioning machines, ducts, water pipes and cables. The I-shaped construction of steel beams, and the regular gaps between adjacent beams, leave plenty of space to hide these away. Steel structures are also lighter than their concrete equivalents, so the foundations can be smaller as well.

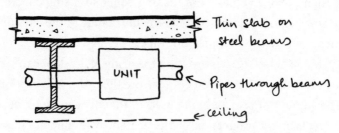

Arranging the steel beams and concrete floors for an office building.

On the other hand, residential buildings and hotels have floors that are sub-divided into flats and rooms, so you're not under as much pressure to create huge open spaces. You can hide concrete columns in walls to support flat concrete slabs. Concrete floors are thinner than steel ones, so you can fit more storeys into a concrete building of the same height. There are fewer cables and smaller ducts to run, and these can be attached to the bottom of the slabs. Concrete also absorbs sound better, so you get less noise transfer between floors – this doesn't matter so much in an office where you, hopefully, don't sleep.

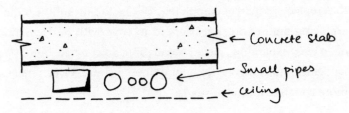

Arranging the concrete floors for a residential building.

Since The Shard has offices on its lower storeys and a hotel and apartments on its higher ones, we used different materials in different places. The lower storeys are made from steel columns and beams to create space in the offices; the higher storeys from concrete to create privacy. While it may seem obvious to use the right material in the right place, it's actually quite an unusual thing to do, and only a handful of structures globally have so far adopted this design. One possible reason is that it's arguably logistically easier (and possibly cheaper) to use the same material throughout, but I'd counter that by saying you achieve a better design for the long term, and it's more sustainable because you use less material. Another reason is that multi-use buildings simply aren't as widespread as single-use buildings. But with the construction of more and more multi-use buildings, I expect the multi-material method will become more common.

Using the materials we have in an efficient way is good engineering. We often think of concrete as being old-fashioned because of its ancient roots, but it's still very much part of the future too. Scientists and engineers are working on new super-strong mixes, and trying to figure out how to make concrete more eco-friendly. Perhaps one day we may find a new material that replaces concrete completely. But in the meantime, cities are being built at breakneck speed to cope with the demands of an ever-expanding, global population. So concrete buildings will grace our horizons for a long time to come. Which means more concrete for me to stroke.

SKY

Over the years I've worked on a range of projects, from the steel footbridge in Newcastle and concrete apartment blocks in London, to the refurbishment of the brick railway station at Crystal Palace. But skyscrapers have become one of my specialities – which is ironic, because I have no head for heights.

Don't get me wrong: I won't freeze up and go bulgy-eyed, like James Stewart at the beginning of *Vertigo*. I don't collapse into a blubbering mess when I look down from a great height, even if my legs have turned to jelly. But there's no doubt that it makes for some uncomfortable moments at work. Most days, I'm safely sat at a desk inside an office (reassuringly low down on the ninth floor). But sometimes I have to don the classic clobber of my profession – hard hat, hi-vis jacket, steel-toed boots – and climb up a structure I've been designing.

So it was with a mixture of excitement and anxiety that I got off the train at London Bridge in May 2012, took a right out of the station and walked up the street towards a plyboard door painted bright blue – a part of the site hoarding, ignored by the thousands of commuters on their way to work. This was once the entrance to The Shard: a sharp contrast to the gleaming glass and white-steel construction that welcomes you today.

'Spire' made from steel →

Internal concrete spine for stability

Hotel made from concrete →

Offices made from steel →

Foundation made from concrete →

The Shard is now a landmark in London, England.

Moving past the plyboard portal, I entered a maze of plastic barriers and wove my way through, slightly worried that I'd get lost, as the fenced pathways were arranged differently from the last time I'd visited. Eventually I stepped tentatively into a cage-like elevator – a hoist – that was inclined slightly to match the angle of the tower. It shuddered and groaned then shot up rapidly, while my eyes stayed glued to the building, not daring to look down. (Knowing that The Shard's elevator was the first inclined hoist ever to be stuck to the outside of a tower was cool, but it did nothing to lessen my discomfort.) When the elevator finally ground to a halt, I emerged halfway up the building. It was quiet and deserted, and its skeleton was bare: rust-coloured steel columns towered above a firm, blotchy-grey concrete floor. Resisting the urge to stroke it, I tried to picture what this place might look like when it was full of people, furniture and activity. On that day, it was quiet.

I willed myself back into the hoist, this time rising to the highest level it accessed – the 69th floor. Here it felt completely different. The structure was open to the elements. Metal barriers protected the edge of the building, as the glass was not yet installed. The solitude of the lower levels was replaced by a flurry of activity – workers shouting instructions, pieces of steel clanging, cranes beeping as they lifted beams, and concrete spewing out of quaking pumps. Above me rose the crown of the tower – its elegant spire – which I had worked on. Another eighteen flights of stairs led to the highest floor. It suddenly hit me that this was the first time I'd been able to go there, as it hadn't been finished on my previous site visits. Today was truly special.

At the top step, though, I had to stop. The tapering shape of the tower meant that this level – the 87th floor – was relatively small. Even standing at the staircase, which ran through the centre of the floor, I seemed close to the edge. My stomach churned. I suppressed a rising feeling of fear. Fresh, chilly air entered my lungs as I took calming breaths with my eyes closed. When I felt less dizzy, I opened my eye (that's right, just the one).

I was at the intersection of the sky and humanity. After months of making models, doing calculations and creating drawings, I was finally seeing the project made real. It felt so much larger and more tangible than the sketches on a piece of paper or drawings on a computer screen. This phase of construction is a thrill: a moment when the niceties of false ceilings and floors are missing, there isn't the restriction of a facade, and the general public has never crossed the threshold. To me it felt like having a backstage pass for the rehearsal of a big rock concert – a privileged glimpse of all the stuff that will soon be hidden away and embellished, but which forms the backbone of what we will finally see. Visiting the site filled me with awe for the object we had created. It motivated and refreshed me, and reminded me why I love the creative process of design and construction, particularly for skyscrapers.

*

If you were to draw a graph of humanity's tallest buildings over time, which is exactly the sort of thing I might happily spend an evening doing, you would see that it suddenly shoots skywards around the 1880s. For millennia, the Great Pyramid of Giza (at 146m) held the record as the tallest human-made structure in the world. It wasn't until medieval times that this record was surpassed, by Lincoln Cathedral (160m), which held the title from 1311 until 1549, when a storm snapped its spire. This made St Mary's Church in Stralsund in Germany (151m) the tallest building – until it, too, lost its spire, to a lightning strike in 1647. It was replaced by Strasbourg Cathedral (which was a mere 142m, but by now the Great Pyramid had eroded so much it didn't reach 140m). The real quest for height began in the nineteenth century, when the first skyscraper was erected in Chicago in 1884. Admittedly, at 10 storeys – a mere 42m – it's hardly what we think of as a skyscraper today, but it was the first tall building to be supported by a metal frame. In 1889 the Eiffel Tower became the first building to hit the 300m mark. Since then our ambitions, and our buildings, have soared. It took nearly 4,000 years to beat the height of the pyramids – shaky spires notwithstanding. But in the past 150 years, our structures have grown from about 150m tall to over 1000m.

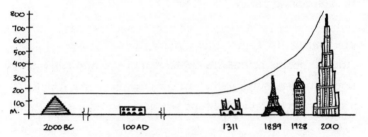

Plotting the heights of the tallest buildings over time demonstrates how technical innovations over the past century have accelerated how high we can build.

Isaac Newton famously said that 'If I have seen further, it is by standing on the shoulders of giants.' Standing at the top of the tallest tower in western Europe (310m), and aware of all the material and techniques that had gone into its making – the clanging steel and beeping cranes to name a couple – I was vividly reminded of how we got here, of the key people in our history who helped unlock the sky. Newton, of course, was one of them: without his third law of motion, for example, I wouldn't be able to calculate the forces at work in an arch. But there are others who pushed us to think outside the box (of simple, single-storey dwellings) and who created the cranes and elevators without which we would still be stuck at ground level or thereabouts. The Shard is built not just on innovative foundations but on a legacy of historical ideas and advances that revolutionised construction and made our skyscrapers possible. For a start, to get a tall building off the ground we have to get things off the ground. Before cranes, the difficulty of this task seriously limited our construction ambitions – until, that is, Archimedes (287–212 BC) invented the compound pulley.

*

The pulley itself pre-dates Archimedes. In approximately 1500 BC people of the Mesopotamian civilisation (in what is now Iraq) used single-pulley systems to hoist water. A pulley is a suspended wheel with a rope wrapped around it. One end of

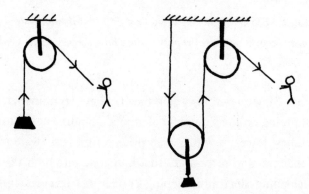

Simple (above left) and compound pulleys (above right).

the rope is tied to the heavy object that needs to be lifted – like a bucket – while a person pulls on the other end. It was a very practical tool, because you could lift objects while standing on the ground and pulling downwards, using gravity to help you. Until the pulley was invented, you had to find a level that was higher than your object's destination and pull upwards. The pulley changed the direction of the force, which meant we could move larger loads.

Archimedes, however, had a restless imagination that he applied to mathematics, physics and even weapon-making, as well as engineering. He improved the pulley by wrapping the rope around not one wheel but several. With one pulley, the force you have to exert to lift a load of a certain weight is equal to that weight. So a 10kg mass needs a force of 10kg × 9.8m/s² (the gravitational pull), which equals 98N. (The N stands for *newtons*: named after the scientist, and another reminder of how key a figure he is for engineering – without his law of universal gravitation I wouldn't be able to make this calculation.) The amount of *energy* you expend is the force you've applied multiplied by the distance. With a single pulley, if you want to lift this weight by 1m, you have to pull the rope 1m as well, so the energy you've used is 98N × 1m = 98Nm (i.e. newton metre).

If you use two pulleys, however, while the energy you expend must remain the same (since you're moving a fixed weight by a fixed amount), you halve the force needed. The reason for this is that the weight is now supported by not one but two sections of rope. Each section of rope needs to move by 1m to lift the weight by 1m, which means you have to pull the rope by 2m. Since the energy is the same, but the distance is doubled, the force you apply is halved. The same principle applies for three pulleys, or ten.

Archimedes made a radical claim to his ruler, King Hiero II, that *any* weight could be moved using his compound pulley system. Unsurprisingly, Hiero was sceptical and demanded that Archimedes prove it. One of the largest cargo ships from the king's arsenal was heavily loaded with people and freight. Hauling it to the sea with ropes normally took the full strength of dozens of men, but Hiero challenged Archimedes to do it alone. Watched by the king and an assembled crowd, Archimedes set up an arrangement of pulleys, wrapped a rope around them, attached one end of the rope to the ship, and pulled on the other. According to *Plutarch's Lives* (biographies believed to have been written in the early second century), 'he drew the ship in a straight line, as smoothly and evenly as if she had been in the sea.'

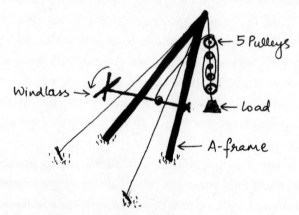

A Roman crane using a five-pulley system.

The Romans recognised the multiple pulley's potential and developed it further by incorporating it into their cranes. Two staves of wood arranged in an inverted V formed the crane's skeleton. The top ends of the staves were fixed together with an iron bracket and the base was anchored to the ground. Between these two staves a rod would be set horizontally (creating an A shape) to act as a windlass: i.e. a rope could be attached to it and then raised or lowered by rotating it, just like the apparatus used to operate a bucket in a well. Fixed to the top of the crane was a two-wheel pulley block; a rope was threaded from the windlass through this and into a third pulley positioned just above the load. At either end of the windlass were four handle-like spikes that could be used to turn it, thereby raising or lowering relatively big loads with ease. If the Romans had to lift something larger they added more pulleys and more rotating sections, and replaced the four turning spikes with a large wheel called a *treadwheel*.

Using a crane with pulleys, a Roman labourer could lift loads 60 times heavier than an ancient Egyptian could handle. And although they are much bigger, the cranes we use today still work on the same principle. Long, square hollow pieces of steel are assembled into a frame to form a very tall tower, and a long arm or *jib* is attached. The jib holds the all-important multiple pulley system, and the human muscles and spike-handles of the Roman version are replaced by petrol power. The jib moves right and left, through 360°, carrying multiple tonnes of steel or glass, attached safely to the modern version of Archimedes' invention.

*

By understanding the potential of cranes and arches, the Romans were able to build bigger. But their abilities were matched by their ambition: they were prepared to think bigger as well. As their empire grew, and the population along with it, the Romans found their towns expanding into large cities. To fit everybody in they built *insulae:* the ancient equivalent of apartment buildings, up to an unprecedented 10 storeys tall. (The pyramids were of course much taller, but you certainly couldn't live in them.)

Spreading across an entire block of the city, the *insulae* were surrounded on all sides by roads (appropriately enough, *insula* means 'island'). Instead of a central atrium for light and air, which was typical in most private homes at the time, the *insulae* had windows facing outwards at the city: in effect they were turned inside out. The first storey was built by installing many columns and then spanning relatively shallow arches between them. Concrete was laid over the curved tops of the arches to level them off and create a floor. Without the arch, far more columns would have been needed to support the floor beams, which would have created even tinier, more obstructed rooms.

To go higher, the Romans layered on more columns and arches. For the first time, they had to consider the design of foundations to ensure that their large, heavy structures didn't sink into the ground. After studying the type of earth present under the proposed building, they constructed foundations made from stone and concrete to hold the structure up.

The most expensive, sought-after apartments were on the ground floor. The higher you went, the smaller and cheaper the dwellings became – which is of course the opposite of today: the height of luxury (literally) is a penthouse that will cost you a small fortune. The *insulae* were rather harried places: without elevators, residents had to trudge the stairs to the upper floors. Since water couldn't be pumped that high, they had to lug clean water up with them, and drag their waste back down (although many would simply throw it out of the window). En route you might even encounter an animal: a cow is said to have wandered up to the third storey of such a block.

The *insulae* were noisy: even after glass windows were invented and replaced shutters, they couldn't keep out the constant commotion of Roman street life. Before dawn, the bakers were out clanging their ovens. Later in the morning, teachers would be shouting out their lessons in the squares. All day you could hear the constant hammering of the gold beaters, the jangling coins of the money changers, the cries of beggars and of vociferous shopkeepers trying to strike a bargain. At night, dancing, drunken sailors and creaking carts added to the din. But worse than the noise and lack of sanitation was the fear that your building might collapse or burn down, as happened to a number of poor-quality blocks. The emperor Augustus instituted an early form of planning restriction, limiting the maximum height to about 20m (later adjusted by Nero to just under 18m), but these regulations were often disregarded. Despite the discomforts, by AD 300 the majority of Rome's population lived in *insulae*. There were over 45,000 such buildings, and in contrast, fewer than 2,000 single-family homes.

For the first time in history, practical tall structures for hundreds of people, spread over many storeys, were built. It was a revolutionary idea – although it must have been a disconcerting experience for the first inhabitants, rubbing shoulders with their neighbours, and a bizarre sight for outsiders unaccustomed to this new way of living. This, though, was the future.

This idea – humans living in layers on top of one another – was the start of what would eventually become the skyscraper.

*

Archimedes took the Mesopotamians' pulley and improved it. Similarly, the Romans took Archimedes' innovation and applied it in new ways, creating heavy-duty cranes in the process. But advances in engineering don't come just from picking up a tradition or innovation and taking it forward. Sometimes they are about breaking with tradition and thinking the impossible. I admire Leonardo da Vinci (1452–1519), for example, who envisioned flying machines, mechanical knights and even a famous concept for a bridge (made from short ladder-like units that could be assembled and disassembled quickly). Another such thinker was Filippo Brunelleschi (1377–1446), who singlehandedly – and, as you'll see, single-mindedly – created one of the most famous domes in Renaissance architecture, and revolutionised construction in the process by building it without a supporting framework. Not bad for a man after whom people shouted, 'There goes the madman!'

By Brunelleschi's time, work on the Cattedrale di Santa Maria del Fiore in Florence had already been under way for more than 100 years. An edict of 1296 had proposed the construction of an edifice 'so magnificent in its height and beauty that it will surpass anything of its kind built by the Greeks and the Romans', and building began that same year, following designs by Arnolfo di Cambio (who was also responsible for two other great Florentine landmarks, the Basilica di Santa Croce and the Palazzo Vecchio). Despite the edict's grandiose assertions, enthusiasm and civic energy – not to mention cash

– waxed and waned in the following decades, and as a result it wasn't until 1418 that the cathedral was finished – except for its dome. During construction, little thought had been given to how someone might place a dome on what was, for the times, a massive hole of 42m.

Brunelleschi's Duomo in Florence, which caps the Santa Maria del Fiore cathedral in this Italian city.

Brunelleschi grew up close to the building site and its unfinished cathedral. Construction had been going on for so long that one of the streets by the site was now called Lungo di Fondamenti: 'Along the Foundations'. As an apprentice, he learned to cast bronze and gold, forge iron and shape and form metals. He later moved to Rome to study the techniques of his ancestors, the ancient Romans. Brunelleschi had always been drawn to engineering and made two resolutions as a young man: to revive architecture to the greatness of ancient Roman times, and to provide a dome for the cathedral. The chance to fulfil both resolutions presented itself when the authorities in charge of the structure ran a competition to find a suitable candidate to build the dome. But Brunelleschi was unlikely to win unless he could overcome the hostility his radical ideas engendered in lesser imaginations, and diplomacy was not his strong point. (On one occasion a committee reviewing his designs had him forcibly ejected from their presence and thrown into the piazza, which is what earned him a reputation as a madman.)

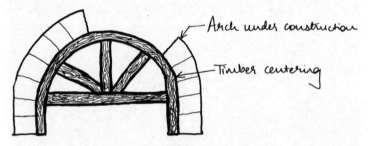

The construction process of building an arch whereby a timber centering allows the stones to be placed in position, finishing with the all-important keystone.

It's perhaps easy to understand why people denounced Brunelleschi's claims that he had a new method of construction. For thousands of years, arches – and domes – had been built in the same way. Carpenters made a timber template or *centering* to match the shape of the underside of the arch. Stonemasons or bricklayers carefully added material around this template, often gluing the masonry together with some form of mortar. They started by laying brick or stone from the base, working their way slowly towards the centre of the arch. The final stage was crowning the arch with a *keystone*. Until the keystone was placed, the curved arms that sprung up from the base remained disconnected. The timber centering supported them; without it the arch would simply have collapsed. Once the keystone was placed, the pathway for the compression loads was complete, and the arch was stable. The centering could then be removed and the arch would remain standing. The construction of domes followed the same process, but used a hemispherical timber centering.

Everyone believed this was the only way to build a dome. Brunelleschi disagreed. He presented a model to the committee that was 2m wide and almost 4m high, made from 5,000 bricks, which he said had taken just over a month to complete and had been built *without* using centering. The claim was met with scepticism, especially since he refused to tell anyone how he had done it.

The panel of judges tasked with choosing the dome's final design repeatedly asked him to reveal his methods, but Brunelleschi refused. At one of the judging meetings, where a number of experts were present and also bidding for the commission, he asked for an egg to be brought into the room. If any of his rivals could make the egg stand on its end, he said, they should win the competition. One by one people took the challenge, and failed. Brunelleschi then tapped the egg hard on the table and left it standing where it was (with a partially broken shell). When the others protested that anyone could have done that, had they known they could break the shell, he countered: 'Yes, and you'd say the same thing if I told you how I intend to build the dome.' He won the contract – though possibly only because there were few other practical solutions. (One person had even suggested filling the cathedral with earth to support the dome during construction. After the dome was completed, the earth would be cleared by small boys eager to get hold of coins deliberately mixed in at the outset.)

I visited Florence when I was a physics student. With the Ponte Vecchio, Giotto's Campanile, the Baptistery and Santa Felicita, it's like an open-air museum of medieval and early Renaissance engineering. *Il Duomo*, as the city's cathedral is affectionately known, is of course one of its centrepieces. I stood outside for a while, taking it all in – the neat symmetry of its three doorways, separated by four tall columns (with another two up above), and a series of very intricate carvings of Mary and the apostles just below the largest of the rose windows. Circles, pointed arches, triangles and rectangles, with coloured bands of stone, came together in pleasing geometric chaos. Eventually I passed through the doorway and my eyes were immediately drawn to the underside of the dome, high above me.

The base was an octagon, and each side had a circular stained-glass window letting in shafts of light. More light entered through an *oculus* at the top of the dome. Above the stained-glass windows were spectacular frescoes depicting the Last Judgement – choirs of angels, saints and personifications of the virtues vied for attention amid layers of painted cloud. It was all lovely, but the scientist in me wanted to know how it worked, to see the dome behind its beautiful embellishments.

The best view of the dome is from Giotto's bell tower, which stands in the piazza near the western corner of the cathedral. The 414 stone steps tested my fitness, but eventually I found myself at the top, looking out at the bank of deep red terracotta tiles and a few of the eight white ribs that define the dome's shape. It was a thrilling viewpoint, and a fitting tribute to Brunelleschi's genius. For me, it's Brunelleschi's unconventional thinking, coupled with the courage to make it a reality, that makes him relevant to modern engineering. It's by thinking beyond the orthodoxy and imagining the 'impossible' that we move engineering forward.

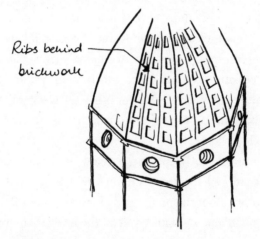

The skeleton of the Duomo that lies between the two layers of brickwork, Brunelleschi's innovation.

Brunelleschi drew the ribs in characteristically detailed sketches. The ribs were made from stone, acting as arches that landed on the eight corners of the hole. These arches supported the edges of the octagonal dome. Between the main eight stone ribs were a further sixteen designed to resist the force of the wind. I couldn't see these from the outside, because Brunelleschi hid them away in the hollow space between two layers of brick skin. By creating this hollow space, not only was he able to hide the secondary ribs, he could also reduce the weight of the dome to half of what it would have been if it was solid. This reduction in weight helped him build the dome without centering.

Brunelleschi had gone back to basics. Brick structures are traditionally built in layers, comprising brick, then a layer of mortar, then another layer of brick, and so on. Imagine a simple garden wall and you've got the idea. Say, however, that you need this wall to curve in towards you (unlikely, I know, but bear with me). At that point, the problems begin: as the wall curves and becomes taller and heavier, it's in danger of overloading and cracking. Mortar is usually weaker than brick, so the continuous layer of mortar, rather than the bricks, is most likely to fail first.

To counter this, Brunelleschi asked his bricklayers to do something they had never done before. He directed them to lay three bricks horizontally, and then to place bricks vertically, like bookends, at either side of the horizontal group. The next layer again alternated three horizontal bricks with vertical bricks at each end. It was a painstaking process: 4 million bricks were laid; workers patiently waited for the mortar to dry on one layer before they started on the next. The layers created a 'herringbone' pattern, so-called because it supposedly looks like the bones of a fish. As an engineer, I admire this idea because of its simplicity. Since continuous lines of mortar were the weak link, Brunelleschi broke up the lines with vertical bricks, making the curving wall far stronger.

A herringbone brick-laying formation in which the vertically laid bricks add strength.

A similarly innovative approach drove the construction of The Shard. While designing its spine (or core), the team of engineers I worked with devised a unique method to build it. To save time on the construction programme, we decided to work in two directions: digging down to form the basement and at the same time constructing upwards. Usually when you want to make a basement, you dig an immense hole with concrete or steel walls holding up its sides. Piles – long shafts of concrete – are installed at the bottom of the hole to support the future building. Then slabs are poured at each basement storey until you get back up to ground level. It's only at this point that anything can be built above ground.

But we did something unprecedented. We asked for the piles to be installed at *ground* level, and huge steel columns to be plunged into the piles. First, the ground floor slab was built, with a giant hole in it. This hole gave workers access to the soil, then diggers removed earth to expose the concrete piles with steel columns inside them. While digging continued downward, a special rig was attached to the newly exposed steel plunge columns; this rig could build the central concrete core.

As the core rose, the basement and foundations were finished. At one point, twenty floors of the huge concrete spine were being held up just by the steel columns – there was no foundation in place. It was a structure on stilts.

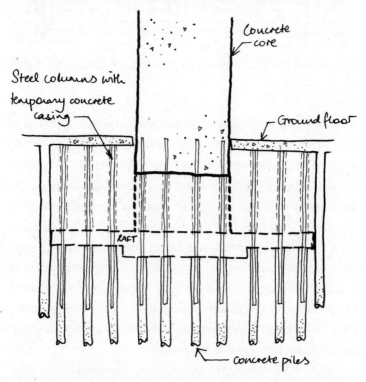

The top-down construction method, which was employed during the building of The Shard, London.

This method, called 'top-down' construction, had been used previously to hold up columns and floors in small structures. But it had never been used on a core, let alone one of this size. It was an engineering first. Our ability to think beyond standard practice saved time and money – we solved a real-world challenge with creativity. Others are now using our idea in their projects – as always, building on existing ideas leads to innovation, whether it's in one of the most famous cathedral domes in the world, or one of the tallest buildings in Europe.

*

On that site visit to The Shard in May 2012, as I shot up the tower in my cage-like hoist to the 34th and then the 69th floors, my eyes glued to the building rather than looking out and down, I couldn't help reflecting on how, without elevators, The Shard – indeed, any skyscraper – simply wouldn't exist. Part of the reason Roman *insulae* stopped at ten storeys was because climbing up and down any further was impractical. Today, we're so used to pressing a button and summoning a mobile cubicle to whisk us up and down our multi-storey towers that we don't give it a second thought. But before the 1850s, elevators in this form didn't exist. And although we started to build skyscrapers fairly soon after the invention of the elevator, such a device wasn't originally designed with buildings in mind, but as a safer way to move materials around a factory.

Like Archimedes, Elisha Otis had a restless and creative imagination. While working in a variety of jobs – carpenter, mechanic, bedstead manufacturer, factory owner – he invented an automatic turner that made the production of bedsteads four times faster; a new type of railway safety brake; and even an automatic bread-baking oven. In 1852 he was hired to clear a factory in Yonkers, New York and, frustrated by the effort involved in transporting materials between floors manually, he turned his attention to how best to accomplish the job mechanically. Methods for moving people and materials from one storey to another had been around for centuries: Roman gladiators, for example, rose from the pits of the Colosseum up into the fighting arena on a moving platform. The problem, however, was that they weren't safe: if the rope shifting the platform up or down suddenly snapped, the platform fell to the ground, probably killing its occupants. Otis wondered if he could fashion something that would prevent this from happening.

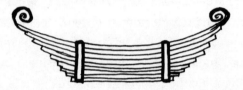

The wagon spring solved the challenges of operating an elevator.

His idea was to make use of the 'wagon spring': a C-shaped spring made up of carefully layered thin steel strips that was commonly used to improve the suspension in carriages and wagons. When it has force on it, a wagon spring is almost flat, but when it's released, it curves. It was this change of shape, caused by force, that Otis planned to use to his advantage. First, he replaced the smooth guide rails (which kept the platform in position during its progress up and down) with toothed or ratcheted rails. Then he created a mechanism in the shape of a goalpost, which had a hinge in the middle and feet sticking out at the base. He attached the spring, then the goalpost, to the rope at the top of the elevator car. When the rope was intact, the spring remained flat and the goalpost square. If the rope was cut, the spring sprung into a C-shape, pushing down on the goalpost and deforming it so that its two 'feet' stuck into the ratcheted rails, bringing the elevator to a halt.

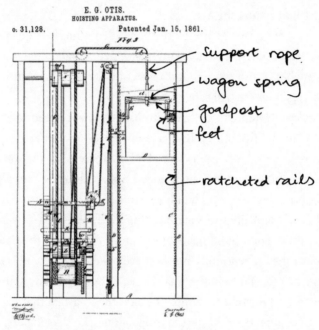

This diagram is included in the patent documents for the Otis Elevator – or 'hoisting apparatus'.

But to bring his invention to the attention of the public, and show them that it worked, Otis needed a big stage – and he found it at the 1853 World's Fair in New York. Entitled the 'Exhibition of the Industry of All Nations', the exposition aimed to show off American technological might, and showcase industrial innovation from around the world. In the vast exhibition hall Otis constructed his elevator with guide rails, ratchets, springs, platform and hoisting machinery, and loaded the platform with goods. When a crowd had gathered, he climbed on top of the platform and had it lifted to its maximum height. As the crowd looked on, he called for the hoisting-rope to be cut, and his assistant swung the axe.

There were gasps as the platform suddenly lurched downwards. And then, just as suddenly, it stopped. It had fallen only a few inches. From the top of it Otis could be heard shouting, 'All safe, gentlemen. All safe.'

Four years later, Otis installed his first steam-powered safety elevator in the five-storey E.V. Haughwout & Co. department store on the corner of Broadway and Broome Street in New York. The eponymous company he founded has continued to supply elevators and escalators to buildings around the world, from the Eiffel Tower and Empire State Building to the Petronas Towers in Malaysia. Such buildings would hardly have been possible without Otis's invention. Until he developed the safety elevator, the height of a building was restricted by how many stairs people were prepared to climb. The elevator smashed that barrier and engineers could start to think about true skyscrapers.

Since then we've been building higher and higher, and we now have the opposite problem: we can't make elevators that travel much further than 500m because the steel cables to hoist them up and down become too heavy for the machinery to work efficiently. It's one reason why elevators often don't go all the way to the top of very tall towers. You go up a number of floors, then change elevators to go up the rest. But engineers are already exploring ways to solve this by using different materials. Replacing steel with carbon fibre – which is stronger but lighter – seems one way forward, but questions remain about how well the carbon fibres can resist fire. As our towers continue to grow, these innovations will be much needed.

Another challenge in super-tall towers is sway. In chapter 1 I talked about controlling the movement of buildings to stop us feeling sick. But there is another reason this control is needed. Elevators run on straight guide rails, and as towers move the elevator shafts and the guide rails fixed to them curve. A small amount of curve is not a problem – the cogs and clasps of the elevator car on the rails have a little give – but too much and the car will grind to a halt, unable to move. The taller buildings become, the more they move and the more curve you experience in the elevator shaft. There are solutions to the problem, ranging from upgrading the elevators themselves, to allowing more give, to stopping elevators running in the worst storms. Ultimately, I'm sure, a modern-day Otis will come up with an ingenious solution. And he – or she – will have to, because the elevator has become an intrinsic part of our everyday life. The equivalent of the entire world's population is moved in an elevator every 72 hours.

*

I was reminded of Elisha Otis during my visit to the Burj Khalifa in Dubai, the world's tallest building (at 829.8m), because his company installed the elevators that were about to take me to the observation deck on the 124th of its 163 floors. It was a more serene journey than my trip up the outside of the tallest tower in Western Europe in a cage-like hoist, although the floor number on the LCD display changed with a bewildering rapidity as we ascended at 36km/h.

(Elisha Otis's original elevator in the E.V. Haughwout Building climbed at just over 0.7km/h.) A minute later I emerged to an unparalleled view. On one side, pure sand extended beyond the buildings to the horizon. On the other, I could see the blue sea and, far away to the left, the cluster of man-made islands that form the famous leaf shape of the Palm Jumeirah. Steeling myself, and feeling protected by

Burj Khalifa in Dubai, the world's tallest building in 2018, which has been made possible partly by the developments in elevator technology.

the floor-to-ceiling glass, I ventured closer to the edge and looked down. Beneath me were a number of tiny, futuristic-looking buildings, like scale models on the set of a sci-fi film. It was a shock to realise that these structures are actually taller than most of the skyscrapers in Europe, and many even in the US. The Burj Khalifa dwarfs everything around it, and plays havoc with your sense of proportion.

'Megatall' skyscrapers like the Burj Khalifa were made possible by a man who started life as a mischievous and lively-minded young boy, born in Dhaka, Bangladesh, in April 1929. Fazlur Khan disliked traditional schooling methods: his inquisitive questions were met with stern responses from teachers; as a result, he didn't take education very seriously (even though his father was a mathematics teacher). Fortunately, his patient, forward-thinking dad realised that his son needed a broader education, and was determined to further his intellectual curiosity while fostering a sense of discipline. He set Fazlur problems similar to those in his school homework, but which made the boy consider solutions far beyond what the homework asked for; he also challenged him to solve the same problem from multiple perspectives. When the time came for Fazlur to choose whether to study physics or engineering at university, his father guided him towards the latter because, he said, it demanded discipline and would require him to wake early for lectures. (In fact, as I can attest, a physics degree involves a lot of early-morning lectures too.) Khan gained a degree in civil engineering at Dhaka University in 1951, finishing first in his class, and went to the US on a Fulbright Scholarship in 1952. In the next three years he acquired two master's degrees and a PhD, while also learning French and German.

It was Khan who came up with the idea of putting a building's stability system on the outside – a brilliant innovation that has since been used on iconic structures around the world, from the Centre Pompidou and the Gherkin to the Hearst and Tornado Towers. Using large pieces of diagonal bracing to form strong triangles, Khan created a stiff external skeleton, effectively turning traditional skyscrapers inside out. This system is often called a 'tubular system' because, like a hollow tube, the outside 'skin' of the structure gives it strength, although the shape of the skin doesn't have to be cylindrical.

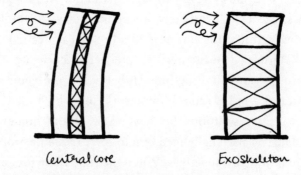

Central core   Exoskeleton

An alternative stability system for buildings is to forgo the conventional central core and instead employ an exoskeleton.

Khan's first commission to employ this concept was the DeWitt-Chestnut apartment building in Chicago. But the real showcase for his novel approach was the completion in 1968 of the city's John Hancock Center which, at 100 storeys (344m), became the second-tallest skyscraper in the world after the

Empire State Building. It is a rectangular cuboid with gently tapering faces, making it narrower at the top than at the base. On each face you can see five giant 'Xs', one on top of the other, that form the bracing for the tower. Fifty years on, its eye-catching design still looks modern and elegant. The pioneering design earned Khan the catchy title 'father of tubular designs for skyscrapers'.

The John Hancock Center in Chicago utilises an exoskeleton to give the tower stability.

The external skeleton was only one of Khan's ideas. He also suggested combining many such skeletons in a cluster. This is like holding a bunch of straws in your hand: each straw is a single tube which by itself is stable up to a certain point; by bunching lots of straws together, however, you can make a much stiffer and more stable structure. The Burj Khalifa employs a variation of this system. Look at a cross-section of the structure and you'll see that it has a distinctive tripartite shape that resembles leaves or petals. (It's become a kind of brand image for the building: as you ascend in the elevator, a light show of row upon row of the shapes dances across the walls in different configurations.) The 'petals' are in fact a series of 'straws' or tubes with exoskeletons of their own which – in their cluster – support one another. This mutual support between the individual pieces means that the tower remains stable despite being so high.

The key to building higher is stabilising the structure externally rather than internally. Perhaps the most precarious experience I can think of was the only skiing trip I ever went on. At first, our instructor wouldn't let us use ski poles, so I had to stop myself from falling over using just my feet. I soon lost count of the number of times I fell over, and the number of bruises I picked up in the process, but once I had managed to stand upright – at least for a short while – I was allowed the poles. And what a huge difference they made: by spreading my arms out and using the poles to stabilise myself, I found I could stay upright longer. Although the poles were much skinnier and less stiff than my legs, by putting them further apart than my feet could reach, I was more stable.

Tall towers with exoskeletons work in the same way: by spreading the stability from a small internal area (analogous to my feet or a building's core) to an outside area (the poles or the exoskeleton), it's possible to create much more stable buildings. Flipping the structure around in this way opened up a number of engineering possibilities: if you built a tower of 50 or 60 storeys like engineers did at the turn of the twentieth century, you could use much less material, making it cheaper. Or if you used the same amount of material as in the older towers, you could build much taller. So, from the 1970s onwards, scores of tubular towers arose, from Hong Kong's Bank of China Tower and the original World Trade Center Towers in New York to the Petronas Towers in Kuala Lumpur, changing the face of our skylines forever and creating the classic modern-city silhouette.

*

With the invention of new building techniques, structural systems, and computing power increasing every year, it's an exciting time to be a structural engineer. Just as the height of buildings has increased on the back of what we've learned from our predecessors, so too has the depth of our knowledge. Today, I can design structures that brilliant thinkers like Leonardo da Vinci struggled with. And in a hundred years, engineers will no doubt find it easy to do things that I struggle with now. My peers and I are building on thousands of years of engineering gifted to us by Archimedes, Brunelleschi, Otis, Khan and countless others.

With today's technology at our fingertips, I don't believe there is a limit to how high we can build. We've beaten so many physical, scientific and technological restrictions over the past 4,000 years that with strong enough materials, a wide enough base, solid enough ground – and, I suppose, enough money – I see no reason why we can't go as high as we want. The real question is: how high do we want to go? A wide base would probably mean very little daylight in the middle of the vast floors. Strong large columns and beams could mean restricted spaces in which to live and work. And what about the safety and convenience of the inhabitants: how long would you need to wait for an elevator, and how would you evacuate tens of thousands of people from a mammoth building?

Technology can undoubtedly take us there. New super-strong materials like graphene are already being synthesised in labs; cranes are getting larger; and new techniques like top-down construction are constantly being used in inventive ways. Science and engineering are leading to the creation of the mega-skyscraper – the Wuhan Greenland Centre (636m) in Wuhan, China; the Merdeka Tower (682m) in Kuala Lumpur, Malaysia; and the dart-like Jeddah Tower in Saudi Arabia, which will be the world's first building to reach a height of 1km – at an unprecedented pace.

But where does it all stop?

The highest I've lived is on the 10th storey, and I loved the view and the new perspective of the city in which I lived. But I wonder how I would feel living much higher than that. In cities like Hong Kong or Shanghai, living on the 40th floor is common for thousands of people: it's something the residents are used to. Eventually, perhaps, it will be commonplace every-where: people are moving to cities in droves, and building high is a good way to fit all of us into an increasingly limited space.

The rapid growth in the height of buildings in the last century has barely given us a moment to consider if we *like* being so high above ground. But now, rather than racing ever higher, we are now stopping to think about our desires. It's about what we *want* to build, not what we *can*. After a spate of building high towers from the 1960s to the 1980s, architects and engineers are questioning what type of buildings are really best for people and the environment. Cultural factors also play a part: different countries are at different stages in their urban development, and can have very different views about whether onwards and upwards is the best approach. I believe that, at some point in the future, the average height of our towers will plateau. Sure, iconic towers will still be built and they will continue to break records. Ultimately, however, our humanity will hold us back from the mega-tall. We want to live with sunlight and air flowing into our homes, and a connection to the earth and to our roots. We might gaze upwards at our structures and marvel at them, but we also need to feel grounded.

EARTH

Mexico City is built on a lake.

It started off as a small island but gradually expanded. The city now spreads far beyond its original site, but the centre of town, which contains most of the historical Aztec and Spanish buildings, sits on that lake. Twenty-eight metres down, the earth is strong and solid; everything on top of that is loose soil that was added later, and the result is very soft, very wet and very weak. It was described to me as a 'bowl of jelly with buildings on top'.

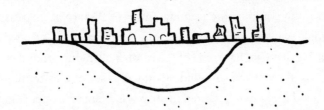

Mexico City, which is built over a lake.

And so the historical centre of Mexico City is sinking. Fast. In the past 150 years it has subsided by over 10m – that's more than a three-storey building.

*

When I was invited to Mexico to give a talk about my career and designing tall buildings, I jumped at the chance, not least because there was so much I wanted to see: the National Museum of Anthropology, the Bosque de Chapultepec, the ancient pyramids at Teotihuacan, and of course the Torre Latinoamericana, once the tallest skyscraper in Mexico City and still one of the best places to appreciate the sheer sprawling vastness of the metropolis. Naturally, I was also keen to explore the unique ground that lies below the city, and the bizarre effect it has had on the buildings there.

In engineering, what lies beneath the surface is just as important as what we can see above it. After all, you can have a well-designed superstructure (the bit above ground) – but if it's not supported by an equally well-designed, stable substructure (the bit below the ground); if the layers and condition of the soil being built on aren't properly understood; if you don't build correctly within that ground – then the structure won't be stable. The end result could be the Leaning Tower of Pisa. (Not the reason I would want tourists flocking to one of my buildings.) Knowing that Mexico City has some of the most challenging ground conditions in the world for building on – plus seismic susceptibility for good measure – I figured my trip was a fantastic opportunity to hear directly from the experts how they keep the city standing straight.

The site of the city was determined by a vision. The Aztecs were told by their god Huitzilopochtli (the God of War and the Sun) that they must move from their highland plateau, and that their new capital must be located where they found an eagle with a snake in its beak sitting on top of a nopal cactus (an image that is now the emblem on the national flag). The Aztecs set off and, after searching for just over 250 years, they found the eagle their deity had foretold. The fact that it was sitting on a tiny island in the middle of Lake Texcoco didn't seem to trouble them (although I can imagine the tribe's engineers cursing under their breath as they surveyed their new, watery building site).

Tenochtitlan, which means 'place of the nopal cactus', was founded in 1325. In its heyday, it was a beautiful city with fertile gardens, canals and massive temples, and its rulers commanded vast swathes of land. To connect the island city to the mainland, the Aztecs built three large causeways by pushing wooden logs vertically into the lake, and then creating pathways on top with soil and clay. These causeways are now the main roads that run through the historical centre of the modern city.

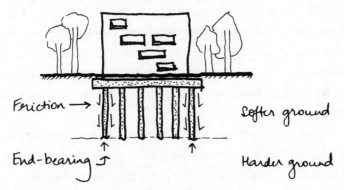

Piles holding up buildings in soft ground.

The logs are examples of *piles*. They come in various shapes and sizes but share a common principle: they are columns put deep into the ground to help support the structure above them. If the ground is soft and not strong enough to support the weight of the structure, piles work to channel that weight in such a way that the soil is not overwhelmed. The ancients generally used tree trunks, but modern piles supporting larger structures are usually made from concrete shaped into cylinders, and sometimes from steel, cast in circular tubes, H or trapezoidal shapes. The foundations of the structure are built at the top of these piles and connected to them through steel bars.

Piles can channel forces into the ground in two ways: by means of friction between the surface of the piles and the soil, or by dumping forces at their base ('end-bearing piles'). Depending on the weight and type of structure being supported, you can have multiple piles, which can vary in length depending on the forces they feel and the type of ground they engage.

Friction piles exploit the friction between the surface of a pile and the ground to carry the load or weight coming from the structure. The more piles you have, the more surface area is in contact with the ground, and the more friction is created. This friction force resists weight – thinking about it in terms of Newton's third law, it is an upward reaction to the downward action of the superstructure.

Sometimes the ground is too loose to create friction against a pile, and then end-bearing piles are used. These are made long enough so that they poke into a deeper, stronger layer of ground. The load in the piles flows into their bases and dissipates into the earth.

In fact, piles don't have to be either friction piles or end-bearing ones: they can be both. Some soils, such as clay, have good friction capacity because they bond to the pile. But say the load is so large, and you're so restricted by available space, that friction alone isn't enough to resist it. In that case you can make the piles long enough to reach a stronger layer of ground. In London, for example, there is a highly compact layer of sand approximately 50m deep that we drill down to for larger structures.

Working out how many piles to use, and how big to make them, is an important part of the engineer's job. The starting point is the soil-investigation report, which tells me what the different layers of ground are, and how thick and strong they are. Then, if I find that a 'pad' of concrete will not be enough to stop the structure sinking, I'll choose to use piles. By consulting the information in the report – and geotechnical engineers – I can calculate how deep the pile needs to be to hit a strong layer, and what the friction properties of the various layers are.

I then have to decide on diameter. A small-diameter pile has the benefit of being cheaper and easier to install, but it may not be strong enough for the job. A larger-diameter pile has a bigger surface area, which increases the amount of friction; the area of the base is also bigger, making it stronger. The calculation is a search for the right compromise. I choose a diameter, calculate how much load a single pile will take based on a chosen length, then divide the total weight of the building by the capacity of one pile to work out how many piles I need. If I can fit that number of piles below the structure, then we can go ahead. If not, I make the pile bigger and repeat the calculation. For a 40-storey tower I designed near Old Street in London, we arrived at a total of about 40 piles between 0.6m and 0.9m in diameter, with some more than 50m long where the loads were greatest. Many modern skyscrapers are held up by piles that work by friction alone (if the ground is good enough so the piles can carry the loads they need to). But the piles in this tower work both by friction and by using end-bearing, as London's clay is relatively weak to quite a depth.

Putting piles in the ground is a big challenge in itself. It wasn't really until modern mechanisation that the huge piles we can now install were possible. Now, piles are often built using a sort of giant corkscrew that twists deep into the ground then reverses out, bringing the soil with it, and leaving a hole that is later filled with concrete. While the concrete is still wet, a steel cage is plunged in to reinforce the pile. For centuries, before mechanisation, most engineers simply pushed piles into the ground, as the Aztecs did at Lake Texcoco. From an engineering point of view their construction was successful, standing firm for the next two centuries.

But then the foreigners arrived.

The Spanish captured Tenochtitlan in 1521, razed it to the ground, and then rebuilt the city on the foundations of the Aztec pyramid temples. They cut down trees around the lake, causing mud slides and erosion that made the lake bed shallower. The water levels rose and the city flooded frequently throughout the seventeenth and eighteenth centuries, causing chaos and devastation (after the flood of 1629 the city was underwater for five years). Eventually, the lake was filled with soil to allow the city to expand, but it still suffered regular flooding because of the high level of water naturally present in the ground.

There is a level in the ground below which natural water flows and saturates the earth: this is known as the *water table*. Dig a hole in an area where the water table is high, and you'll find that the hole fills with water pretty quickly: this is like the original Lake Texcoco. If you fill the hole with earth – which is like Lake Texcoco being filled with soil – then sprinkle on water to simulate rain, eventually water will puddle above the soil (just as our gardens are covered with puddles after a storm because the soil is saturated). This is what happened in Mexico City. The lake was filled in with soil but the water had nowhere to go. Then, the moment it rained, the rain added to the underlying water table and stagnated in the streets of Mexico City. It wasn't until the twentieth century that the flooding was controlled using a huge network of tunnels that led the extra water away. But the legacy of building on such unpredictable, unstable ground can still be seen in the modern city.

*

Standing in the courtyard outside Mexico City's enormous and very grey Metropolitan Cathedral, I scanned the crowds for Dr Efraín Ovando-Shelley, a geotechnical engineer who, according to his photo, wore sunglasses and khakis that made him look a bit like Indiana Jones. The solid, ordered columns of the cathedral were in sharp contrast to the delicate carvings between them, but what really caught my engineer's eye were the cracks in the building. I could see where black space had opened up in the mortar and stone bricks, and the two huge bell towers that flanked the main entrance didn't seem to be completely vertical. But such considerations were cut short when, at exactly the appointed time, Dr Ovando-Shelley appeared wearing his sunglasses, greeted me, handed over a book he had written, and led me towards the cathedral for a very unusual guided tour.

Metropolitan Cathedral, Mexico City.

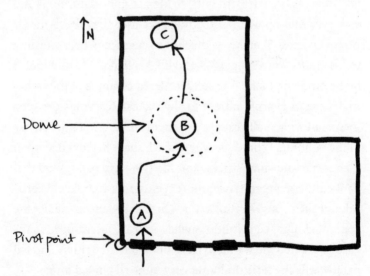

Map of the Metropolitan Cathedral.

As soon as we stepped through the entrance, *(Map, point A)* something felt very odd to me. Swarms of tourists stood rapt by the grandeur of the place, while worshippers sat respectfully hunched in its polished wood pews. But my attention was drawn to the floor. As we moved towards the back of the cathedral, I felt like I was walking uphill. And I was – because of the uneven or 'differential' settling of the ground that has taken place through history, the floor of the cathedral slopes upwards.

Construction of the cathedral began in 1573, on top of the foundations of an Aztec pyramid. The architect, Claudio de Arciniega, knew of the problems with the ground and designed a clever foundation to deal with them. He started by driving more than 22,000 wooden stakes – each 3m to 4m long – into the ground, to 'pin' the soil together and compact it. Imagine a box of sand with lots of kebab skewers pushed into it in a grid pattern. If you shake the box, you'll find that the sand moves around far less than if the skewers aren't there. The stakes performed a slightly different function from piles, since they weren't designed to take the weight of the cathedral, but rather to strengthen the soil.

Following this, the builders erected a massive masonry platform above the stakes. It measured 140m by 70m – about the same width as a soccer pitch but one and a half times longer – and was about 900mm thick. Huge beams were laid on top of this platform in a grid pattern – a bit like a waffle – in such a way that the columns and walls of the cathedral could sit on top of them. The tops of the beams would eventually form the floor of the cathedral, spreading the weight of the columns onto the masonry platform, which in turn would spread the weight over the ground. This sort of foundation (with or without the large beams) is known as a 'raft' foundation.

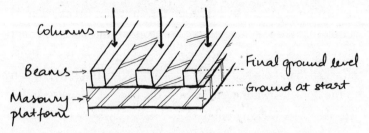

Columns →
Beams →
Masonry → platform

Final ground level
Ground at start

The layers that form the raft foundation of the cathedral.

It does what its name suggests, which is to 'float' on top of the ground. When building on soft ground, the key is not to put large concentrated loads on the soil. If you do, it's like standing on mud in stiletto heels. As many summer wedding guests will know, a sharp heel sinks into the ground because the pressure it exerts on the ground (calculated by dividing force by area) is high. Flat shoes, however, don't sink as easily because the same force is spread over a much larger area – the snowshoe is based on this principle. So the masonry platform in the cathedral acted like a flat shoe on top of mud, spreading the weight of the building over a large area. The trouble, however, is that sometimes the ground is so soft that even spreading the weight of a structure across a large area and avoiding concentrated loads is not enough.

It's probably worth noting here that friction or end-bearing piles were not used to support the weight of the structure. Perhaps because of the pyramid foundations below it, or perhaps because the engineers of the time realised that anchoring piles to the solid layer of earth might cause the opposite problem, making the cathedral rise. In fact, the Angel of Independence victory column in

Mexico City (built in 1910) is supported on piles, and in the 100 years that have passed since it was built, 14 steps have been added to its base as it has become taller relative to its surroundings. Engineers in Mexico City agree that it's best to allow the city's structures to slowly, steadily and uniformly sink.

When it was built, the top of the masonry platform was made level with the ground outside. On top of it were the 3.5m-deep beams, and on top of them was the floor of the cathedral itself. Thus the floor was originally constructed 3.5m above the ground, showing that the engineers knew the structure would sink, and planned that by the time they had finished it would sink just enough to bring the floor of the cathedral down to ground level. The hope was that the structure would sink uniformly, and wouldn't necessarily be damaged. Despite de Arciniega's efforts, during construction, as heavy stone was laid on top of heavy stone, the structure started to sink in a non-uniform way. The south-western corner of the structure (the front-left corner in the diagram) sank more than the north-eastern corner. To compensate for this unsettlingly uneven settling, the builders actually increased the thickness of the 900mm masonry platform on its southern side.

The structural reason why the platform settled unevenly is because soil comes with baggage. It's not enough to meet the soil, ask how it's feeling on the day you start building and then assume it doesn't have any emotions from its past that will affect how it behaves. It has a history and a character that an engineer must consider. The Aztecs had built their pyramid in exactly the place where the cathedral was sited, adding layers to it over time, partly for spiritual reasons, and partly to cover the damage caused by settlement. This construction had affected the physical state of the soil: some areas had already experienced lots of pressure and become consolidated and compacted, while others, which hadn't been weighed down, remained light and less dense. Where new foundations were built on top of consolidated soil they didn't sink much, but the portion built on less-dense soil moved much more.

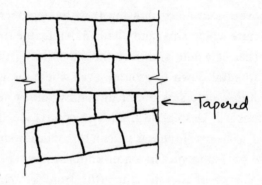

An attempt at realignment.

Even after the Spanish builders had finished the foundations, the structure continued to move unevenly. They tried to compensate for this differential settlement by changing angle as they worked up. Dr Ovando-Shelley pointed out areas where the courses of stone (which would normally be laid flat and uniform) had been cut to a taper. This helped the builders come back to a level line after the layers of stone they had already built had tilted. Other adjustments had been made to counter the continuing subsidence: a column at the southern end of the structure was almost a metre taller than the columns in the north. The cathedral was finished 240 years later, but throughout this time, and beyond, it continued to move erratically.

Dr Ovando-Shelley and I walked along one of the aisles *(see Map on p. 152, point B)* and stopped directly below the central dome. From here hangs a giant, missile-shaped pendulum (or *plumb line*) made of gleaming brass and steel that shows how far the cathedral has shifted. You can simulate this with a string, a small weight and a clear plastic box. Attach the weight to the string, suspend it from the centre of the roof of the box and lay the box on a level table top. You'll see that your makeshift pendulum hangs exactly above the centre of the floor of the box. If you tilt the box slightly, however, the pendulum will move away from the centre. Tilt the box by 45° and the pendulum will hang over the edge of the floor. The Metropolitan Cathedral's pendulum works in the same way: as the foundations tilted, the pendulum stayed vertical. By noting where the pendulum was centred at various intervals over time, the tilt of the cathedral has been monitored.

In 1910 measurements were taken to compare the levels of the two extreme corners. The engineers established that, since 1573, the floor had tilted so much that one corner was a staggering 2.4m higher than the other. It's difficult to imagine a structure tilting by such an extreme amount; not surprisingly, it had a damaging effect on the cathedral's integrity. By the 1990s its bell towers were leaning precariously and in danger of collapsing.

A major restoration project started in 1993; Dr Ovando-Shelley was one of the large team of engineers that worked on it. They accepted that it was almost impossible to stop the structure sinking altogether, but reasoned that if it sank uniformly it would suffer less damage. However, before they could even think about ensuring it settled evenly, they needed to pivot the entire cathedral so it was relatively flat.

As my tour continued, we walked away from the dome to the back of the cathedral *(see Map on p. 152, point C)*. Here, the shimmering Baroque magnificence of the golden Altar of the Kings extended towards the ceiling, covered by a mass of intricate hand-carved figures – an opulent wall of worship designed to assault the senses, to impress, and to arouse reverence. It certainly inspired a feeling of awe.

I, however, was completely transfixed by a tiny metal stud on a column just to the left of the altar. It was relative to this point that the team measured and compared the levels of the floor to establish exactly how much the cathedral needed to be pivoted. The chosen pivot point (the point that wouldn't be allowed to sink any further) was the south-west corner, because this had sunk the most over time. The metal stud was at the northern end of the cathedral, which needed to be pushed down by metres. Just thinking about it made my head spin. And it didn't stop spinning as Dr Ovando-Shelley explained the technique they used to achieve it. Have you seen the sci-fi blockbuster *Armageddon*, in which Bruce Willis and his team must drill a hole in an asteroid and pack it with explosives to prevent a collision with Earth? The plan devised by the cathedral's engineers seemed about as unlikely and difficult to achieve: they would burrow beneath the cathedral and settle the soil. The thought of *removing* earth from underneath a structure to *stabilise* it might seem totally counter-intuitive. But for these exceptional ground conditions, exceptional engineering was needed.

As I said before, though, soil isn't just soil: you have to under-stand its history before you can predict how it will behave in the future. Dr Ovando-Shelley and the team performed a variety of soil tests all over the site to find out exactly how strong or weak the soil was, and how consolidated (or squashed down). Feeding this information into a computer model, they drew a 3D map composed of layers of different colours that undulated and overlapped depending on the strength and type of soil at a particular depth. The model also simulated all the historical events that had affected the soil – from the building of the Aztec temple and the Spanish cathedral to the changes in water level and so on – and created a profile of the ground.

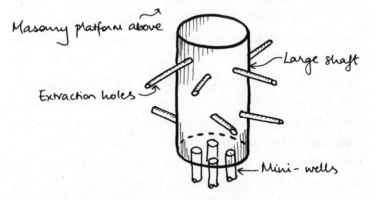

Extraction holes radiating from the large shaft.

The team then bored 32 cylindrical access shafts, 3.4m in diameter and between 14m and 25m deep, through the original masonry raft of the cathedral and into the ground. These were dug laboriously by hand (accessing this confined space with diggers would have been both difficult and dangerous). At each stage of the descent, concrete was cast in a ring around the edge of the hole, creating a tube to keep the soil in place. When the shaft was finished, a second layer of concrete was cast inside the tube to stop the hole from collapsing in on itself. At the base of each shaft the engineers sank four mini-wells from which they could pump out the excess groundwater that would otherwise rise and flood the shafts.

These, though, were not the shafts that were going to save the cathedral. They just provided the means for drilling about 1,500 holes, slightly inclined from the horizontal, with a diameter the size of a fist and between 6m and 22m long, through which soil could be extracted. The plan was that, after the soil had been removed, these holes would naturally close up over time, causing the foundation of the cathedral to settle.

Since the north side of the cathedral was the highest and needed to come down the most, the largest amount of soil was extracted in that area, while much smaller amounts were taken from the south-west corner. More than 300 cubic metres were removed from one north-eastern shaft, whereas only 11 cubic metres were taken from another in the south-west corner. In total, through this vast warren of shafts and tunnels burrowed deep beneath the historical cathedral, and with nearly 1.5 million extraction operations, 4,220 cubic metres of soil were removed from underneath the structure – enough to fill about one and a half Olympic-sized swimming pools.

As you might expect, this soil removal was done carefully and cautiously, in stages, over a long period (four and a half years). All that time the levels in the cathedral were strictly monitored to make sure that any movement stayed within the limits of what the engineers wanted. The arches and columns inside the cathedral were supported with steel beams and props to prevent any damage from sudden, unexpected or large movements. Meanwhile, soil samples were continually taken out of the ground to be tested for stiffness and water content, and were compared with the computer model to make sure reality matched prediction.

The difference in floor level between the north-east and the south-west had been more than 2m, but in 1998, once the north end had settled down by just over a metre, the process was suspended. Even though this left the foundation slightly tilted, the engineers had become concerned about damaging the structure. The lean of the towers had been brought back to an amount that was deemed safe – and so, for the time being, work has stopped.

The large cylindrical access shafts have been left open. They are now flooded with groundwater, but if they are needed in the future – if the cathedral starts tilting again – the water can be pumped out, and more soil removed. For now, the cathedral has been left to the mercy of the soil – but this time it is being watched.

Positioned at strategic points around the cathedral are four pendulums encased in glass boxes that send data wirelessly to a lab in Italy where engineers monitor how the structure is behaving. Pressure pads monitor the loads in the columns, checking they aren't changing too much. A change in load would suggest the structure is tilting again, causing some columns to be more squashed than others. Dr Ovando-Shelley described the cathedral as a laboratory, in which data has been collected for nearly twenty years. It has become a place of science as well as a place of worship.

Since the 1990s, the cathedral has been sinking at a rate of about 60mm to 80mm per year – a slow and steady settling in comparison to the past and, most importantly, an almost uniform one. The movement will continue in the future, but it might slow down over time. This Indiana Jones of engineering had saved his relic, and succeeded in his mission. No Armageddon for Mexico City's Metropolitan Cathedral.

The team of engineers' groundbreaking work has been a subject of study all over the world. In 1999 they worked with engineers in Italy, replicating their methods below the Leaning Tower of Pisa. In Mexico City the engineers were faced with an extreme situation – the decidedly poor condition of the soil, its variability and the sheer size of the cathedral. But the upside of the challenge they faced is that we now have an invaluable body of knowledge that can be used by engineers in the future, particularly those fighting to save our heritage, and those attempting to build in harsher and harsher conditions as our population expands and the climate changes.

Our technical tour done, Dr Ovando-Shelley and I left the cathedral in search of a restaurant for lunch, crossing the Zocalo Square, which was framed by other elaborately designed and decorated buildings that had settled unevenly. He waited patiently as I stopped to take photos of door frames that had skewed from rectangles into parallelograms.

On a terrace overlooking the Zocalo, a waiter served us frozen margaritas. 'Soils have no word of honour,' said

Dr Ovando-Shelley, clinking my glass, 'and neither do geo-technical engineers.' He laughed uproariously. But to me, he had nothing but honour. He, and the team of engineers, had saved the biggest cathedral in the Americas from ruin. And he bought me chicken *mole* for lunch.

HOLLOW

Usually, our homes are an amalgamation of materials – we gather stuff and assemble it, creating something from nothing. But there is a place, with sparsely grassed steppes as far as the eye can see, where shelter was formed the other way round, in an absence of material – where nothing was created from something.

Naturally, I had been very curious to see this, which is why one day I found myself doubled over, surrounded by blackness, craning my neck and straining my eyes, trying to work out where I was. I knew I was deep underground: I had walked down hundreds of winding and incredibly steep stone stairs, past ancient living rooms, kitchens – and death traps – to get there.

I could just about make out that I was in a tiny, coffin-shaped passage, as wide as my shoulders as I crouched, and as wide as my feet at floor level. I wasn't even sure there was enough space for me to turn around and backtrack to the entrance. I could see damp beige stone just ahead of me, but the bright beam of light from my phone torch barely penetrated the darkness beyond. I carefully felt my way along the passage, trying not to bump my head. After what felt like a very long time (though it was probably only a few minutes), I emerged into a small lit cave and felt relief, until I saw the long rectangular recesses carved into the floor – which had once held the remains of those unlucky enough never to find their way out.

I was in Derinkuyu, one of the deepest and largest of the mysterious, warren-like ancient underground cities in the heart of Anatolia in modern-day Turkey. These cities were made possible by the area's three volcanoes – Erciyes, Hasan and Melendiz Daglari – which erupted violently around 30 million years ago. They spread a 10m layer of ash across the region, on top of which flowed lava, which consolidated and hardened the ash, turning it into what is known as *tuff*. The local climate, with its heavy rains, sharp changes in temperature, and melting snow in the spring, gradually eroded the soft tuff until only columns of it remained. The harder lava layer on top of the softer tuff degraded more slowly; now large pieces of lava rock sit precariously on top of the thin ash pillars, giving them a surreal, mushroom-like appearance – and

Fairy chimneys, the local name given to the thin ash pillars and the harder lava layers that sit precariously on top.

their local name: 'fairy chimneys'. The strange landscape acts as a kind of taster for the even stranger things going on below ground.

Geographically, Anatolia stands at the intersection of East and West, and throughout its turbulent history it has been the site of battles between civilisations. The Hittite people occupied the region in around 1600 BC, followed by the Romans, the Byzantines and the Ottomans. The constant warring meant that the locals were always under threat. The Hittites realised that the thick layer of compressed ash beneath their feet was relatively soft, soft enough to carve with a hammer and chisel. They began constructing underground caves and tunnels to hide in while the fighting went on above. Each of the civilisations that followed the Hittites added to these networks, in effect establishing cities in which up to 4,000 people could live for months at a time. Over a period of nearly 3,000 years, hundreds of underground cities were created in the region. Most of them were small, but about 36 had at least two or three storeys.

As I could see at Derinkuyu, the system of caves in these underground spaces was structured like an ant-house: the rooms were not stacked one on top of the other, as happens in our buildings, because that would cause the ash to weaken and collapse. Instead, the rooms were carved out randomly in space, spread out across a large area. The arched ceilings over the rooms and passages were the perfect shape to keep stone in compression, and stable, ensuring that the ground would not cave in on them. A number of ventilation shafts, starting from the surface and running for up to 80m underground, brought in fresh air. The cities were designed to protect against infiltration by the enemy – with huge rolling stone doors to keep them out, deep pits for them to fall into, and cubby-holes behind doors where the residents could hide and ambush their pursuers. The inhabitants even created narrow tunnels up to 8km long to connect adjacent cities, in case their enemies managed to get past all their carefully laid traps.

I'm glad that I'd never had to spent months at a time in Derinkuyu, fearing for my life, but come to think of it, I do actually spend an inordinate amount of time underground. In fact, since I began working, I've spent a total of over five months of my life deep inside London's clay, as I take underground trains – the Tube – to work. Alongside millions of other people, packed into carriages like sardines, it's an uncomfortable reminder that, in my city, space is at a premium. The streets can't accommodate homes, offices, pedestrian paths, trains, trams, cars and cycles – not to mention water pipes, sewers, electricity and internet cables. And why should they? After all, we live in three dimensions and should use all of them, building up and down rather than simply sprawling sideways. The city beneath our feet is brimming with hidden engineering, but these arteries would not have been possible had it not been for the humble tunnel. In Derinkuyu, space was plentiful; tunnels provided safety. In London and many other metropolises, there is a lack of space, and tunnels provide the solution.

*

In the early 1800s, the only river crossing in the entire city was London Bridge – an immensely impractical and laborious situation in a metropolis that was spreading out rapidly on both sides of the Thames. The time taken to navigate the busy city, the wait to make the perilous and excruciatingly slow journey across the choking bridge, and the cost incurred in tolls were all sources of great frustration. In 1805 a company was set up to try to circumvent this by directly connecting the docks at Wapping and the factories at Rotherhithe.

Although the two points were only a tantalising 365m apart across the river, this distance was large enough to make building a bridge impractical – which meant that to get from one to the other, people and goods had to make an arduous 6.5km journey via London Bridge. Besides, putting a new bridge between the docks and factories would have stopped tall ships from reaching higher up the river, causing major problems for the thriving trade the city hosted. The only remaining option was to create a passage under the river. The problem was that canal builders, mining experts like Richard Trevithick, and other inventors had already tried to tunnel without success. The new company's efforts to bore a tunnel under the river were also unsuccessful, until an engineer came up with a solution inspired by a shipworm.

Marc Brunel was born in Normandy, France, in 1769. As a second son he was expected to become a priest, but he showed more interest in drawing and mathematics than in scripture, and entered the navy instead. Fleeing France in 1793 during the French Revolution, he went to America, where he eventually became Chief Engineer of the city of New York. He then moved to London in 1799, to try and persuade the Admiralty to purchase a new system he had invented for producing pulley blocks. He worked on various projects for the armed forces, developing apparatus for mass-producing soldiers' boots, and sawmill machinery at the Chatham and Woolwich dockyards. But he came to the attention of the Thames Tunnel Company (after vigorously lobbying its bosses) because of the tunnelling machinery he had invented.

Brunel carried a magnifying glass in his pocket. While working at Chatham Dockyard, he picked up a damaged piece of timber that had been removed from the hull of a warship, and scrutinised the actions of *Teredo navalis* (the naval shipworm) at close quarters. The worm had two razor-sharp, shell-like 'horns' on top of its head, and as it moved, wriggling and rotating its horns, the wood directly in its path was ground into a powder. The little shipworm ate the powdered wood and wriggled a few millimetres forwards into the space it had just created. The powdered wood travelled through the worm's digestive system and mixed with enzymes and chemicals in its body. The worm then excreted this mixture, creating a thin paste that lined the small tunnel left behind. When exposed to the air inside the cavity, the excretion hardened, shoring up the tunnel. Slowly but surely the worm moved forwards again and again, munching through the wood while creating a strong, lined passageway behind it.

Fully aware of the previous attempts to create a tunnel under the river, Brunel put his genius to work, and came up with a new plan. He realised he could succeed where everyone else had failed by adapting the process he had just observed. He would build his own shipworm: a machine that could tunnel forward and line the hole behind it. But his 'worm' would be made from iron. And it would be colossal.

Brunel's idea was that the device would have two blades, just like *Teredo navalis* – but that these would be twice as tall as a person. The blades would sit at one end of an iron cylinder lying on its side (and looking a little like a fan we might use in the summer to keep cool, but without the cage). A team of men would push the blades round so that they ate at the ground. Hydraulic jacks would push the cylinder forward. The soil which had been cut away by the blades would be transported backwards manually, like the shipworm excreting wood powder. As the cylinder moved forward, it would expose a ring of ground. To shore this up, bricklayers would lay bricks in a ring using quick-drying mortar to glue them together, creating a cylindrical shaft behind the blades, much like the worm's waste lining its tunnel. This process – turn fan, remove soil, lay bricks – would be repeated to gradually fashion a strong cylindrical tunnel.

Brunel's shipworm.

Having sorted out his worm, Brunel now had to find a suitable material for it to burrow into. Obviously, some substances are easier to dig into than others. Take dry sand, for instance. Fill a circular cake tin with sand, then try to scoop out half of it to create a semicircle. You won't be able to, because the particles of sand simply collapse into the space you've just emptied. Similarly, if you try doing the same thing with very wet sand, the liquid nature of the material causes it to flow into and fill the space you're emptying. London sits on clay that's 50 million years old. If this clay is nicely compressed under layers of soil, and not too wet, it forms a fairly stable layer of ground. From an engineer's point of view this is good to work with, because you can slice into it quite easily, and it's unlikely to collapse. Put good clay – nicely compressed and not too wet – in a circular cake tin and remove half of it, and you'll be left with a perfect semicircle of material. On the other hand, London's clay can vary considerably: it can be sandy, weak, watery and inconsistent. For Brunel's invention to work, he had to find good clay.

He hired two civil engineers to investigate in detail what the ground was made of. Paddling around in a boat, they plunged a 50mm-diameter iron pipe deep into the riverbed, then hauled it back out. They then studied the substances that had become trapped in it, looking to identify the different soils inside, and the thickness of each layer. After months of investigating they submitted their findings to Brunel, who decided that the ground was good enough for his plan to proceed without major problems. Before his shipworm could be let loose, however, he needed to burrow deep into the ground.

On 2 March 1825, the bells of St Mary's Church in Rotherhithe pealed as throngs of people made their way to Cow Court, ready to witness a very unusual sight. In the middle of the yard lay a huge iron ring 15m in diameter and weighing 25 tonnes. A brass band began to play as well-dressed ladies and gentlemen appeared, looking out of place in this rather squalid part of London. Amid cheers from the crowd, Marc Brunel arrived with his entire family, and was presented with a silver trowel, with which he laid the first brick on top of the iron ring. Brunel turned to his son, Isambard, who laid the second. Then followed speeches, drinking and toasts to the arts and sciences to mark the inauguration of the Thames Tunnel. But the joyful crowd had no idea just how much the sciences would be challenged in the months ahead.

The iron ring the crowds could see was like the sharp end of a cookie cutter. Two rings of brick separated by a layer of cement and rubble were laid on top of the iron ring, creating a cylindrical tower just under 13m high. On top of this the builders placed another iron ring, which was linked to the bottom one using iron rods sandwiched between the two brick walls. A steam engine was attached to the top of the 1,000-ton structure to pump away water and remove the excavated soil.

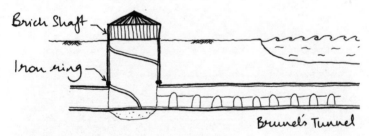

Tunnelling under the river Thames, London.

To use a cookie cutter, we apply the strength in our arm muscles to push it down into the dough. But Brunel's idea was to allow his brick cutter to sink into the ground under its own weight: it was so heavy that it would naturally move through the soft soil. Slowly but surely, the shaft began to sink a few centimetres a day. As it sank, diggers removed soil from the middle of the cylinder, much as you would remove dough from the middle of a cookie cutter.

After getting stuck once, the brick shaft arrived at its final destination. To create foundations, the diggers dug another 6m below the bottom iron ring. In this space, bricklayers filled in three sides of the shaft and the floor, leaving one face open to the ground. This is where Brunel's 'worm' would be deployed to burrow the tunnel.

While all this was happening, Brunel realised that – unlike a shipworm, which could easily turn its blades – humans didn't have enough strength to rotate the blades of his tunnelling machine. He couldn't think of a way to attach a steam engine to provide the power, so instead he came up with a new idea. His solution was to divide the device into smaller sections – 36, in fact – with a single person working in each. He called this enormous machine 'The Shield'.

Working The Shield, the enormous machine used by Brunel and his men to excavate underground.

It had 12 iron frames, each 6.5m tall, 910mm wide and 1.8m deep. Each frame was divided into three 'cells', one on top of the other. The 12 frames were placed side by side to create a big grillage of 36 cells, each housing one worker, and these workers would operate The Shield. At either side of each man in his cell was a set of long rods, spaced at regular intervals from floor to ceiling. These held in place 15 or so planks of wood, stacked one above the other directly in front of the worker, and shoring up the ground in front of The Shield.

Operators in alternating frames (say frame numbers 1, 3, 5, 7, 9 and 11) worked simultaneously. Their task was to remove one wooden board by drawing back the two iron rods holding it in place, and dig out exactly 4.5 inches of earth and put the board at the rear of this new, slightly deeper cavity. They would then push the rods into place to support the board. The next step was to remove the subsequent plank and repeat the process, continuing like that until all the wooden planks in all the 18 cells had been fixed into their new positions. Now that these miners had excavated the section of ground in front of them, jacks at the rear of The Shield propelled their cells forward by 4.5 inches.

At this stage, the odd-numbered frames would be 4.5 inches ahead of the even-numbered ones. It was now the turn of the workers in the even frames to go through the whole process of adjusting rods, removing boards, digging into the earth and repositioning the boards. When they had finished, the even frames were pushed forward. The entire Shield had progressed by 4.5 inches – the exact distance needed to fit one layer of bricks.

Behind The Shield was another flurry of activity. 'Navvies' (as the labourers who built the canals, roads and railways were known, after the word 'navigator') removed the excavated soil in wheelbarrows. Bricklayers stood on wooden planks and carefully laid bricks in the 4.5-inch gaps created as The Shield moved forwards. They used pure Roman cement, which dried very quickly and was incredibly strong – so strong, in fact, that when Brunel tested it by building a block of bricks and dropping it from a height, the cement didn't crack. He even had his workmen attack the block of bricks with hammers and chisels; while the bricks cracked, the cement stood unyielding. Brunel then decided to use this cement throughout the tunnel, despite its great cost (remember that a lot of energy goes into producing pure cement powder, which can be lessened by adding aggregate).

I try to imagine what it must have been like working in the tunnel. Before I'm allowed to set foot on a construction site, I have to pass exams, be trained in health and safety, and put on protective clothing. I walk around doing my job without worrying that I might not leave alive. Conditions in the Victorian tunnel were starkly different: the smell of the workers' sweat, the tallow smoke and the gas fumes made breathing very difficult – workers often emerged from the tunnel with a ring of black deposit around their nostrils. Flammable gases trapped in the soil were suddenly released and, if lamps were inadvertently brought near them, could catch fire and explode. The air was damp and the temperature rose and fell by 30°, sometimes in the space of a few hours. It was also incredibly noisy – bricklayers shouting for more bricks, iron rods clanging, wooden boards thudding and hobnailed boots echoing through the tunnel. Brunel himself became very ill from over-exhaustion, and was prescribed the only treatment that would work: being bled by leeches on his forehead.

Brunel's son, Isambard, who was only in his early twenties at the time, became indispensable on the project as the main engineer running the site. (Sophia, Brunel's elder daughter, was nicknamed 'Brunel in petticoats' by the industrialist Lord Armstrong because Marc Brunel, unconventionally, taught his daughter about engineering. When they were children, Sophia showed more aptitude than her brother in all things mathematical and technical – and in engineering – but it was her misfortune to be born at a time when women had no such career possibilities. She is the great engineer we never had.) But Isambard, like his father, was often taken ill. And things were getting worse: the soil conditions were unexpectedly deteriorating, and funds were running out. At one point the whole operation was shut down and the tunnel bricked shut with The Shield inside it. It took six years for the Brunels to convince the Treasury to put more money into the project. The company directors meddled with Brunel's methods, refusing to obtain equipment he wanted to make the work safer, and pressuring him to work faster despite the risks. The biggest problem, however, was the flooding. The 'good' clay that Marc had been hoping to tunnel through was not consistent, and sometimes it disappeared completely, especially as the workers dug directly below the river.

The Thames was basically a huge sewer; all of London's waste (and many of the city's corpses) were deposited into it. The soil at the base of the river was very wet and of terrible quality, and the tunnel was being dug only a few feet below the river, right into this base. As The Shield moved forward, digging away at the ground, the soil was often displaced more than it should have been. There was also a weak point in the riverbed between The Shield and the brick tunnel, and if the soil was particularly bad it simply collapsed, sending river water coursing through the passageway.

The first time this happened, Isambard fixed the problem by contacting the East India Company and borrowing a diving bell (a chamber containing a couple of people that could be lowered underwater). In it he went to the bottom of the river, found the leak, and laid a bed of iron rods across the gap, with bags of clay piled on top to seal the hole. Once the water had been pumped away, the digging work could restart.

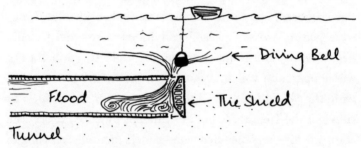

Innundation of the tunnel and the use of a diving bell to seal the breach.

This, though, was only the first of four major floods in which many men died. Isambard himself only narrowly escaped drowning, suffering his first (but not last) haemorrhage as a result, and being forced to leave the site for a few months' convalescence.

Despite the setbacks, however, in 1843, after 19 years' work, the tunnel was finished. Penny-paying pedestrians descended the spiral staircase in the shaft to the tunnel, which in its finished form was spectacular. A line of pillars down the centre supported immense brick arches. Gas lamps lit the passageway and an Italian organ powered by a steam engine played music. Hawkers sold refreshments and souvenirs from little alcoves in the brick walls. In 1852 the first Thames Tunnel Fancy Fair was held, featuring artists, fire-eaters, Indian dancers and Chinese singers.

But only a decade later, as the railways entered everyday life, the tunnel had fallen into disrepute. People no longer wanted to walk through its damp interior, choosing instead to take the flashy new trains. The tunnel became seedy and desolate, the haunt of drunks. In 1865 it was handed over to the East London Railway Company, and by 1869 rail tracks had been installed on the floor and steam trains began chugging through. Today, the London Overground line runs through it. The Rotherhithe shaft, which Marc Brunel managed to excavate so imaginatively, was recently opened to the public and has become a popular tourist attraction. Enter the stumpy circular tower and you find yourself in a cavernous underground chamber containing the remains of spiral staircases, and blotchy, scarred and weathered walls with mysterious black pipes feeding into and out of them. It's an incredibly atmospheric backdrop to the concerts and theatre performances that take place there.

Taking nearly 20 years to build, and then becoming obsolete just over 20 years after it was finished, the Thames Tunnel might not seem like a success. But thanks to Marc Brunel's imaginative engineering, we gained access to the underground parts of our cities. The London Underground – the first underground train network in the world – was made possible because of the work of Marc and Isambard Brunel, who showed us how to build structures in very fluid soil.

*

To dig their tunnels, the engineers building Crossrail (London's new train line) have been using a modern version of Marc Brunel's first and unsuccessful idea. Brunel couldn't get enough power to rotate giant blades, but electricity has made this simple for us. Instead of a manually operated machine, we use 'tunnel boring machines' (TBMs) – which are, of course, anything but boring.

Each of Crossrail's TBMs – described as 'giant underground factories on wheels' – is as long as 14 London buses end-to-end. The front has a huge circular cutter that spins, eating into the ground in front of it. An intricate jacking system pushes the machine forwards. Conveyor belts transport the excavated soil to the back of the TBM and out of the tunnel. A laser guidance system makes sure that the tunnel stays on course. Behind the TBM, a complex array of arm-like devices fix concrete rings in a circle (steel could also be used) to create the tunnel lining.

There's an endearing tunnelling tradition which proclaims that the TBMs must be named – with female names – before work can start. Crossrail ran a competition to name its TBMs in pairs, since the machines work in twos, radiating in opposite directions, starting from a point. One pair is named after the monarchs of the great railway ages: Victoria and Elizabeth. Another after Olympic athletes: Jessica and Ellie; another after the women who wrote the first computer program and drew the beloved London A–Z maps: Ada and Phyllis. Perhaps most fitting of all, though, are the names of the final two TBMs: Mary and Sophia, after the wives of the great tunnel builders themselves, Isambard and Marc Brunel.

PURE

It thrills me to see tourists taking pictures of buildings in a city, because it means that they love engineering – even if they don't realise it. They admire and respond to the ambition and the imagination that have gone into the design – curved canopies, tall silhouettes and unique facades are carefully selected, framed and frozen in time as the dramatic backdrop to countless photographs taken on phones mounted on selfie sticks. This architectural drama is the romantic side of engineering, and not to be underestimated. Nevertheless engineering is ultimately a response to very practical considerations; often it is less immediately exciting things like soil, materials or the law that are the driving force. A building or bridge might look spectacular; in fact, much of what shapes it can be decidedly unaesthetic.

One of the most influential of these considerations is water, which is such a fundamental requirement for humans that we can't survive much longer than three days without it. The structures I design are skeletons: until they have water, they are merely uninhabitable shells. I work with other engineers (mechanical, electrical, public health) to make provisions for the skeleton to support its circulatory system: creating pathways through it and making sure that its foundations, core walls and floors are strong enough to carry the weight of pumps and pipes. It's only when the arteries of water come to life that we create a building fit for the living.

But even though our planet is called 'the Blue Planet' because of the amount of water it contains, the shimmering, salty swathes of sea that cover most of the Earth's surface are not potable. We humans need easily accessible fresh water if we are to survive. But here's the problem: we don't actually have much of this. If all the water on our planet was represented as an area the size of a soccer pitch, then the freshwater lakes on the planet's surface would be the equivalent of the cushion I have on my sofa, while the surface rivers would fit inside the coaster I use under my tea.

Finding water is hard enough – and that's why many of our ancient towns were founded on the banks of a river – but as they grew into cities, as fields growing crops became vast, and as we migrated to live further and further from water sources, *moving* water became a challenge. It's no wonder, then, that in ancient times humans developed extremely inventive ways to track down and transport fresh water. Even today, engineers work hard to create solutions for this technically challenging process, and in parts of the world it is still a huge hurdle to be surmounted.

*

Like many others of the times, the ancients in Persia struggled to find fresh water. In the centre of Iran there is a large, dry, arid plateau that only receives a tiny amount of rain – less than 300mm – each year. As you fly over the country, desert stretches out below you, bleached of colour by the relentless glare of the sun. Occasionally, though, near small villages and towns, or even in seemingly uninhabited patches of the desert itself, you'll notice 'holes' in the sand. From a vantage point high in the sky, they look like the little crab holes that pepper the beach in Mumbai where I grew up. (I used to sit and stare at them for ages, waiting and hoping for a scuttling creature to appear.) But these holes are neatly arranged in straight lines, and are in fact much larger. Thankfully, they weren't the work of some giant crab, but were dug by humans, over the past 2,700 years. And throughout that time they have been essential to the survival of the people that live there.

These holes are part of the *kariz*, as it's known in Persian (or *qanat*, in Arabic): the system used by the ancient people of Persia to bring their life force – water – from below the ground.

To see how they were built, let's transport ourselves to the desert of two and a half millennia ago. The *muqanni* or worker looks near a hillside or slope for signs of the presence of water – a fan of deposited soils, perhaps, or a change in the type of vegetation. At a promising location, he takes a spade and digs a cylindrical well just over half a metre in diameter. To move the dirt he uses a windlass to haul a leather bucket full of soil up and down. Under the blazing sun, he keeps at it, hoping to find damp soil – a possible sign that the water table is close. Sometimes, he goes down as far as his tools will let him, but he doesn't find anything. At other times, he finds water hiding very deep, more than 200m down. Once in a while, he need only dig down 20m before he finds moisture. That's on a good day.

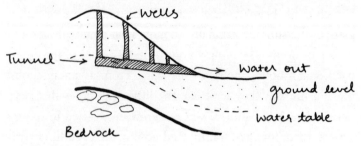

An ingenious kariz.

But the *muqanni's* work has only just started: it's still possible that all he has found is a tiny bit of water that will quickly run out. He needs to make sure that his discovery is the real thing. So he leaves his bucket in the new shaft and, over the next few days, checks how much water, if any, has collected in it each morning. If he wakes up every day to a full bucket, he knows he has struck gold – or, rather, something even more valuable: he has found the face of the *aquifer* (an underground layer of permeable rock that contains water). He and his fellow *muqanni* then dig wells, one after the other, in a straight line down the slope of the hill.

Using a plumb line to measure depth, the *muqanni* dig each of these wells slightly deeper than the previous one. It may seem strange to dig a line of wells like this, but here is where the ingenuity of the *muqanni* lies: their village contains 20,000 people, and trekking up the hillside, drawing water and carrying it back would be a laborious task. Of course, this is done in many places around the world, but here the terrain – the hilliness and type of soil – means the *muqanni* can make the villagers' lives easier.

The wells finished, the workers start to dig a tunnel horizontally from the base of one well to the base of the next, creating a conduit about 1m wide and 1.5m high – just big enough for them to walk through so they can build the next phase.

This tunnel slopes gently, joining up the bottoms of the wells, and will bring the water out of the mountain. The slope of the tunnel is important: if it is too steep, the stream of water will be too strong and fast, eroding the soil and eventually causing it to collapse. If, on the other hand, the slope is too gradual, water will not flow easily, and will stagnate.

The *muqanni* light an oil lamp and place it at the mouth of the tunnel. And as they march into the mountain, they watch the flame so they can make sure they're working in a straight line. Noxious fumes may emerge from the ground to suffocate them, so the oil lamp not only acts as a beacon but as a kind of warning light: if the flame burns steady and bright, there's enough oxygen around. If it burns a different colour or goes out, it shows there are other gases present. There are other hazards too. Loose or crumbly soil could cause the tunnel to collapse, so where required the *muqanni* make hoops of baked clay and push them into the tunnel. The hoops act like two arches joined together: the weight of the loose soil pushes on to the hoops and puts them into compression. Clay is strong in compression, so the hoops reinforce the tunnel and stop it caving in.

There is a final hazard to be broached when the workers reach the head well (the first well, with its base at the face of the aquifer). They have to break through the aquifer very carefully, otherwise a jet of water might burst through and drown them.

Managing all this safely depends on the *muqanni*'s experience being passed from generation to generation: the techniques used to build *kariz* today haven't changed a great deal since ancient times.

The length of the conduits varies hugely, from 1km to over 40km. Some produce continuous water while others are seasonal. To maintain the system, the *muqanni* use the extra wells they dug. The frequent build-ups of silt and debris can be removed using the windlass to lower buckets into the wells. With regular repairs they can last a very long time.

There are said to be over 35,000 *kariz* in Iran – networks of hundreds of thousands of underground conduits all built by manual labour and still providing an important source of water. The city of Gonabad houses the oldest and largest known example in the country. It is 2,700 years old and its 45km conduit provides water for 40,000 people. The main well is deeper than The Shard is tall.

*

Digging down to an aquifer was one strategy the ancients had for supplying their citizens with water. But with water sources, terrain and tools differing across civilisations and eras, other ingenious solutions were invented, including many we still use today. By the end of the eighth century BC, the two canals providing water for Assyria's capital city, Nineveh, were no longer adequate to serve the burgeoning population. King Sennacherib (who reigned between 705–681 BC) – had previously used his engineering skills to dig canals through Babylon to flood and destroy it. Now, he was forced to find an additional source of water and channel it to Nineveh. He started nearly 50km away, at the watershed of the River

Atrush. From here he constructed a canal to the headwaters of the River Tebitu to increase the amount of water the Tebitu received. The river had earlier been dammed to create the reservoir that had supplied most of Nineveh's water. This extra water would flow to his city through the two existing canals, increasing its supply.

There was, however, one problem. To get from the river to the canals that led to Nineveh, Sennacherib's new conduit had to cross a small valley and, without a water pump, there was no way to push water up the far slope. Undeterred, Sennacherib conceived a structure that could carry water across the valley – what we know as an *aqueduct*. We think of the Romans as the foremost engineers of aqueducts, but the Assyrian king's edifice predates their efforts by several hundred years, making it one of the oldest such structures in the world. You can still see its remains at Jerwan in northern Iraq.

Technically, the word 'aqueduct' refers to any artificial channel used to transport water from one place to another: it can be a canal, a bridge, a tunnel, a siphon (a pressurised pipe), or any combination of these systems. The Nineveh aqueduct bridge was the greatest construction of Sennacherib, a master builder who also created much of Nineveh's civic architecture, including the legendary 'Palace Without a Rival'; he may even have been responsible for the Hanging Gardens of Babylon. Over 2 million cubes of stone went into the aqueduct's construction, each about half a metre wide. The end result was 27m long and 15m wide, made from pointed *corbelled* arches (a curved shape supported by projecting pieces of stone) that were over 9m high. A channel on top of the bridge allowed water to travel across the valley. The channel was lined with a layer of concrete to prevent the water from leaking away.

Load travels down not around.

A corbelled arch.

Incredibly, the new canal and aqueduct bridge were completed in only 16 months in 690 BC. When the structure was nearly complete, Sennacherib sent two priests to the upper end of the canal to perform religious rites. Before the allocated time for the ceremony, however, the gate holding back the water suddenly opened, releasing the river into the channel. The engineers and priests were terrified of the reaction this might provoke from the king, as Nature had defied his wishes. But the king decided this was actually a good omen, because the gods themselves were so impatient to see his great work completed that they had caused the gates to fail. He went to the head of the canal to inspect the damage, had it repaired, and rewarded his engineers and workmen with brightly coloured cloths, golden rings and daggers.

*

Finding and transporting water are two of the engineer's big challenges. But once you've got it, you have to know what to do with it: the third, equally important challenge, is storing it, ready for use. The Romans, who took aqueduct engineering to an impressively sophisticated level, came up with suitably ambitious storage solutions, such as the Basilica Cistern, situated in – or, rather, under – the centre of Istanbul in Turkey.

Basilica Cistern, Istanbul.

The Romans didn't invent the cistern: since at least the fourth millennium BC people in the Levant region (modern-day Syria, Jordan, Israel and Lebanon) had been building structures to hold water. Cisterns might seem like simple things to make, but in truth the biggest ones are impressive feats of engineering.

The Basilica Cistern, for example, has immense walls – up to 4m thick – to resist the pressure from multiple gallons of stored water. To stop water leaking out, the Romans carefully sealed the walls with a coating of lime plaster about 10mm to 20mm thick. Since the roof of the cistern supported a public square, it had to be strong enough to support the weight of buildings, roads and pedestrians above.

When I visited Istanbul, the sun had pushed the thermometer to a stifling 35° Celsius, and I was grateful to descend the old stone steps into the cool air of the cistern's vast underground space. Uplighters emitted an orange-red glow and soothing music played in the background from speakers I couldn't see. I stepped onto raised wooden planks built recently to allow tourists to walk around. Below me, there was a pool of crystal-clear water a few inches deep in which grey, ghostly carp silently swam. I stood watching them, until I was jolted out of my daze by drops of water falling on my head and arms.

I looked up to see a roof made from beautiful red Roman bricks – the flat kind – with thick layers of mortar between them. Large arches spanned between numerous columns to create a grillage. Between these arches stood quadripartite vaults (domes which are divided into quadrants by four ribs). The breathtaking structure was held up by 12 rows of 28 columns, 9m high, all made from marble and arranged in a regular grid pattern. The tops of the columns varied – some had classical Greek and Roman designs on them; others were plain and bare – they had been salvaged from temples or other ruined structures. Some of the columns had split over time and were strapped together with flat pieces of black iron. A couple had the head of the Greek Gorgon Medusa carved at their base, the venomous snakes of her hair curled menacingly around her face. Her gaze was said to turn people instantly to stone, but here one of the carved heads lay upside-down while the other was on its side – a haphazard arrangement that somehow negated the deadly effect of her gaze. One column, known as the peacock column, was engraved with a curious pattern of circles and lines: these represent the tearful eyes of hens, and apparently the column was built as a homage to the hundreds of slaves that died during the cistern's construction.

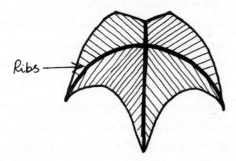

A quadripartite arch.

The Basilica Cistern was built by Emperor Justinian in AD 532. Lying beneath the Stoa Basilica, the large public square on the first hill of what was then called Constantinople (after the Emperor Constantine, who in AD 324 made the city the capital of the Roman Empire), it was capable of holding 32 Olympic-sized swimming pools' worth of water. The cistern received its water via an aqueduct that was connected to natural springs near the region of Marmara. It serviced the Great Palace, the residence of the Roman emperors, until they moved away, and it was subsequently forgotten about. In 1545, a scholar called Petrus Gyllius was talking to local residents as part of his research into Byzantine antiquities. After a little persuasion and coaxing, he discovered they had a mysterious secret – they could lower buckets through holes in their basement floors and miraculously haul up fresh, clean water. Sometimes, they even found fish swimming in their buckets. They had no idea why or how this happened – they were just glad to have a source of clear water (and sometimes even food), and until Gyllius came along, they had kept the secret to themselves. Gyllius realised that their homes must be above one of the famed Roman cisterns, investigated further, and found it.

I for one am glad he did – the place has a dramatic magic of its own, and has captured the imaginations of many people, including the thousands of tourists who have visited since it was refurbished and reopened in 1987. And, of course, the director of *From Russia with Love*, who filmed James Bond and Kerim Bey punting stealthily among the columns in sharp grey suits, on their way to spy on the Russian embassy.

*

It's incredible that something as big and impressive as the Basilica Cistern could simply be forgotten. It's incredible, too, how cavalier the Romans appeared in their attitude to water. Many historians believe that the rainwater they received was enough to live on, and that the aqueducts were for their baths and fountains. It seems extraordinary to perform such ambitious feats of engineering just for luxury and indulgence, particularly because in many parts of the world, then as now, water was in short supply and it took every ounce of an engineer's ingenuity to make it count.

In 2015 I visited Singapore to stay with a friend in her flat on the fourteenth floor of a tower block with wonderful views over the city. I checked with her that the tap water was safe to drink (of course it was) and that she had hot water available for a shower after my long flight. She warned me not to waste water, to turn off the shower when I was soaping myself and to make sure no water was dripping when I had finished.

I was impressed at her efforts to preserve water and be eco-friendly, but a longer conversation we had after my shower made me realise why this was. From a young age, it had been drilled into her by her parents, her school and her college that water is a precious resource not to be wasted. This is because Singapore has no natural aquifers or lakes. There are a few rivers that have been dammed to create reservoirs, but the country basically has no natural sources of water. Throughout its history, whether under British rule or as an independent nation, supplying its inhabitants with enough water has been a constant challenge.

The earliest sources of water in Singapore were streams and wells, which served the country adequately when the population was a mere 1,000. But after 1819, when Sir Stamford Raffles made the country part of the British Empire, the numbers greatly increased. By the 1860s, 80,000 people were on the island, and the rulers began building reservoirs to store water. In 1927, an agreement was reached with neighbouring country Malaysia, enabling the Singaporeans to rent land in Johor, from where they could pipe untreated water from the Johor River. In a reciprocal arrangement, another pipe from Singapore to Johor enabled the islanders to return some water once it had been treated. During the invasion and capture of the island by the Japanese in the Battle of Singapore (in 1942), the pipes were destroyed, leaving the people with enough water for just two weeks. 'While there's water, we fight on,' declared the region's commanding officer, Lieutenant-General Arthur Percival – but on 16 February he was forced to surrender.

This dire situation stayed in the minds of the people long after the Japanese left – to be replaced once more by the British – until 1963, when the country became, briefly, part of the Malaysian federation. So when Singapore gained full independence on 9 August 1965, water self-sufficiency was one of the government's top priorities.

In 1961 and 1962, Malaysia signed agreements to supply water to Singapore, one of which expired in 2011; the other is set to expire in 2061. For Singaporeans, it's a vulnerable position to be in, particularly in our water-dependent, high-consumption modern world, and I imagine they are concerned about their autonomy, given that they depend heavily on a neighbour for such a fundamental resource. If, for example, the whole area were to experience a drought, Singapore might end up at the mercy of another country. So for Singapore, water is as fundamental to its national interests as medicine or spies are to others.

As a result, Singapore is busy engineering a solution to its somewhat precarious situation. The Public Utilities Board (PUB) has developed a strategy called 'Four National Taps'. This refers to the four sources of water it will harness as efficiently as possible to provide a high degree of self-sufficiency for the country.

The first National Tap is rainwater. Singapore's location and exposure means it receives over 2m of rain every year. To conserve it effectively, engineers have created water catchments: areas of land where rainwater is collected rather than being allowed to drain away into the sea. A network of canals and basins has been built to trap the rain and channel it into dammed streams or reservoirs for storage. This has involved a massive clean-up operation, as over time many of the country's streams had become polluted by discharge from homes and businesses. So the PUB relocated polluting businesses and set about legally protecting the water stores from contamination. Rainwater is now being collected and stored in two-thirds of the island's land area. A few streams remain to be dammed – mainly those close to the sea, which have slightly salty water (which wouldn't be usable without some treatment). But once the engineers have finished, a massive 90 per cent of the land will be used, making Singapore the only place in the world that collects and conserves virtually all of its rainwater.

The second National Tap is water from Malaysia, which Singapore will continue to import until the agreement runs out. The third National Tap is recycled or reclaimed water. Although the practice of recycling waste water is not new – Los Angeles and other parts of California have been doing it since the 1930s – it is still far from commonplace.

Singapore first started thinking about recycling waste water in the 1970s, when the appropriate technology was still too costly and relatively unreliable. Eventually, however, it improved to the point where the project became viable, so now waste water is collected from homes, restaurants and industry and subjected to a three-stage purification process, using the latest in membrane engineering.

The first stage is *microfiltration*, during which the water is passed through a semi-permeable membrane. This is typically made from synthetic organic polymers such as polyvinylidene fluoride, which allow certain atoms or molecules to pass through but not others, and filter out solids, bacteria, viruses and protozoan cysts. Essentially, the membranes are microscopic versions of a colander, holding onto solids but allowing liquid to drain through. The water that escapes still has dissolved salts and organic molecules in it, so the second stage of recycling is designed to remove these, using a process called *reverse osmosis*.

Osmosis is the movement of a solvent (a substance that can dissolve others – the most common example is water) from a less concentrated solution to a more concentrated one, until the two concentrations are equal. It is an important part of our natural world – the means by which plant roots absorb water from the soil, for example, and by which our kidneys extract minerals such as urea from our blood. You can see the process in action for yourself, using an egg, vinegar, and treacle or corn syrup. First, soak the egg in vinegar for a couple of days, to dissolve the calcium in the shell and leave what is in effect an osmotic membrane. Then put the egg in treacle or corn oil.

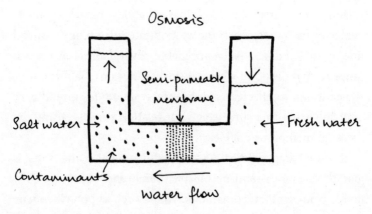

The process of osmosis.

Over the next few hours wrinkles will appear in the surface of the egg as water leaves through the membrane, dehydrating the egg in the process. Remove the shrivelled egg and put it in fresh water, and you'll see the process reverse, as water goes into the egg via the membrane, plumping it back up.

Osmosis happens naturally: fresh water filters through to mix with salty water easily. But if you want to produce more fresh water, you need to use pressure to 'push' the salty water through the membrane, which blocks the salt, bacteria and other dissolved matter. The pressure you apply needs to be bigger than the natural osmotic pressure, so you can force fresh water molecules through the semi-permeable membrane. This is reverse osmosis.

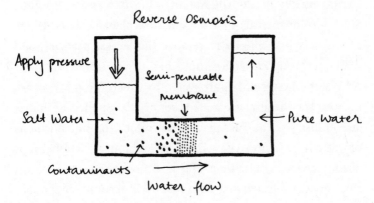

The process of reverse osmosis.

Reverse osmosis can remove up to 99 per cent of dissolved salts and other contaminants. So while the water coming out of this process is already of a high quality, there might be a few bacteria or protozoa still in it. As a backup, the water is disinfected using ultraviolet light to kill off any remaining microorganisms, and then it is ready to be distributed.

In 2003, after years of testing, NEWater – which is what the recycled water is called – was introduced to the public. During the parade of Singapore's 37th National Day, the Prime Minister, Goh Chok Tong, the founding Prime Minister, Lee Kuan Yew, and the thousands of people attending all opened a bottle of NEWater and sipped it while the cameras rolled. No one got ill. In fact, NEWater is used mostly in industrial estates and fabrication plants that require water of an even higher quality than drinking water. NEWater has passed over 100,000 tests and actually surpasses the World Health Organization's requirements for water that's fit for human consumption – even if its origins make you squirm.

And finally, the fourth National Tap is seawater. In 2005 Singapore opened its first desalination plant in Tuas, where seawater is first filtered to remove the largest particles, and then put through reverse osmosis in much the same way as for NEWater. The result is pure water, to which the minerals we need to stay healthy are added, before it's supplied to homes and industries. The Tuas plant can produce 30 million gallons of water (130,000m³) a day. The third and fourth National Taps already produce more than 50 per cent of the country's needs. By 2060 it's projected that the scheme will account for about 85 per cent – a spectacular and potentially life-saving transformation brought about by clever planning and engineering.

*

That Singapore collects most of its rainwater for reuse and is planning for long-term water sustainability demonstrates how engineering can solve critical, real-world problems. It's an age-old challenge, involving the most basic and essential of molecules, but one which is now being addressed using some of the most advanced technology available. As time goes on and our global population increases – and with it the demand for water – engineers and scientists across the planet will have to confront the escalating challenges of locating this precious liquid, creating new pathways to channel it, and enhancing the science to purify it.

Otherwise, we will not survive.

CLEAN

My visit to Japan in 2007 was one of the most memorable and inspiring trips I've been on. My mum and I wandered the streets of Tokyo marvelling at the vending machines that dispensed eggs, fruit, ramen and even puppies, and we ate at sushi restaurants where enthusiastic chefs and waiters shouted out everyone's orders in a harmonious chorus.

I was also intrigued by the toilets, which played music, and which featured buttons that lit up, and cleaning sprays that automatically sanitised, making a normally mundane act an exciting affair. In my experimentation, I did press a few buttons and regretted it pretty quickly – but, hey, I felt cleaner afterwards, if a little violated. When we left Tokyo for more remote locations, we encountered much more basic squat toilets: it was a stark contrast – but nothing compared to medieval Japan.

Long before the Tokugawa shogun regime (1603–1868) was established in the country, solid human waste – euphemistically known as 'night soil' – was being traded. It was loaded onto ships that sailed all around Japan, distributing it. Unsurprisingly, the ships carried a rancid stench with them, and people complained about these fetid vessels being docked alongside ships carrying tea. Magistrates, however, decided that the trade was vital, and that people would just have to deal with the stench.

Trading human faeces was important because of the particular challenges this small island nation faced. Because of its topography, Japan had little land for growing crops, yet the population was booming and increasing food production was essential. So the land available for agriculture had to be used intensively to produce enough food, with more than one harvest per year. This meant that the natural nutrients of the soil were rapidly becoming depleted. Traditionally, the Japanese had turned animal waste into fertiliser to replenish the soil, but there weren't many animals on the island, so the inhabitants had to look elsewhere for a solution. They found the answer in their own sanitation: the burgeoning population created a lot of waste. So the Tokugawa shoguns decided to make a virtue out of necessity by removing the waste to ships, and then trading it with farmers looking to boost their crops.

The turd trade was soon big business. During the early years of the Tokugawa shogunate, the country began to depend on one of the biggest cities at the time, Osaka, for fertiliser. Boats laden with vegetables and fruits would arrive in the city and exchange their produce for its citizens' night soil. However, the value of the night soil quickly increased (inflation affects faeces too, apparently) and vegetables were no longer enough to pay for such a valuable commodity: by the early eighteenth century, people were buying it with silver. Laws came into force stating that the rights to faecal matter produced by the occupants of a dwelling belonged to the landlord, though they were generous enough to assign the rights of urine to the tenants themselves. The price of faecal matter from 20 households a year amounted to the same as the cost of grain one person would eat annu-ally. Night soil was by now an integral part of the housing mar-ket: the more tenants that landlords had, the more waste they could collect, so the cheaper the rent.

Eventually farmers, villagers and city guilds were all fighting over rights to buy night soil. By the mid-eighteenth century, lawmakers in Osaka assigned ownership and monopoly rights to officially recognised guilds and associations that would determine a fair price. Even then, the high prices crippled the poorer farmers, and people risked harsh jail terms by turning to theft.

Night soil collection may have become a cause of conflict, but it had some unexpected benefits, too. Because waste was collected so obsessively and carefully, the water sources people used to collect drinking water were less likely to be contam-inated. Other cultural practices helped: the Japanese drank most of their water in the form of tea – boiling the water got rid of many disease-causing microbes. And those who followed the ritual practices of Shinto, had strong views about sources of uncleanliness – blood, death, illness – and 'purified' themselves if they came into contact with anything unclean. All this meant that life in Japan in the mid-seventeenth to mid-nineteenth centuries was more sanitised and hygienic than in many coun-tries in the West, and the Japanese suffered lower mortality rates as a result.

The twentieth century was different. With the constantly growing population and the Second World War causing dev-astation (not least in economic terms), the good quality of life people had enjoyed deteriorated. In 1985 only about a third of communities had modern sewage systems – a lag caused mainly by the success of the pre-modern methods for dealing with waste. In the 1980s the sewage network was modernised, and nowadays the Japanese are famous for their advanced toilets, in extreme contrast to the night soil trade that flourished not so long ago.

Whether in modern times or the distant past, the way in which a city deals with its waste has been an indicator of how successful and enterprising it is. Almost every home in the cities of Harappa and Mohenjo-daro in the Indus Valley Civilisation (around 2600 BC) was connected to a water supply and had a flushing toilet. In our densely packed post-industrial cities, efficient waste disposal has always been of vital importance. As Florence Nightingale (whose hygiene initiatives revolutionised Victorian hospitals and homes) acknowledged in an 1870 Indian Sanitary Report: 'The true key to sanitary progress in cities is, water supply and sewerage.' Those of us lucky enough to have a great sanitation system rarely give a second thought to where our poo goes once the toilet flushes. Those that don't, on the other hand, are all too aware of the disease and death that festering waste can bring. It might be a subject that makes most of us squeamish, but as the population of our planet rockets, adequate sanitation is becoming increasingly important.

*

'The trouble is,' said Karl, 'no one gives a shit about poo.' He stormed off.

At the time, I was working on the design of a small apartment building near Oxford Street in central London. While I was busy arranging columns around the car-parking bays and the swimming pool in the basement, my drainage-engineer friend Karl was working out how much waste water would be produced by showers, sinks and toilets inside the building, and by rainfall outside. Once he had calculated the amount of flow per hour, he had to make sure there were enough pipes to convey it into London's sewage system. From historical records, we knew there was a large sewer adjacent to our structure, but we didn't know exactly how big and full it was, or whether it was in reasonable condition. We wanted to know if we could use it to discharge waste from our structure, but also if digging the basement near this sewer would damage it. Karl had written to a survey company to gather information about the pipe so he could complete his design.

One day, Karl turned up with a DVD and, without much explanation, asked me to feed it into the computer and press play. Almost at once I shrieked and scrambled to turn the thing off. Among my colleagues, in the middle of my office, on my computer screen – which suddenly seemed enormous – were being displayed the results of the sewage survey. I hit the stop button and told Karl I wouldn't watch it – and that's when he told me off and strode away.

Chastened, I sat down, took a deep breath and clicked play. The film was shot by a small camera mounted on a robot on wheels being driven through a sewer wirelessly by someone standing safely on the ground. The brick walls were a deep red colour and looked pretty clean despite the unappetising contents that had flowed through them for the past 150 years. The sewer was surprisingly large – I could have walked through it without crouching – and fashioned into a distorted oval a bit like an egg standing on its narrow tip. This shape helps waste to flow easily – in times of low flow, the speed of the effluent is high since it's in the lowest and narrowest portion of the sewer; at high flows, the larger crown creates space.

The amazement I felt watching this robot moving through a landmark piece of engineering easily overrode any feelings of queasiness I had at seeing what lay at its bottom. Over the next week, Karl and I (having quickly put our faecal fracas behind us) studied the film in detail, and decided that the nearby sewer was intact and in good condition, so waste from the new building could be discharged into it. (We couldn't simply dump it all in when it suited us because of the risk of overwhelming the sewer. So, like in many buildings in London, we created an 'attenuation tank' in the basement where the waste is stored and released into the pipe at an acceptable rate.) It was an exciting moment for me: I was creating a real physical link to the pioneering engineering work done by Joseph Bazalgette more than a century ago, when he envisioned and built a vast network of sewers under the capital. At that time London sorely needed such a system, for in the early nineteenth century, living in London was a very disgusting experience.

*

Originally, the plains of London were served by a number of tributaries that provided plentiful water and fish on their path to the River Thames. But as the population of the city increased considerably in the mid-thirteenth century, the quality of the water deteriorated. Things got worse until, eventually, the tributaries were nothing more than open sewers and dumping grounds for animal and even human corpses. By the fifteenth century, 'water carriers' made a livelihood for themselves collecting water from wells in two barrels tied to a stick across their shoulders, but the rivers were in such a state that even going upstream didn't help. The water the citizens of London were drinking was contaminated with their own waste, and dead bodies.

The city also housed 200,000 cesspits – cylindrical pits, often lined on the inside with bricks in an attempt to keep them watertight, about 1m in diameter and twice as deep, with a sealed base and a lid at the top. Their purpose was to store human waste: people would take the chamber-pots they had used to relieve themselves, and empty them into these tanks. It was then the job of the 'nightmen', 'rakers' or 'gong-farmers' ('gong' was apparently a medieval term for a latrine) to periodically clean these out, carrying the waste in buckets to fields. This was better than having the waste in the streets, but it was still decidedly unhygienic, given that the fields weren't especially far from central London. Cleaning the pits was obviously unpleasant work, but it was also dangerous – spare a thought for Richard the Raker, who in 1326 fell into a cesspit and was asphyxiated and drowned in a putrid mix of urine and faeces.

Attempts by the Commissioner of Sewers to pass Acts to build new sewers in the 1840s remained inadequate. Introducing 'water closets' (or the modern-style water-flushed toilets) only made the situation worse: the leaky cesspits had been barely adequate to hold concentrated waste, but now litres of water were being emptied into them, flooding them. In 1850, to try and overcome this, the pits were banned, but as a result the sewers (which were only designed to take away surface water from rain) became completely overwhelmed. All waste – human and other – ended up in the Thames, which was still used by people for washing, cooking, and drinking.

The vile mixture of waste and water in London led to severe and devastating cholera epidemics. They usually struck in late summer or autumn, and half of the people who contracted the disease died. The outbreak in 1831–2 killed more than 6,000 people; it was followed by two more major outbreaks in 1848–9 (just over 14,000 dead) and 1853–4 (another 10,000 fatalities). The common belief at the time was that cholera was airborne and that you contracted it by inhaling a poisonous 'miasma'. But during the 1854 outbreak Dr John Snow (1813–1858) monitored the health of people drawing water from a contaminated pump in Soho, and collected evidence that this was not the case: cholera was, in fact, spread by contaminated drinking water.

Thomas McLean's etching 'Monster Soup commonly called Thames Water' of 1828 was a grotesque satire on the city's water supply.

That London's waste was ruining the capital became particularly apparent during the unusually hot summer of 1858, which warmed up the banned-but-festering cesspits and the sewage-filled Thames and its tributaries, so that the city smelled even more horribly pungent than usual. And so the 'Great Stink' (as it was called) began. It became so unpleasant that people soaked their curtains in a lime chloride mixture to try and hide the stench. The smells were so noxious that ministers working in the House of Commons and the lawyers at Lincoln's Inn were unable to work, and they made plans to abandon the city.

The only upside of all this was that, having been affected by the awful conditions first-hand, the government finally became determined to get rid of the stench and the cholera that came with it. In 1859, after years of rejecting plans from engineers to solve the sewage problem in London, officials finally approved the works proposed by Joseph Bazalgette.

Bazalgette was described as having an indifferent temperament but a pleasant and genial smile. He was considerably below average height but his long nose, keen grey eyes and black eyebrows gave him the impression of being a powerful man. He was born in Enfield on the outskirts of London in 1819 and pursued a career as a civil engineer. A taxing stint working on the rapid expansion of the railways led to a nervous breakdown in 1847, after which he became a surveyor for the Metropolitan Commission of Sewers, tasked with solving the problem of drainage in London. He was later appointed to the Metropolitan Board of Works, whose job it was to devise a solution to London's problems with waste disposal.

Bazalgette's plan made use of the Thames' old tributaries, which were now basically sewers, and which had been diverted to flow along brick culverts or channels. The diversions helped to satisfy the demand for more housing: restricting the rivers to narrow culverts allowed people to build homes close to the edge of the water. These culverts were often buried under roads, freeing up even more space. Their highest point was away from the river, and they flowed down in a north–south direction until they reached the Thames (which flowed west to east), where they deposited their putrid water.

Joseph Bazalgette decided he would intercept these culverts and their horrible contents. He did this at various points, creating a web of new sewers that sat below the old culverts. Inside these old culverts, to partially block the water flow, he constructed weirs (a form of water barrier) that were half as tall as the height of the culvert. Then, in front of these weirs, he bored holes through the floor so that most of the waste water would be redirected into his new sewers below. Hold up your left hand with your fingers spread out, then put your right hand below it at right angles to the left, and you've got a good representation of Bazalgette's system. Your left hand is the series of old tributaries flowing through their culverts; your right is Bazalgette's new sewers.

North of the river, Bazalgette installed sewers below the culverts at three points. The first was far north, where the culverts were relatively high (for those familiar with London, this branch runs from Upper Holloway through Stamford Hill and Hackney down towards Stratford). About halfway between this 'high-level' sewer and the river, he installed a 'mid-level' sewer, running from Bayswater, below now world-famous shopping area Oxford Street, and Old Street. This collected more of the wastewater as it hit the weirs and poured down through the holes in the base of each culvert. Finally, very close to the river, he put in a 'low-level' sewer to capture the remaining water. South of the river he did something similar but used only a high-level sewer (running from Balham through Clapham, Camberwell and New Cross to Woolwich) and a low-level one (going from Wandsworth through Battersea, Walworth and on to New Cross). This was because there were fewer people living there, and the extent of the city south of the river was less than in the north. In total, this system, end-to-end, would have measured 160km.

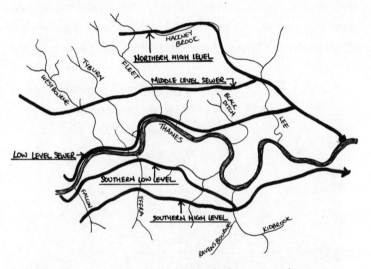

Bazalgette's main sewer network that reached across London.

The Victoria, Albert and Chelsea embankments in London are all products of his work. These contain the low-level sewers that run alongside the River Thames. Just as engineers before him had restricted the width of the tributaries of the Thames by putting them in culverts, Bazalgette narrowed the mighty river itself with these embankments. His new underground routes not only housed the new sewers, but also created space for the first underground railway: the London Tube.

When designing the five main sewer pipes and the hundreds of offshoots, to calculate the size required Bazalgette made a generous allowance for the amount of waste produced by every one of the 2 million inhabitants of the city. Then, figuring that these sewer constructions would only be done once, he doubled the size. His five sewers were at their highest point where they began in the west, and they sloped two feet for every mile as they travelled east towards two new pumping stations. Designed by Bazalgette and the architect Charles Henry Driver, these were Crossness (which served

The interior of the Victorian pumping station with its decorative ironwork at Crossness Sewage Treatment Works, Erith, London.

the two southern sewers) and Abbey Mills (which served the three northern ones). Solid, imposing and cathedral-like, both pumping stations are masterpieces of late-Victorian architecture. The real surprise is the interior of Crossness, in which the vast pump machinery is surrounded by gleaming brass and extravagantly ornate, colourfully painted wrought ironwork. In fact, the stations have appeared on screen several times, notably in films such as *Batman Begins* and *Sherlock Holmes*.

Having travelled through the sewers and reached the pumping stations, the waste had to be lifted back up to a level high enough that it could flow naturally to large sewage storage tanks further east. North of the river, the waste was stored at Beckton, while south of the river it was stored in a tank next to the Crossness pumping station. The reason the waste needed to be raised was so it could flow under gravity into the Thames when the river flowed out towards the sea on its ebb tide. At this time, the contents were still being dumped into the river untreated.

Bazalgette was told to make sure his tanks were far enough east so that, in the worst case, if they *had* to be emptied during an incoming tide because they were full, the back-flowing sewage wouldn't come as far west as Westminster – the ministers didn't want a repeat of the smells they experienced in 1858. In fact, by narrowing the river, Bazalgette inadvertently caused its tidal range to go further back than before, so occasionally the smells did get quite fusty.

Although the idea behind Bazalgette's sewage system was in essence quite simple, executing it was not, as building the new sewers meant digging up London's roads. It must have been an incredibly invasive and complex piece of work, digging down to the right level, constructing egg-shaped brick sewers and the connections to the culverts, then filling the hole and redoing the roads. But it was worth it, because life in the capital slowly began to get better.

The quality of water in central London improved dramatically. Bazalgette's sewers (2,100 kilometers of them, made up of more than 300 million bricks) were finally completed in 1875. By that time, the ravages of cholera in London were a thing of the past, in large part due to Bazalgette's practical, efficient and imaginative piece of engineering.

*

Bazalgette took the sewage from central London and deposited it outside the city into the River Thames, which ultimately took it out to sea. The waste was not treated, so the system basically moved the disease-causing elements from a populous area to a deserted one. If this sounds to you like a somewhat old-fashioned approach, then it may come as a surprise to learn that we use exactly the same system today.

Nowadays, in new waste systems, the water collected from rainwater goes ideally into pipes that are separate from those that pick up sewage from homes and offices and industrial waste from factories and restaurants. The idea is that the rainwater, which is not polluted, can be discharged into seas or rivers, while the sewage and industrial waste are taken to treatment plants.

At the plants, the polluted waste is broken down into its base chemicals, using a series of physical, chemical and biological processes. 'Physical' could mean filtration: passing water through membranes to remove impurities. 'Chemical' is the addition of substances to the waste, which react with it to break it down. 'Biological' is a similar process, but using bacteria to break down the waste. The aim is to create 'treated effluent' – an environmentally safe fluid for disposal – or 'sludge', solid waste that can also be disposed of or used as an agricultural fertiliser.

That, at least, is the theory. In practice, it rarely works like that. Shockingly, estimates by UN-Habitat (an agency monitoring places where people live) state that, globally, 90 per cent of waste water is released into the environment untreated or after only primary treatment. And at the moment, London is no exception. This is because Bazalgette's sewers are 'combined sewers', which means that they carry everything – rain, sewage, industrial effluence. Bazalgette was incredibly forward-thinking in designing the sewers for the waste of 4 million people (twice the population of Victorian London) plus rainwater. Now, though, the population in London is 8 million and we are still using this system, which is nearly 150 years old. The reason it still works most of the time is because the sewers are big enough to cope with the 1.25 billion kilograms of poo they receive each year. But since the system is working at relatively full capacity, it can't cope with rain as well, so even if it rains just 2mm in a day (which is a common occurrence in dear, damp London), these combined sewers fill up and overflow.

Dotted along the sides of the River Thames are 57 pipes that discharge this overflowing waste directly into the river. You can see the exit point for one of these at Battersea, where there is a large, reinforced iron door set into the river bank; there's another under Vauxhall Bridge. The Vauxhall one alone currently releases 280,000 tonnes of waste a year. Some of these exit points were built in Bazalgette's times, while others were added later. In 2014, excess flow had to be discharged into the river more than once a week, equating to 62 million tonnes of untreated sewage released into the Thames every year. That's the equivalent in weight of more than 8,500 blue whales plunging into the river *every week*. If we do nothing, that will almost double by 2020. Such statistics are liable to make anyone breathe uneasily. Fortunately, however, between now and 2023 a huge project to deal with this problem will be under way, under the feet of unsuspecting Londoners: the Thames Tideway Tunnel.

I made an appointment to see Phil, one of the directors of the project to create a new 'bowel' for the capital. We settled down in a large canteen to chat about urine and faeces – or, more specifically, how they will now be removed in a more modern manner.

'Our scheme is an extension of Bazalgette's legacy,' Phil explained. 'One that, I believe, he would have done himself if London's population had grown to such extents during his lifetime.' The premise of the project is simple: 150 years ago, Bazalgette intercepted the decaying tributaries. Now, the Tideway Tunnel will intercept Bazalgette's sewers: instead of the wastewater from his sewers overflowing into the river, it will overflow into a new network of tunnels.

The scale of the project is impressive. At 21 sites around the city – including one at the Vauxhall discharge point – new vertical, cylindrical shafts will be dug up to 60m deep to collect the excess sewage. Most of these will be built at the edge of the river. The first step is to build a large *cofferdam*, a watertight enclosure where the construction site can be set up. Within this area, a new shaft will be installed close to the existing sewage discharge point. Chambers will then be built to connect the existing discharge point to the shaft. So, instead of flowing into the river, the sewage will flow through the chambers into the new shaft. As Phil pointed out, while it's fine providing a new system, it's also extremely important that it's invisible, both to sight and smell (I pictured living next to a large toilet). Acres of public gardens and parks will be developed on top of these shafts. So in a few years' time, you'll be sitting on a bench by the river sipping your cappuccino, surrounded by grass and trees, while literally tonnes of sewage per second pour from Bazalgette's sewers into the shaft below you. When the waste reaches the bottom of the vertical shaft, a pipe will carry it through to the new tunnel.

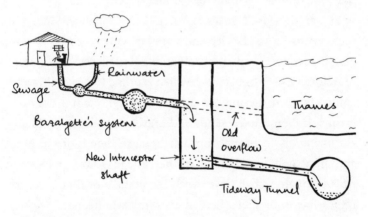

Intercepting sewage via the planned Tideway Tunnel; the future of the sewage system within London.

This main artery is 7.2m in diameter: large enough to contain three double-decker buses side by side. It starts at Acton in West London and falls 1m for every 790m that it runs east. By the time it reaches the pumping station at Abbey Mills, the tunnel is as deep as a 20-storey building is tall. From Abbey Mills, the sewage is pumped to the Beckton sewage treatment works.

The majority of this tunnel runs below the River Thames in central London, which is a really interesting engineering strategy. It's an excellent idea to do this, because running new infrastructure under a busy city is difficult at the best of times. But London in particular has a large underground tunnel network and thousands of buildings with deep foundations. By running the tunnel below the water itself, it passes under only 1,300 buildings (which might seem a lot until you consider how many more it would have been if the tunnel had run under land instead). It also goes below 75 bridges and 43 tunnels, including the Tube tunnels, as it burrows under the city.

The ground itself poses another huge challenge. Since the tunnel runs across the city, and slopes downwards from west to east, it encounters different soil at different places. At the start in Acton it goes through clay, which is prone to expanding and contracting. In the middle section, through central London, it runs through mixed sands and gravels, which are problematic materials to tunnel through because they move around and aren't cohesive. Finally, in the east, in Tower Hamlets it runs through a chalk layer with big chunks of flint in it. It's impossible to predict where all this flint will be, and, because it's difficult to cut through, it can cause delays as the tunnel boring machines (TBMs) struggle to slice their way through the ground. The tunnel needs to be strong, especially at the junction between two different types of soil, because one type of soil might be much more cohesive or drier than the other, and apply different forces to the tunnel as it expands and contracts. Five TBMs will work at the same time in different parts of the city, moving in different directions, to form tunnels that will eventually join up to create the 'super sewer'.

The aim of this mind-boggling project is to bring down the number of discharges into the river from 60 a year to 4, reducing the amount of wastewater from 62 million tonnes to 2.4 million tonnes a year. I asked Phil why the discharges couldn't be completely stopped, and he explained that these four discharges would happen only when there is very heavy rain: during such storms, the sewage is diluted considerably as the storm water mixes with the waste, so the discharge into the river is not toxic. The oxygen levels in the river would not really be affected by these diluted overflows because of the natural biological processes in the water that maintain its ecosystem. To reduce the discharges to zero, the Tideway Tunnel would have had to be twice as big.

Engineers often need to compromise in this way: the ideal solution is not always the most practical one. Ideally, we would have separate sewer pipes for rainwater and for waste, but this would mean having to more or less shut down London and dig up all the streets to put in a brand-new system. Even more ideally, we wouldn't discharge into the Thames at all, but this could actually be worse for the environment. Creating a tunnel of the necessary size would mean removing twice as much soil from the ground, resulting in a much longer construction process with bigger machines and much more energy. This method would also reduce the amount of water in the river itself, as the flow of the natural tributaries would be completely shut off.

The Thames Tideway Tunnel project will clearly have a momentous effect on the quality of the river: no longer will swimmers and rowers have to worry about sloshing through human waste. But what made me even happier was when Phil pointed out that the project will incorporate new treatment plants. We've come full circle from Bazalgette's solution, and are adding another network of shafts and tunnels to his system to meet the needs of the modern city. But this time we will be decontaminating our waste so we don't contaminate our seas.

Today, we pay homage to Bazalgette for having the skill and imagination to create a sewer system we are still able to use, 150 years later. Hopefully, the current expansion of the system will serve us just as long and, in a century's time, city dwellers will be thanking us for giving London a new bowel.

That's probably enough about poo.

IDOL

When I walk into a room for a meeting, I'm often the only woman there. Sometimes I keep a tally – 11 men and me, 17 men and me. The most, I think, was 21 men and me. Surrounded by men, I conduct my business, bemused when one of them swears, then looks sheepish and apologises directly to me (they've clearly never seen me driving my car in heavy traffic). I have opened countless work-related letters addressed to 'Mr Agrawal' – after all, if you can't tell my gender from my name, you have a greater than 90 per cent chance of being right if you go for male. Because, much to my frustration, I am in a minority in my profession.

Working in a man's world can be challenging in all sorts of ways, sometimes comical, other times trying. It's hard to keep a straight face and conduct professional conversations about finite element modelling or soil strength profiles when I'm in a site office surrounded by pictures of naked women. On one occasion a builder asked me if I wanted my picture taken in my 'costume', in other words, the hard hat and hi-vis jacket I wear regularly for all the site visits that are part of my job. I've heard stories from other women in the industry about how they've been (illegally) asked in job interviews when they plan to get married and have children.

Thankfully, these are mostly occasional occurrences. And ultimately I love what I do and believe that anyone can succeed in my field with persistence and resilience. I acknowledge that being in a minority can even have advantages – people tend to remember me after a meeting because I've spoken knowledgeably about concrete and cranes while wearing a chic dress and shoes. And it has provided some unusual opportunities to be a spokesperson for engineering, such as fashion and make-up shoots.

My engineering idol: Emily Warren Roebling.

I admire many engineers – I've talked about many of them in this book – but Emily Warren Roebling holds a special place in my heart. She understood technical concepts as well as any of the male engineers churned out by universities that wouldn't admit women, yet she never trained as an engineer: she simply learned when she had to. Her brilliant communication skills earned her the respect not only of labourers on site but also of the highest-ranking politicians of the time. What's more, pioneering innovations in engineering were implemented under her watchful eye.

Being in a minority and working in construction has its difficulties in the twenty-first century, but Emily did all this at a time when most believed that womens' brains were not even capable of understanding the complex mathematics and engineering she mastered. Her masterpiece, the Brooklyn Bridge, remains one of the most iconic symbols of New York.

*

From a very early age it was clear that Emily was extremely intelligent and had a keen interest in science. Despite a 14-year age gap, she shared a very close relationship with her oldest brother, Gouverneur K. Warren. He entered West Point military academy at 16 and then joined the Corps of Topographical Engineers, surveying for future railroads and mapping areas west of the Mississippi. He went on to fight with distinction in the American Civil War (a statue to him stands at the entrance to Prospect Park in Brooklyn). Warren was Emily's hero. When their father died he assumed responsibility for his family, encouraging Emily's interest in science and arranging for her to be enrolled in the Georgetown Visitation Convent, a preparatory school for women. There, she further explored her passionate interest in science, history and geography and became an accomplished horsewoman. In 1864, during the American Civil War, Warren was posted far away, but Emily made the arduous journey to visit him and, during her stay, met Warren's friend and fellow soldier Washington Roebling. Contrary to her usually balanced and sensible nature, she fell in love at first sight. Six weeks later, he bought her a diamond ring.

Throughout the rest of the war, Emily wrote long affectionate letters full of details of her life. But Washington destroyed them soon after he read them, saying that the letters made their separation much more painful to him. Emily, on the other hand, saved everything he ever wrote to her, and in less than a year she had more than 100 letters containing all his thoughts, fears and affections. While he was away fighting in the war she visited his family, and they took a great liking to her. Finally, after 11 months of correspondence, Emily and Washington Roebling were married on 18 January 1865, and Emily stepped seamlessly and gracefully into the role of a typical Victorian housewife: tending to house and family in the shadows of her husband.

Washington's father, German-born John Augustus Roebling, was an accomplished engineer, and Washington planned to follow in his father's footsteps. In 1867 John sent Washington to Europe to study building methods, one of which was inspired by the Romans.

*

The relatively light and small structures that the Romans built in the early years of their empire didn't really need foundations because the ground was strong enough to carry them. But as they mastered the techniques of construction, their structures increased in size and weight, and the Romans learned that foundations were a crucial part of the design of their creations, which would otherwise move or sink. It was relatively easy to build foundations on land by digging out the soft earth and laying strong stone or concrete on the firmer, deeper layers of earth. Doing the same in a river, though, was – as you might imagine – more complicated. But being the inventors they were, the Romans came up with a solution.

They sometimes supported their structures by driving piles made from logs into the ground. They inserted the piles using *piledrivers*: machines made from inclined pieces of timber connected together in a pyramid shape, and about two storeys tall. Pulleys and ropes attached to the apex of the pyramid allowed men or animals to raise a heavy weight. A wooden log was pushed into the ground as deep as could be achieved manually, and then the rope holding the weight was released, dropping it on top of the log and pushing it further. The process was repeated until the log was completely submerged.

To create foundations in water, the Roman engineers started by using piledrivers to install wooden piles into the riverbed in a ring around the position of their intended foundation. They inserted two concentric rings of piles, and then packed the space between them with clay to seal it. The water within the ring was bailed out, leaving a dry area in which they could work. This sort of construction is called a *cofferdam*. It's a technique still used today (in the Thames Tideway Tunnel, for example, as we saw in the previous chapter), but using large steel piles shaped like circular tubes or trapezoids.

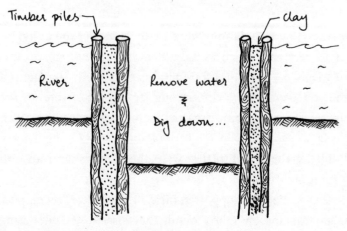

Building foundations in water the Roman way.

Inside the dry cofferdam, Roman workers dug out mud until either they hit rock, or the cofferdam started leaking. On top of the strong ground they built a pier of stone or concrete in layers. (Using their special pozzolanic cement they were able to make solid concrete even in damp and soggy environments.) Once the pier was built, they piled rocks against the base to stabilise it further, then put mud back into the hole to its original level. The base of the pier or column and the pile of rocks were buried in the riverbed. The timber piles were then removed and water flooded back in. The workers continued to build the pier as high as it needed to be to support the bridge structure above.

*

The Roman cofferdams worked in places where the water was not too deep. But Washington Roebling wanted to know how to go deeper. Driving piles wouldn't work because they would be really tall and wouldn't be able to resist the push of the water. So he studied *caissons*.

A caisson is a chamber with a watertight top and an open base, which penetrates into the mud of the sea- or riverbed. (You can picture this by pushing an upside-down tumbler into a pot of water that has sand at the base: the tumbler rim pushes into the sand while the sealed top keeps water from coming in.) One chute from the surface provides access to workers so they can descend into the chamber, and another is the passage through which they can take materials in and out. But if you want to really go deep into the water, there is another challenge. The further you descend, the greater the pressure of the water, and the harder it pushes against the walls of the caisson.

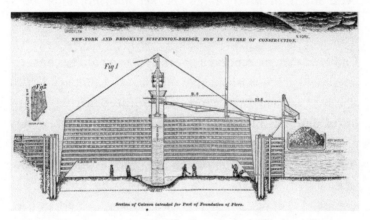

The immense caisson used during the construction of the Brooklyn Bridge.

To counteract this pressure, you can use a *pneumatic caisson*. These are 'normal' caissons with an added feature: compressed air is pumped into them. The pressurised air stops water from coming in and also balances the push of the water on the sides. An airlock gives workers access to the chamber. Engineers started to use these literally groundbreaking innovations to install foundations for bridges around the middle of the nineteenth century, and Washington Roebling was fascinated. He even considered using explosives in the confined space – a technique that, for obvious reasons, hadn't been tried before.

Emily began to assist her husband's research, studying caissons alongside him, and using the scientific methods she had learned at the Georgetown Visitation Convent to understand bridge engineering. Little did she realise at the time, that the dangers of working in the highly pressured environment of a caisson would eventually lead to a catastrophic change in their lives, one from which Emily and her husband would both emerge very different people.

*

Until the late nineteenth century there was no bridge connecting Brooklyn to the island of Manhattan, and although ferries shuttled back and forth across the East River, they were often halted during winter when the river froze over. So there was great pressure on the government to improve the situation. A bill was passed chartering the New York Bridge Company to do exactly that, and in 1865 John Augustus Roebling was appointed to design and make cost estimates for a bridge over the East River. They were to arrange for funds, which were to be split between the City of New York and the City of Brooklyn (which at the time were separate cities), along with private investors. Two years later, John Roebling began to lead the entire project.

The central section of the bridge he designed took the form of a suspension bridge, which has some similarities to the cable-stayed form I used for the Northumbria University footbridge: both employ tall towers to which cables are attached. And in both types the cables are always in tension, which holds up the deck. However, the two bridges differ in the way the tension force channels itself into the ground.

The journey of the forces in a cable-stayed bridge is direct. As the deck pulls down on the cables, putting them in tension, these cables, which are connected directly to the towers, compress the towers. In a suspension bridge, however, the weight of the deck pulls on cables that in turn pull downwards on *another* cable – a *parabolic cable* – which is suspended from the tall towers at each end. (*Parabolas* are curves with a particular shape – for the mathematically minded, if you plot a graph of $y = x^2$, you get a parabolic curve.) The parabolic cable is anchored to foundations on the opposite side of each of the towers. The parabolic cable exerts a downward force on each tower, putting them into compression and channelling the forces into the foundations. This is the difference between the two types of bridge: cable-stayed bridges don't have parabolic cables.

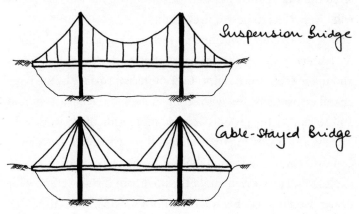

Suspension versus cable-stayed.

Work on the Brooklyn Bridge began in 1869, but disaster struck almost immediately. An accident on site left John Roebling with tetanus and he died a few weeks later, without even seeing the first stone of his spectacular structure laid.

Washington Roebling was the natural successor to his father, and took on the role of Chief Engineer on the project. To sink the piers for the bridge, he made use of the caissons that had caught his imagination in Europe. But his were larger than any that had been used before, and he was also going much deeper under water. With layers of heavy stone on the lid, he drove two huge chambers – each 50m wide by 30m long – into the river, one on the New York side and another on the Brooklyn side.

While this looked to be a reasonable engineering decision on paper, reality soon sank in. During the first month of excavation, progress was so slow that the engineers questioned whether they should give up and start again with a different approach. As columns of black smoke emanated from steam engines, and tar barrels, tools and stacks of stone and sand cluttered the site, reports began to surface from the workers about what it was like to be in the caisson.

It was incredibly loud in the restricted space, and shadows darted everywhere; the pressure affected the workers' pulses and made their voices faint. Every internal surface of the large chambers was covered with slimy mud, and the air was humid and warm. As the ground became more difficult to work with – constantly throwing up boulders through which the caisson couldn't cut through – Roebling began experimenting with explosives. He worried about the quality of the air and how his design would affect his workers, not knowing at the time that his own health would soon be ruined.

Over the next few months, having spent hours deep below the surface, Washington suffered exhaustion, temporary paralysis and deep pain in his joints and muscles. He had even hired a doctor to be on hand to supervise the condition of the men working in the Brooklyn caisson, which was deeper than the New York one. Without a full understanding of the health issues that he and his workers were facing, Washington shrugged off his symptoms and continued working. But even though the pain was temporary, the feeling of numbness in his extremities was not. He became a victim of *caisson disease*, in which nitrogen is released into the blood, causing acute pain (liable to make the sufferer double over, which is why the condition became known as 'the bends') and even paralysis or death. Now, of course, we understand the dangers of moving from high- to low-pressure environments too quickly – divers, for example, ascend at a rigorously controlled rate so the gas can be expelled. In 1870, however, caissons were a relatively new innovation and, although the engineers knew something about the dangers of working at depth, they had yet to determine the mechanism for avoiding injury.

Washington was in constant pain – in his stomach, his joints and his limbs – and severely depressed. Ravaged by headaches, he was losing his eyesight and was upset by the slightest noise. But only he had the knowledge and ability to oversee the project in his father's place. Nevertheless, Washington's physical condition made it impossible for him to be actively involved; even normal day-to-day tasks were now a struggle. His mental state left him loath to speak to anyone except Emily. It seemed as though all the years of design and planning that the Roeblings had put into the bridge, and all the personal sacrifice they had endured, were going to count for nothing. Emily, however, had spent a long time with her husband and father-in-law, hearing about bridge design and engineering, and even helping with the technical research. Slowly, she began to get involved. It was, however, a huge step. The idea of a woman being involved in the project, and perhaps even leading it, was unprecedented. Apart from the doubts and mistrust everyone would likely feel for Emily – from the builders on site to the investors – did she herself have the confidence and resolve to act as a liaison between her husband and the site, let alone take over the role of Chief Engineer?

With some background in science, but no detailed knowledge of bridge design, Emily began by taking extensive notes from her husband. She feared that he would not live to see the bridge completed. She then took over all correspondence on her husband's behalf, regularly writing to the offices of the company. With unwavering focus, she started to study complex mathematics and material engineering, learning about steel strength, cable analysis and construction; calculating catenary curves, and gaining a thorough grasp of the technical aspects of the project. Emily was determined to see her family's legacy built.

She soon realised that these skills alone were not enough for her to successfully lead the project: she had to communicate with the workers on site, and the powerful stakeholders. So she began visiting the site every day, instructing the labourers on their work and answering their questions. She supervised the build and relayed messages between her husband and the other engineers working on the project.

As Emily grew in confidence, she relied on Washington less and less. Her gut instincts guided her decisions and her blossoming skills helped her anticipate problems before they happened. Records of all work on site and responses to letters were diligently filed, and she tactfully represented her husband at meetings and social events. When bridge officials, labourers and contractors visited looking for her husband, she intercepted, answering their questions with authority and confidence. Most of them left satisfied, and many of them addressed all future correspondence to her – and in their minds she became the true authority. (At one point during the build, there were investigations into the honesty of some of the suppliers. In 1879, representatives of one of the contractors, the Edge Moor Iron Company, keen to allay suspicions, wrote a letter addressed to 'Mrs Washington A. Roebling' that made no mention of soliciting opinions from her husband.)

Yet Emily conducted her work in Washington's name. Rumours circulated that she was the actual Chief Engineer and the real force behind the bridge. News outlets made oblique references to her: the *New York Star* commented archly about 'a clever lady, whose style and calligraphy are already familiar in the office of the Brooklyn Bridge'. During the entire period of construction, the Roeblings kept their home life strictly private – no magazines or newspapers were permitted to interview them.

Despite Emily's careful handling of the project, problems started to proliferate. Costs mounted. Twenty men died from accidents and the caisson disease. Washington's health showed no signs of improvement. The so-called 'Miller Suit' had been filed. Warehouse owner Abraham Miller sued the cities in charge of the bridge, demanding that they remove the structure in its entirety; claiming that it would divert trade to Philadelphia; challenging the cities' ability to fund the project; and presenting a number of shipmasters, shipbuilders and engineers who would testify against the safety of the steel cables used in the bridge. Only the determined efforts of Senator Henry Murphy, a long-time supporter of Washington's father, led to the suit finally being settled. Even the Roeblings didn't escape accusation – it was claimed they had conducted questionable transactions with steel manufacturers, and they were investigated for accepting bribes, before eventually being cleared. The board of trustees overseeing the build changed and political dogfights broke out between new and old members. And then, in 1879, the Tay Bridge in Scotland – one of the biggest and most famous bridges in the world at the time – collapsed in a gale, killing 75 people. A headline in the *New York Herald* wondered: 'Will the Tay Disaster Be Repeated Between New York and Brooklyn?'

In 1882, despite Emily's skilful and assured work on behalf of her husband, the Mayor of Brooklyn decided to replace Washington Roebling as Chief Engineer on the basis of physical incapacity. He passed a motion with the Board of Trustees to dismiss Roebling, calling for a vote at their subsequent meeting. After much argument, political wrangling, and reporting in the press, they gathered, debated, and cast their ballots.

By a majority of just three votes, the men agreed to let Washington Roebling continue running the project until its completion. Nearly half a lifetime later, when Roebling was asked what part Emily had played in building the bridge, he answered 'her remarkable talent as a peacemaker' among the divisive personalities involved in the bridge's construction. I like to think of her as the polished negotiator: patiently listening to every side of the numerous arguments, offering tactful words of caution to the men, and smoothing difficulties in a highly-charged political atmosphere. Emily was clearly instrumental in ensuring her family's legacy remained intact.

Before the bridge was opened to the public, one final test had to be conducted: checking the effect of a trotting horse on the structure. Even at that time, the dangers of resonance – movement caused by users of a bridge – were well understood, so precautions were taken to establish that the bridge was stable and safe for different modes of transport. Carrying a live rooster as a symbol of victory, Emily was the first person to ride across the bridge in a horse-drawn carriage.

A few weeks later, on 24 May 1883, she was given the honour of accompanying President Chester Arthur's procession as he officially opened the bridge, while her husband watched on proudly through a telescope from his room. The day – which came to be known as 'The People's Day' – was declared an official holiday in Brooklyn. Fifty thousand residents spilled into the streets, celebrating and hoping to catch a glimpse of their President and their new bridge. Numerous speeches revered the bridge as a 'wonder of science', and an 'astounding exhibition of the power of man to change the face of nature'. Or, in this case, the power of woman. During the ceremonies, Abram Hewitt, one of Washington Roebling's competitors, stated: 'The name of Emily Warren Roebling will . . . be inseparably associated with all that is admirable in human nature and all that is wonderful in the constructive world of art,' and called the bridge '. . . an everlasting monument to the self-sacrificing devotion of a woman and of her capacity for that higher education from which she has been too long disbarred'.

The official opening ceremony of the Brooklyn Bridge.

Today, on one of the towers supporting the bridge there is a bronze plaque dedicated to the memory of Emily, her husband and her father-in-law. Placed there by the Brooklyn Engineers' Club, it reads:

'The Builders of the Bridge
Dedicated to the memory of
EMILY WARREN ROEBLING
1843 – 1903
Whose faith and courage helped her stricken husband
COL. WASHINGTON A. ROEBLING. C.E.
1837 – 1926
Complete the construction of this bridge
From the plans of his father
JOHN A. ROEBLING. C.E.
1806 – 1869
Who gave his life to the bridge
"Back of every great work we can find
The self-sacrificing devotion of a woman"'

Emily Warren Roebling was technically brilliant and liked by just about everyone she ever worked with. She was held in high esteem and shown great respect by the forces behind the bridge, regardless of their role or aspiration for the project. That she, as a woman, could traverse every social circle, and was welcomed by politicians, engineers and workers, her opinions heeded and instructions followed, was in itself proof of her exceptional skills, in an age when a woman's presence on a construction site was unheard of.

The commemorative plaque to the Roebling family on the Brooklyn Bridge.

As a young structural engineer at a similar age to Emily when she was working on the bridge, I am well aware of the challenges and pressures involved in constructing a key architectural landmark in a major world city. But I came to my greatest engineering challenges after years of structured technical training, experience, guidance and support – gaining my chartered engineer's qualification on the way. Emily did it without any formal training; she was not even a qualified engineer. Tragic circumstances forced her into a situation in which she never expected to find herself, yet she excelled and triumphed. This was not just any bridge – its 486m span made it by far the longest bridge of its time. It was the first to use steel wires for suspension cables, and the first to employ caissons of such enormous size, and explosives within them. It was a pioneering structure that has persisted to this day.

In my research I have been surprised to see the disparity in the way Emily's contribution is acknowledged by commentators. In some places, she is highlighted as the true force behind the project. In other sources, there is absolutely no mention of her at all. But, compared to equivalent women of her time, her contribution has received some recognition at least. I am delighted that her name endures on the commemorative plaque. She is an inspiration to me because, despite the monumental challenges she faced, she delivered the most advanced bridge of its time, using every skill an engineer needs – technical knowledge, the ability to communicate with labourers and persuade stakeholders, and tenacity – at a time when women were expected to be silent and inconsequential.

BRIDGE

'Flirtman called again. Managed to get rid of him in only 3 minutes and 23 seconds.'

At a party, I had been introduced to a man who chattered away at me, altogether too suave and flirtatious for my liking – or rather, the type that considers himself suave but isn't really. Eventually I extricated myself and was careful to steer clear of him for the rest of the evening. But not quite careful enough – somehow we ended up swapping phone numbers.

Over the next few weeks, he called me a couple of times. The first time, my mum had just arrived from India, so I fobbed him off with a polite, 'Sorry, my mother has just arrived, I can't talk now.' The second time I got rid of him in just over three minutes, and proudly emailed a friend to tell her so.

But Flirtman – as he'd become known to me and my friend – was persistent. He called and emailed a few more times (the conversations began to extend past three minutes). Finally, I agreed to go on a date with him. It was then that I found out something unexpected about this young man – he was a complete geek. We talked about physics, programming, architecture, history; and I discovered that he spent hours reading

Wikipedia, and that his brain had an uncanny capacity for interesting but essentially useless facts. I left dinner hiding the little flutter I felt.

I can't think how it happened, but over the course of that dinner Flirtman spotted that I too am a bit of a nerd, and he developed a cunning strategy to get my attention. The morning after our first date, I opened my emails to see a message headed: 'Bridge of the Day no. 1'.

'An example of why you should do a proper damping analysis,' read the email: it was the Tacoma Narrows Bridge in Washington which collapsed dramatically in 1940 in a relatively light wind. Each morning after that I'd log on, still half asleep, and a grin would spread across my normally grumpy face as I saw that a new Bridge of the Day had appeared. In fact, for a whole week, he found and sent me a Wikipedia link and a picture of a bridge: one which had a funny story, a unique design, had suffered a catastrophic failure or just looked beautiful. Was I that obvious? Surely it couldn't be that simple to win me over . . .

Even though I still thought of the sender of the emails as slightly trying, I enjoyed his bridge stories, and learned about examples I hadn't even heard of. After a week of such offerings, I at least had to acknowledge that he'd pulled off a pretty good chat-up line. It's not every day you get serenaded by a series of bridges. And so, in honour of Flirtman, here is my version of Bridge of the Day. I've chosen five of my favourite examples from around the world – but unusual or obscure ones that, hopefully, you haven't heard of. Each bridge is made from different materials, ranging from silk to steel. I've chosen them from different periods in history, and they demonstrate different methods engineers had for building. One bridge moves because it's designed to, one is unintentionally bouncy, and one was made by an ancient king. Each has its own unique engineering feature – offering a glimpse into the thousands of creative ways that humans have crossed valleys and rivers through the ages.

No. 1: Old London Bridge

Old London Bridge: that was often falling down.

This isn't a bridge I've seen, because it was finally demolished in 1831. With its tumultuous history it holds an air of mystery for me: it's the legendary bridge – built thanks to the passion and perseverance of one person – that spanned the Thames for more than 600 years. What fascinates me above all is that for centuries it served the people of London faithfully – but, ultimately, badly. Despite its impressive longevity, Old London Bridge failed as a structure.

The Romans, as you might expect, were industrious and efficient bridge-builders. But after their western empire declined in the fourth and fifth centuries AD, very few bridges were built until the 1100s. At that point, the Church started to fund and construct a large number of bridges. Many of these had chapels where one could pray for safe passage, and contribute financially to its upkeep. There is a legend that Saint Bénézet (who was inspired by a vision to build the famous Pont d'Avignon) founded the Fratres Pontifices or 'Brothers of the Bridge', who built bridges wherever they were needed for religious or community purposes.

Spurred on by this development, Peter of Colechurch, curate of a small chapel in London, decided to raise funds to build a new bridge over the River Thames in 1176. He collected donations from the king, peasants, and everyone in between, in order to build the first stone bridge in London. Previously there had been a wooden bridge that had been variously destroyed by storm, fire, military strategy or simple neglect. Building this structure, however, would prove to be a big challenge for Peter, as it was the first time anyone had proposed a bridge with stone foundations in a tidal river. The Thames is not an easy stretch of water to bridge: it moves up and down by almost 5m, has a very muddy bed, and contains fast-flowing water, making it extremely difficult to build the foundations and piers to support the deck. Even getting materials to the site promised to be a struggle, as stone had to be bumpily transported over the poor-quality cobbled roads provided for travellers. Undaunted, Peter took on this mammoth task.

People in medieval London must have been dumbfounded by the elaborate construction of their first stone bridge. They would have heard the ear-splitting thuds of the piledrivers, mounted on barges, which slowly wound up a large weight, then dropped it to whack piles into the riverbed. They would then have seen artificial islands called *starlings* built on top of the piles. Each was shaped like a rowing boat, and was constructed by amassing stones and rocks of different sizes. The starlings – and the piers or columns that rose from them to support the deck of the bridge – were huge and irregular in size, ranging from 5m to 8m in width. The populace watched as carpenters attached wooden skeletons shaped like arches to the piers. These were centering, on top of which carved stone was placed (after it had been perilously lifted from barges) to create the arches. The people of London had to wait an entire year to see just one arch completed.

In 1209, 33 years later, the bridge – which was 280m long and nearly 8m wide – was completed, but Peter of Colechurch did not live to see it. He died after 29 years of service to the structure, and was buried in the crypt of its chapel.

The finished bridge was extremely crude. It had 19 arches of different shapes and sizes, made from randomly cut stone in the pointed Gothic style. Although the pointed arch inspired by Islamic architecture was all the rage in buildings and churches at the time, it was not an efficient shape to use for a bridge. Certainly, such arches allowed medieval churches to be taller than ever before – but the bridge didn't need to be tall, it needed to be the right height to link both sides of the river. A more traditional semicircular Roman arch would have been more appropriate, but it looks like the engineers were going for style over substance. At its centre was a drawbridge to allow tall ships to pass through, and each end was surmounted by a defensive gatehouse.

The River Thames rises and falls with the tides. By blocking nearly two-thirds of it, the overly broad starlings and piers of the bridge severely restricted the natural flow of the river. So, when the tide turned, the water was much higher on one side of the bridge than the other, because it couldn't flow past, and the choked water created deadly rapids. All but the most foolish sailors avoided passing under the bridge during those times, for fear that their boats would overturn, casting them into the river. But hundreds died. Maybe their lives would have been saved had they paid heed to the saying, inspired by the bridge, which cautioned that it was made 'for wise men to pass over, and for fools to pass under'.

To make matters worse, houses began appearing on the bridge. Now, I like the idea of living on a bridge – watching the river change as the day went on and enjoying spectacular sunsets would undoubtedly have been an uplifting experience. This has worked beautifully on the Ponte Vecchio in Florence in Italy, where the carefully planned and executed houses and shops create a feeling of peace and civic order. By contrast, the houses on London Bridge only added to the chaos.

Squashed between the carriageway and the edge of the struc-ture, numerous three- and four-storey houses and shops were built, until there were over a hundred such dwellings. Temporary stalls were set up in front where shopkeepers sold their wares. Public latrines overhung the sides of the structure, discharging waste directly into the river below. The bridge had not been designed for the weight of the buildings, and the buildings themselves were not safely separated from one another, creating a huge fire risk. The bridge really was an accident waiting to hap-pen. Most of the houses were destroyed by a fire in 1212, along with thousands of unfortunate people who had crowded onto the bridge to watch the flames take hold at one end – but then strong winds carried embers to the opposite end and started a new fire, trapping them in the middle. More than 3,000 bodies were found severely or partially burned, and many more were reduced to unidentifiable ashes. In 1381 and 1450, revolts and rebellions again laid waste to many parts of the bridge.

By the fifteenth century, the buildings on the bridge had doubled both in number and in height. These tall, overhanging structures created dark and dismal passages through which carts, wagons, cattle and pedestrians fought their way. At peak times it could take an hour to cross. Between the overloading of the bridge by the houses, the effects of fires, and the wearing away of the supporting piers by the rapids flowing between them, some portion of the structure was always crumbling and collapsing into the water.

In 1633 a third of the homes were destroyed by yet another fire, although this was perhaps a blessing in disguise, because it created a gap between houses on the bank and those on the bridge. This probably saved the structure from disaster in 1666, when the Great Fire of London couldn't spread across this void. It was, quite literally, a narrow escape, but it seems that the residents and shopkeepers didn't learn their lesson. Another fire in 1725 destroyed over 60 houses and two of the arches.

*

The houses were finally demolished in 1757, and the bridge survived past the turn of the century until 1832, when the new London Bridge (designed by civil engineer John Rennie) was constructed alongside it. But the original bridge is still firmly embedded in our culture – when I was little, my mother taught me the nursery rhyme inspired by its precarious history, singing, 'London Bridge is falling down, my fair lady' in her slightly accented, out-of-tune voice. It's a rare song about engineering. It teaches the future engineers among us about the perils of bad design before we can even walk.

No. 2: The Pontoon

THE BRIDGE OF BOATS OVER THE HELLESPONT, USED BY XERXES.

The Pontoon: bridging the sea with boats.

When we think of a bridge, we usually picture something high up in the air, neatly straddling the obstruction it needs to avoid. My second bridge, however, defies this image. Seeking revenge, the ancient Persian king Xerxes built an immense 'bridge' to cross nothing less than the sea. But instead of flying over the water, he used its buoyancy to create a unique bridge, known as a *pontoon*.

Xerxes' father, Darius I, was one of the greatest emperors in history, ruling unopposed from the steppes of central Asia to the tip of Anatolia. His empire was far larger than Alexander the Great's (and, under his successors, it grew larger still). Between 492 and 490 BC he decided that the tiny Greek city states must fall under his rule, and he marched to Marathon to battle an army from Athens and Plataea. His surprise defeat there marked the end of the first Persian invasion of Greece.

Darius had planned a second attempt but died before he could fulfil his plans. Xerxes never forgot the humiliation his father had faced at Marathon, and was determined to fulfil Darius' dream of bringing the Greek states under the heel of the Persian Empire. Xerxes spent years training soldiers, planning and accumulating supplies before he attacked, and once again, while most of the Greek states submitted to him, he faced resistence from the Athenians and the fierce warriors of Sparta.

A challenge arose in 480 BC when the Persian army needed to march into Thrace via the Hellespont, the strait (now known as the Dardanelles) that separates modern-day European and Asian Turkey. After the first attempt at a crossing failed when a violent storm destroyed the bridges the Phoenicians and Egyptians had built, Xerxes ordered that the waters be given 300 lashes for their insolence. And he had the engineers who built the two failed bridges beheaded.

The replacement engineers, presumably in a bid to save their necks, built a much more substantial structure. The Persians had to travel 1.5km across a deep strait – at the time a huge distance to span, and very difficult to do using the traditional bridging technique, which was to build foundations underwater on solid ground and then span material between supports. Instead, as Herodotus tells us in *The Histories*, they gathered 674 ships (a combination of *penteconters*, Greek ships with 50 oars, and *triremes*, low, flat ships with three banks of oars) and arranged them side by side in two lines. Laid above each row of boats were two cables of flax and four cables of papyrus. These extremely heavy cables tied the boats together, and created the base of the deck.

The engineers cut long planks of wood from tree trunks and laid them edge-to-edge on top of the taut cables. The planks were tied together and evenly covered with a layer of broken twigs and branches, after which soil was thrown on top and stamped down to create a surface that the army could walk on. The engineers also laid heavy anchors upstream and downstream of the bridge: those to the east stopped the boats being pushed down the strait by winds from the Black Sea, while the others resisted the winds from the west and the south. Fencing was installed along the sides of this wide walkway to prevent the horses from seeing the water and being spooked.

Once this bridge of boats was ready, Xerxes offered prayers for safe passage. He threw his cup, a golden bowl and a Persian sword into the straits, possibly as an offering to the sun, or possibly as a form of appeasement to the sea. The army then began to cross this monumental pontoon bridge en route to the Greeks at Thrace. It is said that it took seven days and seven nights for the Persians – including Xerxes' elite fighters known as 'The Immortals' – to cross from one side of the strait to the other.

Despite this feat of engineering, the military side of the story is less epic. Xerxes was defeated at the battles of Salamis and Plataea and, after losing large numbers of men to war or starvation, he retreated back to Persia. Although he managed to subjugate Nature, Xerxes couldn't do the same to the Greek people.

*

Floating or pontoon bridges are believed to have originated in China sometime between the eleventh and sixth centuries BC, when engineers used boats with boards on top to cross large rivers. Use of the pontoon bridge continued through ancient Roman and Greek times – a notorious example was supposedly assembled by Caligula so he could show off his clothes in parades. During the World Wars, soldiers often used this technique because it allowed them to assemble and dismantle a path across water quickly and efficiently. Floating bridges are a great option when water is deep, the span is long and time is short. But storms and currents in the water affect them badly: there are many examples (such as the Murrow and Hood Canal Bridges in the USA) which have failed in strong storms. If one of the boats fills up with water, it drags down the others, until the whole line sinks. Fortunately, engineers no longer face the same fate as those that once served Xerxes.

No. 3: The Falkirk Wheel

The Falkirk Wheel: a rotating bridge.

As it bounced up and down with the waves, and sideways with the currents, walking across Xerxes' boat-bridge would have been disconcerting. We don't like our structures to move perceptibly: it scares us into thinking that they are not safe. But what if a bridge is designed to rotate? Many bridges allow land vessels to cross water, but one of my favourite bridges enables water vessels to cross land.

The Celtic doubled-headed axe was a formidable weapon, with a blade on either side of its shaft so that, in battle, a brave warrior could swing it right or left with equally destructive results. Unlikely as it may seem, this menacing tool is the inspiration for one of the coolest and most unusual structures in the world – the Falkirk Wheel.

The low-lying canals of Scotland were once a flurry of activity. The Union Canal, opened in 1822, went from Falkirk to Edinburgh as a way to bring coal into the capital and feed the new industries that were setting up factories in the city. The Forth and Clyde Canal (opened in 1790) served the same purpose for Glasgow, at that time a small town that was rapidly growing into the industrial heartland of Scotland. However, once the railway network began to develop in the 1840s, these canals, like so many others, became redundant, because it was quicker to transport minerals by train. The canals gradually fell into disrepair – by the 1930s they were in such a state that a portion of the canal system was filled in. A former transport artery was sealed off for good – or so it seemed.

At the end of the twentieth century, architects and engineers conspired to reopen the canals by creating a new waterway-based link between Glasgow and Edinburgh, specifically between the Forth and Clyde Canal and the Union Canal.

Making the 200-year-old waterways usable again offered environmental and economic advantages to the communities that lie along them. But doing this presented some technical challenges, foremost among them a large steep slope that had to be crossed. The traditional way canal-builders dealt with a slope was by means of a *lock*. Between the lower and higher sections of canal, they constructed a long, narrow, tall-sided chamber with a door (or pair of doors) at each end, which could seal or 'lock' in the water. Bargemen ascending the canal would manoeuvre into the chamber and close the lower doors behind them. They would then lift the 'paddles' (shuttered openings) at the other end of the lock, allowing water to flow in from the higher canal. Gradually the lock filled, to the point where the water level was the same as the higher section of canal. At this point the bargemen could open the upper doors and float on their way. A bargeman descending followed the same process in reverse. Originally, this journey between Edinburgh and Glasgow meant a wearying day-long passage through 11 locks, opening and shutting 44 lock gates along the way. Hardly an easy task – and in any case the locks had since been removed. So the engineers had to do some smart thinking.

Today, if you travel west from Edinburgh along the Union Canal towards the Clyde or Glasgow, you eventually reach a place where the land drops sharply away on either side, leaving you chugging along an aqueduct that thrusts out boldly into seemingly empty space. This is the end of the Union Canal. At this point, your boat is 24m up in the air, roughly as high as the top of an eight-storey building. To get from this elevation down to the lower basin and float off along the Forth and Clyde

Canal, your boat must now enter the embrace of an exceptional piece of engineering, a modern take on the Celtic axe.

An immense vertical wheel (like a Ferris Wheel) 35m in diameter lies in front of your boat. The wheel has two axe-shaped arms that rotate through 180°. Each arm houses a sort of 'gondola': a vessel large enough to carry two canal boats and 250,000 litres of water. A hydraulic steel gate stops the water from the high-level canal pouring out. When the wheel's gondola is aligned with the end of the aqueduct, the gate at the end of the canal and the gate at the end of the gondola open, and the boat can manoeuvre straight into the gondola. The doors are resealed – and the arms start to rotate.

At a funfair, as the Ferris wheel turns, you'll have noticed that your seat also moves, so that you remain sitting vertically. As you travel from the bottom to the top and back again, your orientation stays the same. In a similar fashion, a complex system of gears and cogs makes sure that the gondolas on the Falkirk Wheel always remain horizontal as the arms swing through the air. To complete one 180° turn needs little power – the same amount of electricity as boiling eight kettles of water. This is largely thanks to Archimedes and his famous principle, which states that when an object is placed in water, it displaces its own weight. If, for example, you have one boat in one gondola, but no boats in the other side, the two gondolas will still weigh the same. The boat will have displaced an amount of water from its gondola equal to its own weight. So as long as the water levels in both arms are equal, you only need minimal power to overcome inertia and start the wheel rotating, and then momentum carries the balanced arms round until they're switched off. The Falkirk Wheel brings boats from the upper basin to the lower basin (or vice versa) in just five minutes, compared to the full day required to negotiate the canal's original system of locks.

*

There are a few examples of boat lifts around the world – such as the Strépy-Thieu in Belgium, the Niederfinow Boat Lift (the oldest working boat lift in Germany), and at the Three Gorges Dam in China (now the tallest boat lift in the world, it moves boats vertically by a colossal 113m) – but there is a particular thrill to watching and travelling on the Falkirk Wheel. Perhaps this is because it taps into our childhood memories of fairgrounds. It's an example of how engineering has an aesthetic and even nostalgic side to it that plays a part in how we respond to a structure.

No. 4: The Silk Bridge

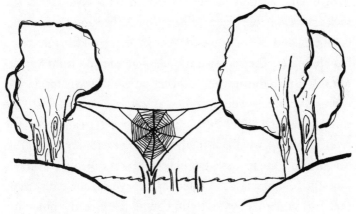

The silk bridge: the longest web-bridge in the world.

One evening I was reading a book with the television on, just letting the reassuring sound of the show's host fill the living room, but not paying any real attention. Until, that is, I heard the words 'strong material' and 'bridge' and, as you can imagine, my ears pricked up like a cat's. The host was talking about one of the most prolific bridge builders in the world – and, exceptionally, the builder is female, and lives in Madagascar.

She's about the size of a thumbnail, has eight very hairy legs and her body is heavily textured like the bark of a tree, which, as David Attenborough went on to explain, is the camouflage that protects her from predators. She also has a *spinneret*, which is the bit of her body responsible for making her the brilliant bridge engineer she is.

Darwin's bark spider can build a bridge up to 25m long (that's 1,000 times her own size), spanning rivers or even lakes. However, unlike most bridge builders, she's not looking for a way to get from one side to the other. She's looking for a way to get food.

Scurrying around in the vegetation on a river bank, she seeks out a suitable place for her project (like any professional engineer) and then releases dozens of sticky silk threads from her spinneret. They spray out – just like they do from Spider-Man's wrists in the movies – and are caught by the natural wind currents that exist above bodies of water in dense forest. The threads are carried in a thin, almost invisible, stream across the river to catch and attach themselves to vegetation there. This line of silk – the first stage of the build – is called the *bridging line*. The bridging line is a catenary, the typical curve of a cable that sags under its own weight. Giving the thread a quick tug to make sure it's secure, the spider uses the hairs on her legs, which are tiny hooks, to reel in the line a little so that it doesn't sag too much.

She tests her bridging line by walking along it, and as she does so she uses more silk and secretions to reinforce the line, making it even stronger. When she reaches the other end, she reinforces the bridging line's attachment to the vegetation by spinning more thread around it. It's important that the connection – which the wind made simply by sticking the line to a branch – will be strong enough to carry the weight of the rest of the structure.

Now the bridging line has to be anchored. The spider searches for bits of vegetation, such as large blades of grass sticking up from the water, and then moves along the line until she is almost directly above them. Slowly, producing more silk, she lowers herself and attaches her anchor point to a blade of grass close to the surface of the river, creating a T-shaped skeleton for her web.

Over the next few hours, the spider effortlessly shuttles back and forth, using the T-shaped skeleton as a base on which to attach more silk threads. From bridging line to anchor thread, she keeps producing and weaving new silk, in a big circular pattern. Some of the silk is not sticky: it functions as part of the structural frame of her construction. The rest is sticky, and will be the part of the web that actually traps her food. Eventually she will have fashioned a giant orb that can be more than 2m in diameter.

Darwin's bark spider is the only known spider that bridges water to trap its food. Its victims are the tasty mayflies, dragonflies and damselflies that zip over the middle of rivers. The large diameter of the web means that small creatures like birds or bats could also potentially get caught.

The sheer size of the web is hugely impressive, but the silk used to build it is even more astonishing – which makes sense: to build such a large structure you need an exceptional material. The bark spider's silk has been tested in a laboratory by connecting it to hooks that are slowly pulled apart. The results show us that these tiny creatures produce silk with an incredible *elasticity*. This is the property of a material to stretch under a load and then recover: if, after the load is removed, a material shrinks back to its original size, it is *elastically deformed*; if it doesn't fully recover to its original shape, it is *plastically deformed*. Tests have shown that the bark spider's silk is twice as elastic as other known spider silks. It is also very *tough*. Toughness is the property of a material that measures how much energy it can absorb without fracturing. It is a combination of strength (how much load the material can resist) and *ductility* (how much it can deform without breaking). In fact, the silk of Darwin's bark spider is the toughest biological material we've found so far – it's even tougher than steel.

Elasticity and toughness are a great combination for building materials. Take, for example, rubber bands: if you have a thin and stretchy band, you can stretch it very far but only with a small load, as it's elastic and ductile but not very strong. A very thick band made from brittle rubber can take more load but might snap suddenly, because it's strong but fragile. The bark spider's silk, on the other hand, has the ideal balance of all these properties. It can absorb large forces, and at the same time it can stretch a long way without snapping. This balance makes it the perfect material for building the world's largest spider webs.

*

I have included the Darwin's bark spider's bridge as a reminder that it is not just humans who create structures: in fact, as this creature demonstrates, we are still catching up with Nature. We are only now starting to build bridges that span as far as this spiders' does compared to our own body size – the Akashi Kaikyo Bridge in Japan currently holds the record for the longest span, at 1,991m. We are already being inspired by Nature (we call this type of design *biomimicry*) – the ventilation system of the Eastgate Centre in Zimbabwe is inspired by porous termite mounds, and the Quadracci Pavilion of the Milwaukee Art Museum has a retractable shade inspired by the wings of birds. But I believe we can learn even more. It would be the dream of any engineer to develop a super-material like spider silk that is incredibly tough and light, and which could be launched across a river or a valley, allowing the air to carry its threads to the other side. Then we too could create a long bridge in a few hours – just as quickly as the Darwin's bark spider does.

No. 5: Ishibune Bridge

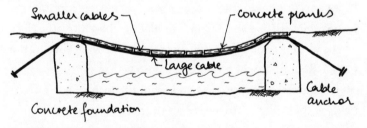

The Ishibune Bridge: a catenary bridge.

At our hotel in Tokyo, my mum and I had been given a piece of paper on which an address had been written in a series of delicate, swirling strokes that looked like little pictures. The writing was beautiful, but illegible to us, so we simply handed the paper to our taxi driver, and hoped it was enough to get us to our destination.

It was raining so hard we could barely see where we were going, but we were aware that we had left the city and were now surrounded by steep slopes covered in dense green forest. Driving higher and higher up a narrow, winding road, we finally reached a red gate with more beautiful characters inscribed upon it. Our driver came to a halt and waved us out of the car – I hoped he would still be there when we came back. I zipped up my jacket and walked along a narrow dirt path looking for the Ishibune Bridge, the perfect example of a simple stress-ribbon bridge – a form that, until I'd planned that particular trip, was unknown to me.

Earlier in the year I had been awarded a travel grant by the Institution of Structural Engineers – my proposal had been to study a special type of bridge. Speaking to my colleagues and doing some research made me aware of the stress-ribbon bridge, a graceful, simple form of which there are less than a handful of examples in the UK. I wanted to learn more about them and understand why they are so rare. My proposal was to travel to Europe and Japan – to places where these bridges are used to great effect – and report back. I went first to the Czech Republic, where engineers showed me a huge range of structures that use the stress-ribbon technique – from bridges spanning motorways to a tunnel built using the same principles. Then, at a German university, I met researchers who had built a 13m-long prototype in a lab, and who were doing tests and experiments on it. I got to do some 'testing' myself – basically jumping up and down on its deck to try to make it resonate.

To make your own mini version of a stress-ribbon bridge, use two tins of baked beans placed a metre apart to simulate bridge abutments, then lay two thick pieces of string over the tins, taping the ends to the table, which represents the ground. To turn this into a stress-ribbon bridge, it needs a deck, which you can make with matchboxes. Poke two holes through the boxes – one in each side – then lay them out on top of the strings. Thread cut pieces of rubber bands through the holes to link the matchboxes. The rubber band will stretch, compressing the matchboxes together.

If you press down on the model bridge in the middle of its span, you'll see the supporting strings tighten (in other words, develop tension); the strings pull at the tape which secures their ends to the table. A stress-ribbon bridge works in a similar way. Steel cables are slung across the gap to be bridged. The cables are thick – with a diameter about the size of a fist – and consist of numerous thin steel wires spun together to form a strong rope, which is protected by a rubber sheath. Concrete abutments at either end support the cables, which are anchored tightly into the ground. The anchors are strong enough to take the force of the cables pulling on them even when the bridge is loaded with lots of people. Planks of concrete (equivalent to the matchboxes), with grooves on the underside, are placed on top of the steel cables and connected to them to keep them in place. The planks have holes running through them, through which smaller steel cables are threaded and tightened to tie the planks together and make the deck stiffer.

The shape of these bridges reminds me of the basic rope bridges made by our ancient ancestors. Like them, and the bark spider's bridging line, a stress-ribbon bridge is a catenary. A stress-ribbon bridge is also very light – the concrete planks are quite thin at about 200mm – and the natural curved shape of the steel cables gives them a slender and satisfying aesthetic. And, just as importantly, as far as I'm concerned, these bridges are practical too, being relatively quick to build. Once the foundations are done, the lifting of the pre-made concrete planks onto the cables is a straightforward and speedy procedure, so building them has less of an impact on the surrounding environment.

The curved red ribbon of my Japanese bridge crossed a deep ravine with a small but rapidly flowing stream at its base. As the rain hammered down, I stepped out onto the deck. It was a little bouncy. I walked up and down several times, varying my speed, then jumped on it, to see what that felt like. The movement was disconcerting, and I realised why – even though stress-ribbon bridges look fantastic and are quick to build – some people don't like them.

Because they are light and rest on cables, there's a large sag in the middle, and the ends of the bridge have relatively steep slopes which can be tricky for people with buggies or those using wheelchairs. And these bridges can be lively – their lightness and flexibility mean that, as you walk across them, they can feel unstable. Even though they are perfectly safe, stress-ribbon bridges usually move. The sag, combined with the bounciness, can give the impression that these bridges are a little dodgy. People in the three countries I'd already visited loved them, but they were used to their movement. Elsewhere, a misplaced perception of instability, and a lack of strong ground in which to anchor the tendons and keep the structure stable, might be reasons why stress-ribbon bridges haven't caught on.

By now I was soaked to the skin, but I spent a long time examining the expert engineering (after all, I had travelled nearly 10,000km to see this bridge that was so unusual back home in Britain). When the bridge shook, I clung to a handrail with one hand, trying to keep hold of my umbrella with the other. I found it difficult to stand in the middle for too long, as the valley's depth, the rapid water gushing through it, and the fact that it moved the most here, unnerved even me.

Nevertheless, like any self-respecting engineer, I made sure that my mum got plenty of snaps of me *in situ* on the structure, before we raced back to the taxi, where the driver lay asleep in his reclined seat. We drove back to Tokyo, still a little soggy from our visit.

The stress-ribbon structures I studied during my travels have stayed with me: I'm inspired by the fact that the simple rope bridge has evolved to incorporate modern technology and materials – and that despite its modernness, this new form has retained the simplicity and elegance of its forebear. New engineering doesn't always have to be big and bold; sometimes it can draw on humbler roots.

*

Bridges are all very well, but no doubt you're wondering how things worked out between me and Flirtman. All I can say is that I came to regret emailing my friend to boast that I had got rid of him in three minutes. Four years later, she read that email out loud in front of hundreds of people. During her bridesmaid's speech. At my wedding.

Yes, dear reader, I married him.

DREAM

Imagine, for a moment, a world without engineers. Abandon Archimedes. Banish Brunelleschi, Bessemer, Brunel and Bazalgette. Forget Fazlur Khan, oust Otis and, yes, get rid of Emily Roebling and Roma Agrawal. What do you see?

More or less nothing.

Of course, there'd be no skyscrapers, no steel, no elevators, no houses and no sewers under London (gross). No Shard. There'd be no phones, no internet and no TV. No cars, nor even carts – which is perhaps just as well since there'd be no roads or bridges either. And we'd be wearing more or less nothing too: there'd be no stitching together of animal skins to make clothes. And no tools for foraging, no fire for safety, no mud huts or log cabins.

Engineering is a big part of what makes us human. Sure, there are crows that can fashion a piece of wire into a hook to retrieve food, and octopuses that carry coconut shells for protection, but – so far at least – we have the edge. Engineering furnished us first with the essentials – food, water, shelter, clothing – and then with the means to cultivate crops, build civilisations and fly to the Moon. Tens of thousands of years of innovation have brought us to where we are today. Human ingenuity is boundless; we will always aspire to manufacture more, to live better, to solve the next problem – and then the next. Engineering has created, in the most literal way, the fabric of our lives; it has shaped the spaces in which we live, work and exist.

And it'll shape our future, too. Already, I can see certain trends in engineering – irregular geometry, technology such as robotics and 3D printing, a quest for more sustainability, the merging of different disciplines (such as in biomedical engineering), a mimicking of Nature – that will once again change the way our landscape looks and feels, and the ways in which we inhabit the planet. Even if some of these seem the stuff of science fiction at the moment.

Computing capacity has made it possible for us to draw complicated, cambering shapes, such as the flowing surfaces of the Spanish Pavilion at 2010's World Expo, the undulating Guggenheim Museum in Bilbao, and the Heydar Aliyev Center in Azerbaijan, which is as intricately shaped as a conch shell. This move towards the geometrically complex takes us away from the traditional square or rectangular building and towards more natural forms. At present, creating these shapes is still expensive because it involves curving steel and shaping it into bespoke contours, or building intricate moulds for concrete. Interestingly, these moulds can account for up to 60 per cent of the total construction budget of a project – only to be binned once the concrete has hardened. In fact, to date, keeping the cost of the moulds (or *formwork*) down is one of the reasons why our columns, walls and beams tend to be rectangular: it's cheap and easy to buy rectilinear pieces of plywood.

So with this emergence of curvy shapes, we need to think smartly about how we're going to build them. (Concrete is a good option since its liquid origin makes it ideal for transforming into any shape.) One way is to use large polystyrene blocks, painstakingly carved by hand or by machine, against which concrete can be poured. But this creates much waste, because the blocks are useless once the concrete has hardened. An exciting idea – which has been around since the 1950s but has so far only been used sparingly – is the flexible membrane mould. Almost any material, ranging from hessian or burlap to light sheets of plastic made from polyethylene (PE) or polypropylene (PP), can be used. These fabrics start off slack and shapeless – but introduce some wet concrete and we're quickly reminded of what a responsive and sensory material it is: concrete interacts with the fabric, stretching and moving it to create a final shape. Two seemingly disparate materials come together in a symbiotic relationship of pressure and restraint.

Spanish architect Miguel Fisac designed the MUPAG Rehabilitation Center in Madrid (opened in 1969), using this technique to create a façade that looks cushion-like and bouncy. At one of the entrances to the Heartlands Project in Cornwall is a wall that looks like a flowing piece of silk suspended from the sky; touch it, however, and you feel solid concrete. I'm sure we're going to see more structures like this, including many on a much larger scale, because using PE or PP sheets as formwork eliminates a huge amount of waste; plus they don't tear easily – and if they do, the tears don't propagate. Moreover, nothing – including concrete – sticks to them, so they can be used multiple times. The internal steel reinforcement skeleton doesn't need to change much; neither does the concrete mix itself. But the challenge so far has been that we're simply not used to working this way. It completely changes the aesthetics of structures: architects and engineers need to catch up, as do the logistics and procurement of construction. But they will, and I bet you that when they do, I won't be the only one caught stroking concrete in public.

Talking about stroking materials: at the University of California, Berkeley, I once got my hands on some 3D-printed modules (which ranged in size from my palm to a dinner plate) that could be assembled to make small installations, walls, facades and shelters. The modules were in a range of colours, and when I asked why I was gobsmacked by the answer. The white ones were salt. The black ones, recycled rubber tyres. The brown and the grey ones were more familiar materials – clay and concrete, respectively – but the purple ones were made of grape skins. That's right: grape skins. A research team led by Ronald Rael is investigating the use of unusual materials (mixed with resins to create a printable paste) to build stuff. I love the fact that, as well as working with traditional materials in a futuristic way – from geometric concrete blocks with irregular perforations to small gorgeously patterned hexagonal clay tiles for use on facades – they are also experimenting with waste materials, including those from the local wine industry. Some of their designs are self-supporting and don't require any additional structure. It got me to thinking about how 3D printing, along with exciting new combinations of materials, could lead to a future where we print these pieces and then assemble our own homes.

And 3D printing is not only being used on a modular scale – in fact, the world's first 3D-printed footbridge was opened in

Madrid in December 2016. At 12m long, it was analysed to find out exactly where the forces were being channelled; material was then deposited only in those sections – meaning minimal material, less waste and a lighter end product. Robots are also being designed to lay bricks and pour concrete on site: manufacturing embraced this trend decades ago, and now its the turn of the construction industry to catch up.

Taking the return to nature in form and material another step further is *biomimicry*, whereby not only do you mimic the *shape* of beehives, bamboo or termite mounds, but also their *function*. A famous example of this technique is the burdock burr that inspired Velcro: we copied its hooks, and its ability to stick to other surfaces. Nature builds simply and with as little material as possible, and we can reflect this principle in our structures. The skulls of birds, for example, have two layers of bone between which is a complex web of truss-like connections separated by large air pockets – in fact, bone tissue forms naturally around the cells that experience high pressure, leaving voids elsewhere. London-based architect Andres Harris conceptualised a curving canopy using a web of cushions around which a structure, similar to the birds' skulls, can be cast. Similarly, the Landesgartenschau Exhibition Hall in Stuttgart gets its inspiration from the sea urchin, which has a skeleton made from interlocking plates or *ossicles*, each of which is sponge-like and lightweight. The exhibition centre is made from 50mm-thick plywood sheets, analysed carefully by software and then fabricated robotically and assembled. If you magically expanded an egg to be the same size as this structure, the plywood would be thinner than the egg's shell.

Nature also heals itself: the human body can detect when something is wrong (often making us feel pain) and then, through a series of steps, fixes the problem. So far, with structures, we have had to intervene and perform repairs – or surgery – when things go amiss. However, a team led by Phil Purnell of the University of Leeds is designing robots that can travel – like white blood cells – through pipes in the road to diagnose defects which can then be fixed before they lead to erosion and collapse. The Institute of Making's Mark Miodownik is leading a team developing 3D printing technology that can be carried by drones to repair potholes and other road defects so we won't need to close down roads to repair them, saving money and easing traffic – the end of roadworks, perhaps? And a team at the Cambridge Centre for Smart Infrastructure and Construction is looking at adding nervous systems to new structures. A single fibre optic cable, tens of kilometres long, with continuous sensing elements, can measure the strain and temperature inside piles, tunnels, walls, slopes and bridges. Data that has never been available before can be collected, and will not only help engineers learn from these designs, but also warn them of impending problems.

If I try and imagine what the world of the future will look like, I imagine these biological forms interspersed with pencil-thin towers and conserved historical structures. Already, towers such as 432 Park Avenue in Manhattan boast of their slenderness ratio (it's 14 times taller than it is wide). A challenge for stability and sway, these ultra-thin skyscrapers usually have dampers. I expect we will see more and more such structures combining offices, apartments, shops and public areas as the battle for space in congested cities intensifies. Many of our historical structures are starting to underperform as time goes on: their water and drainage pipes are often inadequate; lots of heat is lost as they were never well insulated; and beams and floors can be seen to sag. Walk around London and you will notice ornate old facades shooting up into the sky seemingly unaided because the buildings behind them have been demolished. But these facades are being surreptitiously supported by a network of beams and columns that hold them steady until a new building is put in place. Using technology such as lasers to create detailed 3D maps will make it easier for engineers to understand the old and combine it with the new.

And if I really think into the future far beyond my lifetime, I imagine my descendants living underwater in pods made from paper-thin glass that cannot be shattered. Our bridges could span ten times the distance they do today because they'll be made from graphene, the 'super-material' of our future. Perhaps we will even 'grow' our homes from biological material that can be shaped and modified to suit our needs.

But for now, I arrive home every night to the welcoming arms of my old, rectangular, solid-brick Victorian flat. As I turn out the lights (still holding my rather more haggard stuffed-toy cat from New York) and begin to doze, I wonder what the Vitruvius and the Emily Roebling of the future will create. The possibilities are limited only by our imaginations – for whatever we can dream up, engineers can make real.

ACKNOWLEDGEMENTS

Thank you:

Steph Ebdon, who planted the seed of writing a book in my mind, even though I laughed and said it would never happen. I'm ecstatic that it did.

Patrick Walsh, agent extraordinaire, who believed in me and my idea, taught me how to add texture to text, and supported me through every step of the process. Leo Hollis for his support, and that timely introduction to Patrick.

Natalie Bellos, brilliant editor, who saw something in my proposal and guided me through its years of development. Her insights, dedication (even while on leave) and attention to detail are unparalleled. Lisa Pendreigh and Lena Hall for turning it into a real object, for getting it over the finish line. Pascal Cariss for making my 'sentences sizzle' – you breathed life into my words. Ben Sumner for his impeccable copy-editing. The global Bloomsbury team, for nurturing my baby and making it the book it is today.

The brilliant Mexican engineers I met: Dr Efraín Ovando-Shelley (Instituto de Ingenieria, UNAM), who showed me the Metropolitan Cathedral; and Dr Edgar Tapia-Hernández, Dr Luciano Fernández-Sola, Dr Tiziano Perea and Dr Hugón Juárez-García (Universidad Autónoma Metropolitana –

Azcapotzalco), who explained the challenges of the ground and earthquakes. The British Council, for organising a very memorable trip to Mexico City.

Phil Stride of Tideway; Karl Ratzko, Neil Poulton, Simon Driscoll of WSP; Ronald Rael of the University of California, Berkeley: for their time and interviews aiding my research. Robert Hulse at the Brunel Museum for his insight.

Rob Thomas, fountain of book-related knowledge at the Institution of Structural Engineers' library, who found the most obscure sources for me and was always available to listen to my ramblings. Debra Francis of the Institution of Civil Engineers' library for her help.

Mark Miodownik, whose book *Stuff Matters* is my inspiration (it's still on my bed-stand): the nicest guy you could ever meet, you've done so much for me. Timandra Harkness for her support, and her introduction to my wonderful writing friends at NeuWrite, who commented and critiqued.

John Parker, Dean Ricks, Ron Slade; everyone on 'The Shard Team'; the directors at WSP: for an amazing decade of learning and growing. David Holmes and Gordon Kew at Interserve; John Priestland, Mike Burton, Peter Sutcliffe and Darran Leaver at AECOM: all my superbly supportive employers – I know I am an 'unusual' employee.

David Maundrill, Joe Harris, May Chiu, Dr Christina Burr, James Dickson, Pooja Agrawal, Niri Arambepola, Emma Bowes, Chris Gosden, Jeremy Parker, Karl Ratzko and Chris Christofi, dear friends and colleagues (and sister) who read chapters, fact-checked and helped me.

The engineers and scientists, organisations and institutions out there that have inspired me to go out and tell people about what we do. For giving me a platform to speak and write. I am optimistic about the future of our collective profession – its innovation, its impact, its inclusivity.

My family, the whole extended worldwide clan – my grandparents, aunts, uncles, cousins, nieces and nephews, and my mother-in-law – who have always been my cheerleaders, and who waited patiently for this huge project to be finished. My friends, whom I haven't seen much of recently – I will be back. You were always there. My loved ones who are no longer here, I miss you.

My parents, Hem and Lynette Agrawal; my sister, Pooja Agrawal: where do I start? For always telling me I could achieve everything I wanted through 'hard work', for inspiring me with LEGO, science and our worldwide travels, for giving me the best education, for challenging and questioning me, for all your love.

And finally, my *Flirtman,* aka Badri Wadawadigi, who has navigated me – sometimes kicking and screaming – through four years of writing, for reading the words more times than anyone else, for reminding me that I can do it when I didn't believe I could, for telling me off when I procrastinated, for intelligent feedback, for naming the book, for encouraging and pushing me to achieve more than my dreams, and for your love. Let there be many more Bridges of the Day.

Roma Agrawal, February 2018

SOURCES

Addis, Bill. *Building: 3000 Years of Design Engineering and Construction*. University of Michigan: Phaidon, 2007

Agrawal, Roma. 'Pai Lin Li Travel Award 2008 – Stress Ribbon Bridges.' *The Structural Engineer*, Volume 87, 2009

Agrawal, R., Parker, J. and Slade, R. 'The Shard at London Bridge.' *The Structural Engineer*, Volume 92, Issue 7, 2014

Ahmed, Arshad and Sturges, John. *Materials Science in Construction: An Introduction*. Routledge, 2014

Allwood, Julian M. and Cullen, Jonathan M. *Sustainable Materials – Without the Hot Air: Making Buildings, Vehicles and Products Efficiently and with Less New Material*. UIT Cambridge, 2015.

Balasubramaniam, R. 'On the corrosion resistance of the Delhi iron pillar.' *Corrosion Science*, Volume 42, Issue 12, 2000

Bagust, Harold. *The Greater Genius? A Biography of Marc Isambard Brunel*. The University of Michigan: Ian Allan, 2006.

Ballinger, George. 'The Falkirk Wheel: from concept to reality.' *The Structural Engineer*, Volume 81, Issue 4, 2003

Barton, Nicholas and Stephen Myers. *The Lost Rivers of London: Their effects upon London and Londoners, and those of London and Londoners upon them*. Historical Publications, Limited, 2016

British Constructional Steelworks Association. *A Century of Steel Construction 1906–2006*. British Constructional Steelworks Association, 2006.

Blockley, David. *Bridges: The Science and Art of the World's Most Inspiring Structures*. Oxford: Oxford University Press, 2010

Brady, Sean. 'The Quebec Bridge collapse: a preventable failure.' *The Structural Engineer*, 92 (12), 2014 (2 parts)

Brown, David J. *Bridges: Three Thousand Years of Defying Nature.* London: Mitchell Beazley, 1993

Bryan, Tim. *Brunel: The Great Engineer.* Ian Allan, 1999

Clayton, Antony. *Subterranean City: Beneath the Streets of London.* London: Historical Publications, 2010.

Cross-Rudkin, P. S. M., Chrimes, M. M. and Bailey, M. R. *Biographical Dictionary of Civil Engineers in Great Britain and Ireland, Volume 2: 1830–1890*

Crow, James Mitchell. 'The concrete conundrum.' *Chemistry World*, 2008

Davidson, D. 'The Structural Aspects of the Great Pyramid.' *The Structural Engineer*, Volume 7, Issue 7, 1929. (Paper read before the Yorkshire Branch at Leeds on 7 February 1929.)

Dillon, Patrick (writer) and Biesty, Stephen (illustrator). *The Story of Buildings: From the Pyramids to the Sydney Opera House and Beyond.* Candlewick Press, 2014.

European Council of Civil Engineers. *Footbridges – Small is beautiful.* European Council of Civil Engineers, 2014.

Fabre, Guilhem, Fiches, Jean-Luc, Leveau, Philippe, and Paillet, Jean-Louis. *The Pont Du Gard: Water and the Roman Town.* Presses du CNRS, 1992.

Fahlbusch, H. 'Early dams.' *Proceedings of the Institution of Civil Engineers - Engineering History and Heritage*, Volume 162, Issue 1, 1 Feb 2009 (19–28)

'The Falkirk Wheel: a rotating boatlift.' *The Structural Engineer*, 2 January 2002

Fitchen, John. *Building Construction Before Mechanization.* MIT Press, 1989.

Giovanni, Pier and d'Ambrosio, Antonio. *Pompeii: Guide to the Site.* Electa Napoli, 2002

Gordon, J.E. *Structures: or why things don't fall down.* Da Capo Press, 2009.

Gordon, J.E. *The New Science of Strong Materials: or why you don't fall through the floor.* United States of America: Penguin Books, 1991

Graf, Bernhard. *Bridges That Changed the World*. Prestel, 2005.

Hanley, Susan B. 'Urban Sanitation in Preindustrial Japan.' *The Journal of Interdisciplinary History*, Volume 18, No. 1, 1987

Hibbert, Christopher, Keay, John, Keay, Julia and Weinreb, Ben. *The London Encyclopaedia*. Pan Macmillan, 2011.

Holland, Tom. *Rubicon: The Triumph and Tragedy of the Roman Republic*. Hachette UK, 2011.

Home, Gordon. *Old London Bridge*. Indiana University: John Lane, 1931.

Khan, Yasmin Sabina. *Engineering Architecture: The Vision of Fazlur R. Khan*. W.W. Norton, 2004.

Lampe, David. *The Tunnel*. Harrap, 1963.

Landels, J.G. *Engineering in the Ancient World*. Berkeley and Los Angeles: University of California Press, 1978.

Landau, Sarah Bradford and Condit, Carl W. *Rise of the New York Skyscraper 1865–1913*. New Haven and London: Yale University Press, 1999.

Lepik, Andres. *Skyscrapers*. Prestel, 2008.

Levy, Matthys and Salvadori, Mario. *Why Buildings Fall Down: How Structures Fail*. United States of America: W.W. Norton, 2002.

Mathewson, Andrew, Laval, Derek, Elton, Julia, Kentley, Eric and Hulse, Robert. *The Brunels' Tunnel*. ICE Publishing, 2006.

Mays, Larry, Antoniou, George P. and Angelakis, N. 'History of Water Cisterns: Legacies and Lessons.' *Water*. 5. 1916-1940. 10.3390/w5041916.

McCullough, David. *The Great Bridge: The Epic Story of the Building of the Brooklyn Bridge*. Simon & Schuster, 1983

Mehrotra, Anjali and Glisic, Branko. *Deconstructing the Dome: A Structural Analysis of the Taj Mahal*. Journal of the International Association for Shell and Spatial Structures, 2015.

Miodownik, Mark. *Stuff Matters: Exploring the Marvellous Materials That Shape Our Man-Made World*. Penguin UK, 2013.

Oxman, Rivka and Oxman, Robert (guest-edited by). *The New Structuralism. Design, engineering and architectural technologies*. Wiley, 2010.

Pannell, J.P.M. *An Illustrated History of Civil Engineering*. Univerity of California: Thames and Hudson, 1964.

Pawlyn, Michael. *Biomimicry in Architecture*. RIBA Publishing, 2016.

Pearson, Cynthia and Delatte, Norbert. *Collapse of the Quebec Bridge, 1907*. Cleveland State University, 2006

Petrash, Antonia. *More than Petticoats: Remarkable New York Women*. Globe Pequot Press, 2001

Poulos, Harry G. and Bunce, Grahame. *Foundation Design for the Burj Dubai – The World's Tallest Building*. Case Histories in Geotechnical Engineering, Arlington, VA, August 2008.

Randall, Frank A. *History of the Development of Building Construction in Chicago Safety in tall buildings*. Institution of Structural Engineers working group publication, 2002

Salvadori, Mario. *Why Buildings Stand Up*. United States of America: W.W. Norton and Company, 2002.

Santoyo-Villa, Enrique and Ovando-Shelley, Efrain. *Mexico City's Cathedral and Sagrario Church, Geometrical Correction and Soil Hardening 1989–2002– Six Years After*.

Saunders, Andrew. *Fortress Britain: Artillery Fortification in the British Isles and Ireland*. Beaufort, 1989.

Scarre, Chris (editor). *The Seventy Wonders of the Ancient World: The Great Monuments and How They Were Built*. Thames & Hudson, 1999.

Shirley-Smith, H. *The World's Greatest Bridges*. Institution of Civil Engineers Proceedings, Volume 39, 1968.

Smith, Denis. 'Sir Joseph William Bazalgette (1819-1891); Engineer to the Metropolitan Board of Works.' *Transactions of the Newcomen Society*, Vol.58, Iss. 1, 1986

Smith, Denis (editor). 'Water-Supply and Public Health Engineering', *Studies in the History of Civil Engineering*, Volume 5

Sprague de Camp, L. *The Ancient Engineers*. Dorset, 1990.

Soil Survey, Tompkins County, New York, Series 1961 No.25. United States Department of Agriculture, 1965.

Trout, Edwin A.R.. 'Historical background: Notes on the Development of Cement and Concrete', September 2013

Tudsbery, J.H.T. (editor). *Minutes of Proceedings of the Institution of Civil Engineers*

Vitruvius. *The Ten Books on Architecture* (translated by Morgan, Morris Hicky). Harvard University Press, 1914.

Walsh, Ian D. (editor). *ICE Manual of Highway Design and Management.* ICE Publ., 2011.

Weigold, Marilyn E. *Silent Builder: Emily Warren Roebling and the Brooklyn Bridge.* Associated Faculty Press, 1984

Wells, Matthew. *Engineers: A History of Engineering and Structural Design.* Routledge, 2010.

Wells, Matthew. *Skyscrapers: Structure and Design.* Laurence King Publishing, 2005.

West, Mark. *The Fabric Formwork Book: Methods for Building New Architectural and Structural Forms in Concrete.* Routledge, 2016.

Wood, Alan Muir. *Civil Engineering in Context.* Thomas Telford, 2004.

Wymer, Norman. *Great Inventors (Lives of great men and women, series III).* Oxford University Press, 1964.

https://www.tideway.london

http://puretecwater.com/reverse-osmosis/what-is-reverse-osmosis

http://www.twdb.texas.gov/publications/reports/numbered_reports/doc/r363/c6.pdf

http://mappinglondon.co.uk/2014/londons-other-underground-network/

http://www.pub.gov.sg/about/historyfuture/Pages/HistoryHome.aspx

http://www.clc.gov.sg/Publications/urbansolutions.htm

http://www.thameswater.co.uk/

http://www.bssa.org.uk/about_stainless_steel.php?id=31

https://www.newscientist.com/article/dn19386-for-self-healing-concrete-just-add-bacteria-and-food/

http://www.thecanadianencyclopedia.ca/en/article/quebec-bridge-disaster-feature/

http://www.documents.dgs.ca.gov/dgs/pio/facts/LA workshop/climate.pdf

http://www.cement.org/

http://www.unmuseum.org/pharos.htm

http://www.otisworldwide.com/pdf/AboutElevators.pdf

http://www.waterhistory.org/histories/qanats/qanats.pdf

http://users.bart.nl/~leenders/txt/qanats.html

http://water.usgs.gov/edu/earthwherewater.html

http://www.worldstandards.eu/cars/driving-on-the-left/

http://journals.plos.org/plosone/article?id=10.1371/journal.pone.
0026847

https://www.youtube.com/watch?v=gSwvH6YhqIM

http://www.livescience.com/8686-itsy-bitsy-spider-web-10-times-
stronger-kevlar.html

http://linkis.com/www.catf.us/resource/flbGp

http://www.bbc.co.uk/news/magazine-33962178

http://www.romanroads.org/

http://www.idrillplus.co.uk/CSS ROADMATERIALSCONTAINI NGTAR
171208.pdf

http://www.groundwateruk.org/Rising_Groundwater_in_Central_
London.aspx

http://indiatoday.intoday.in/story/1993-bombay-serial-blasts-terror-
attack-rocks-india-financial-capital-over-300-dead/1/301901.html

http://www.nytimes.com/1993/03/13/world/200-killed-as-bombings-
sweep-bombay.html?pagewanted=all

http://www.bbc.co.uk/earth/story/20150913-nine-incredible-buildings-
inspired-by-nature

http://www.thinkdefence.co.uk/2011/12/uk-military-bridging-floating-
equipment/

http://www.meadinfo.org/2015/08/s355-steel-properties.html?m=1

http://www.fabwiki.fabric-formedconcrete.com/

http://www-smartinfrastructure.eng.cam.ac.uk/what-we-do-and-why/
focus-areas/sensors-data-collection/projects-and-deployments-
case-studies/fibre-optic-strain-sensors

http://www.instituteofmaking.org.uk/research/self-healing- cities

PHOTOGRAPHY CREDITS

INDEX

Page numbers in *italic* refer to the illustrations

Index

Index

Index

Index

Index

Index

A NOTE ON THE AUTHOR

Roma Agrawal is a structural engineer who builds big. From footbridges and sculptures, to train stations and skyscrapers – including The Shard – she has left an indelible mark on London's landscape.

Roma is a tireless promoter of engineering and technical careers to young people, particularly under-represented groups such as women. She has advised policymakers and governments on science education, and has given talks to thousands worldwide at universities, schools and organisations, including two for TEDx. She is a television presenter, and writes articles about engineering, education and leadership.

Roma has been awarded international awards for her technical prowess and success in promoting the profession of engineering, including the prestigious Royal Academy of Engineering's Rooke Award.

Built is her first book.

www.RomaTheEngineer.com
@RomaTheEngineer

A NOTE ON THE TYPE

The text of this book is set in Minion, a digital typeface designed by Robert Slimbach in 1990 for Adobe Systems. The name comes from the traditional naming system for type sizes, in which minion is between nonpareil and brevier. It is inspired by late Renaissance-era type.

Dawn T. Steward
2130 Blossom Lane
Winter Park, FL 32789

ENOUGH
IS
ENOUGH

ENOUGH
IS
ENOUGH

The Hellraiser's Guide
to
Community Activism

DIANE MACEACHERN

AVON BOOKS ◆ NEW YORK

ENOUGH IS ENOUGH: THE HELLRAISER'S GUIDE TO COMMUNITY ACTIVISM is an original publication of Avon Books. This work has never before appeared in book form.

AVON BOOKS
A division of
The Hearst Corporation
1350 Avenue of the Americas
New York, New York 10019

Published by arrangement with the author
Library of Congress Catalog Card Number: 94-873
ISBN: 0-380-77335-X

Library of Congress Cataloging in Publication Data:
MacEachern, Diane.
Enough is enough: The hellraiser's guide to community activism / Diane MacEachern.
p. cm.
1. Community leadership. 2. Civic improvement—United States. 3. Local government—United States. I. Title.
JS341.M33 1994 94-873
303.48′4—dc20 CIP

First Avon Books Trade Printing: August 1994

Printed in the U.S.A.

OPM 10 9 8 7 6 5 4 3 2 1

To Maria
a terrific partner, a wonderful friend

ACKNOWLEDGMENTS

There is absolutely no way I could write a book, hold down a "regular" job, and keep track of the rest of my life without the support of my friends, colleagues, and family members.

My husband, Dick Munson, always a terrific sounding board, was a constant source of support. Our children, Dana and Daniel, will be glad this project is over. I am grateful to Gail Ross, my agent and good friend, for all the time and energy she expended in shepherding this guide through the vagaries of the book publishing world. My editor at Avon Books, Jody Rein, is simply the best. I also appreciate the support of editor-in-chief Bob Mecoy, editor Gwen Montgomery, and the rest of the Avon Books staff who worked to make this effort a success.

The staff of Vanguard Communications, including Lottie Gatewood, Jennifer Harrison, Rose Matthews, and Mary Tracy Packard, all lent moral support and an enthusiastic word of encouragement whenever they were needed. Barbara Briggs formatted the manuscript over and over and over again with her usual good humor and cheerful smile.

Neil Grauer, a first-class writer in his own right and the Vanguard wordsmith, researched and wrote some of the case studies included herein, setting a high standard for the rest of the book. Thanks, Professor.

Deanna Troust deserves a special note of unqualified appreciation. As the primary researcher of the case studies that bring this book to life, Deanna spent endless hours tracking down and writing about amazing people and organizations whose innovative tactics and inspiring success stories prove that you can raise hell and win. She filled information gaps and persuaded people to dig deeply into their scrapbooks to lend us their precious photos. There is no way this book could have been written without her help.

I am also indebted to the many activists who gave so graciously of their time in telling us about their stories and sending us newspaper clips and campaign materials. These include Marian Borneman and Kathy Rhodes (Boyertown Community Library), Aurora Castillo (Mothers of East L.A.) and Frank Villalobos (Barrio Planners, Inc.), Herb Ettel (Trial Lawyers for Public Justice), Janet Groat (Jobs With Peace), Matt Peskin (National Association of Town Watch), Terri Schalater (Boycott Colorado, Inc.), Robert Sinclair Jr. (South East Queens Community Partnership), Phil West (Common Cause of Rhode Island), Dave Crandall (Inland Empire Public Lands Council), Karlynn Fronek (League of Women Voters of Minneapolis), Philippa Gamse (National Committee for Prevention of Child Abuse), Jan Garton (Cheyenne Bottoms), David Goldsmith (HandsNet), Shirley Hart (League of Women Voters, Bucks County, Pennsylvania), Lance Hughes (NACE), Bill Leland (Global Action and Information Network), John Cardwell (Citizens Action Coalition of Indiana), Courtney Goodman (American Arts Alliance), Terry Moore (Dump Patrol), Galen Nelson (People for the American Way), John Randolph (Sipsey Wilderness Area), Nancy and Jim Ricci (RID Connecticut), Pat Blumenthal (Colorado NARAL), James Bock (*Baltimore Sun*), Vincent DeMarco (Marylanders Against Handgun Abuse), Bernard Horn (Coalition to Stop Gun Violence), Jef Feeley (*Daily Record,* Baltimore), John Papagni (South East Community Organization), Tracy Arpen (CAP Signs), Sue Bell (Gifts in Kind America), Suzanne Fahey (League of Women Voters of Metro Columbus), Bob Forney and Kate Potter (Citizens for a Healthy Environment), Judy Norsigian (Boston Women's Health Book Collective), Kathleen Prudence (Peaslee, Over-the-Rhine Community Council), Steve Kemme (*Cincinnati Enquirer)*, Kimberly Walker (Mission Possible, Children's Trust Fund of Oregon), Tony Bloome (Common Cause), Josh Feit,

Kirsten Cross and Kelle Louaillier (INFACT), Helen Lichtenstein (Advocacy Institute), Connie Mahan (National Audubon Society), Tim Muck and Todd Putnam *(National Boycott News)*, Northern Rockies Action Group, Tift Pelias (League of Women Voters Education Fund), Tim Wall (Citizens Committee for New York City), Mike Williams (Citizens Clearinghouse for Hazardous Wastes), Anne Zorick (Co-op America), Roy Morgan (Americans for the Environment), Phillip Wallace (Council of State Governments), Steve Coleman (Friends of Meridian Hill, Inc.), Bill Hearn (Maryland Special Olympics), Tom Hall (Partnership for a Drug-Free Florida), Valerie Heinonen (Interfaith Center on Corporate Responsibility), Robert Barnhart *(Third Barnhart Dictionary of New English)*, Arthur T. Keefe III (The Humane Society of the United States), John A. Hoyt (The Humane Society of the United States), Jan Hartke (EarthKind), Brian Trelstad (Center for Environmental Citizenship), LeRoy Moore (Rocky Mountain Peace Center), and Michael Beer (Nonviolence International).

I would also like to thank Karen Sagstetter, Colleen Cordes, Jack Caravanos, Katherine Isaac, Seven Locks Press, the Change America Now Foundation, Karen Menichelli and the Benton Foundation, the South End Press and the *Grassroots Fund-raising Journal* for providing me with their own books and articles. And I appreciate the helpful comments on the manuscript provided by Rich Lombardi, a citizen lobbyist based in Lincoln, Nebraska; Jay Sherman, the grassroots coordinator for the Chesapeake Bay Foundation; Charyn Sutton, a grassroots activist and communications specialist based in Philadelphia; fund-raising experts Jenny Thompson of Craver, Mathews & Smith and Amy Leveen; and Holly Schadler, an expert in federal election law. Furthermore, it was very generous of the Rails-to-Trails Conservancy to allow me to reprint the Special Event Checklist (on page 78) that I developed cooperatively with Peter Harnik when RTC celebrated the conversion of its five-hundredth rail trail.

Finally, in a true case of saving the best for last, I wish to acknowledge the constant support and continued good humor of Maria Rodriguez, the co-founder and executive vice president of Vanguard Communications. The time expended to research and write this book far exceeded our extremely optimistic projections. But as usual, Maria encouraged our company to put principles before profit to do what needs to be done. Thanks for everything, Maria.

CONTENTS

INTRODUCTION

How do you "save" the place where you live?

Thousands of Americans ask themselves this question every day. Crime is getting out of hand and they want it stopped. Another neighborhood kid gets busted for dealing drugs—or worse, gets shot in the crossfire during a gang war or drug bust. A local factory continues billowing noxious fumes and you can't stand by and watch your kids get sick. The biggest employer in town just keeps laying off workers instead of finding creative ways to adapt to a changing marketplace and keep people employed.

What do you do? What in the world *can* you do?

Well, if the people who are featured in this book are any indication, you can do a helluva lot.

Most people start by taking a simple step. They sit down around their kitchen table with their neighbors, share their concerns, and get up the gumption to write a letter or make a phone call to someone they think is more powerful than they are.

But it doesn't take long to realize you need to do more—and that if *you* had the power, you *could* do more.

So you begin to get more people involved, to make plans . . . to organize. The job confronting you seems enormous, almost impossible. But little by little, momentum builds. More and more people join your campaign. The media begin covering your concerns. Politicians take your requests seriously, and donors even give you money so you can keep on organizing.

You don't win every battle, but you win enough of them. And eventually, you make a difference. Waste dumps get closed down, employees go back to work, drug dealers move away.

By now, the realization has hit you. The more "hell" you raise, the more powerful you become. And with that power come concrete improvements in your life and the lives of others in your community.

This book is about "raising hell." It had its genesis in the NIMBY (Not In My Backyard) movement that energized so many environmental struggles over the last two decades. But just like activism itself, the concept for the book evolved from a list of strictly defensive actions into a presentation of preemptive strategies and tactics for true social change. Thus, *Enough Is Enough: The Hellraiser's Guide to Community Activism* offers both ways to fight back and a course to fight for, a tool to empower people so that they not only can protect themselves from undesirable developments but pursue their own constructive objectives.

Beyond NIMBY

The Oxford Dictionary of New Words claims that the term NIMBY was first used around 1980 as a derogatory label for the antinuclear movement and other citizen efforts that were perceived to oppose any large local construction project. NIMBY made British headlines when outspoken environmental secretary Nicholas Ridley protested the construction of a housing development near one of his homes and a walking trail near another. (Ridley's ironic outcry came shortly after he mocked members of Parliament from the Tory party for themselves having NIMBY attitudes toward housing projects.) The press subsequently characterized citizens as suffering from the "NIMBY syndrome" whenever they opposed the construction or operation of a facility that was *thought* to be dangerous or threatening, particularly because it was located near their homes.

The phrase became more popular in the United States in the 1980s, as housewives and blue- and white-collar workers in inner cities, the suburbs, and rural areas began organizing to find and clean up toxic waste sites and other facilities that were believed to pose an environmental or health hazard to them and their children.

As places like Love Canal* became the subject of anxious dinnertime conversation, NIMBY began catching on. The battles started focusing not only on toxic dumps and power plants but also on chemical manufacturing operations, oil drilling facilities, and prisons; the location of crack houses; the shutdown of employment centers; the closing of schools.

The concept of NIMBY receives justified criticism when it is invoked hypocritically to preserve an elite, privileged sanctuary unconnected to the general good, in the manner of British environmental secretary Nicholas Ridley. But NIMBY also properly reflects a venerable American tradition: the right of individual citizens to protect their safety and way of life by uniting in a common cause.

For many people, those concerns have led them to expand the concept of "backyard" to include community and country. To some degree, that accounts for the increasing success activists have enjoyed in rallying people to the larger campaigns they wage—because more and more national and even global issues are being perceived as local. John Muir, the naturalist who founded the Sierra Club a hundred years ago, once sagely observed, "Everything in the universe is hitched to everything else." Today, many people know what he meant. Which is why the nineties' corollary to Muir's statement—"Think globally, act locally"—makes more and more sense. The point isn't to shift responsibility for shouldering the problem somewhere else but to reduce the existence of the problem over all. If none of us allow a toxic waste dump to be located in our own backyard, perhaps we'll get to broader toxics-use reduction nationwide.

I hope this book will give you a sense of hope, help you seize the initiative, and leave you with the firm belief that you can make a difference. It is written particularly for individuals and groups who have no money, no experience, and no power, but who are rich in commitment, determination, grit—and, I hope, a sense of humor.

---

*Love Canal is a neighborhood in Niagara Falls, New York. In 1980, more than 800 families were evacuated from their Love Canal homes after housewife-turned-activist Lois Gibbs demanded that the government protect people from toxic chemicals the Hooker Chemical Company had buried there.

How to Use This Book

The most basic rule about hell-raising is that in order to do it successfully you need to set your sights on a clear goal, then pursue that goal by developing and implementing a logical, step-by-step strategy. Each step involves maintaining a clear vision of the whole process—in other words, you have to know what your end goal is before you can figure out what steps will get you there.

So think of this book as a series of building blocks. The "blocks" you use to build a strong foundation can be found in Part I, "Laying the Foundation: An Overview of Successful Hell-Raising." These include tips on how to pick the "right" issue (Chapter 1), develop an organizing strategy (Chapter 2), take advantage of many available organizing tactics (Chapter 3), build a potent organization (Chapter 4), and raise the money you need to keep going (Chapter 5). A detailed case study sets the stage for the first four chapters; it focuses on the inspiring battles the Mothers of East Los Angeles waged to defeat the construction of a prison, a hazardous waste incinerator, and other unwanted projects in their neighborhood. Chapter 5, "Fund-raising," contains many stories that demonstrate how versatile and lucrative grass roots strategies to raise money can be.

Part II, "The Building Blocks," offers strategies and tactics you can use to build on your organizing foundation and achieve your goals. This section is divided into specific chapters that deal with communications (Chapter 6), lobbying (Chapter 7), and action at the polls (Chapter 8). Each of these chapters begins with a well-developed case study intended to help you understand how to use the "building block" being discussed.

Although each case study is designed to highlight one topic in particular, every story inevitably touches on many other topics as well, since when it comes to organizing, nothing happens in a vacuum. Thus, while the focus of the first story in Chapter 1 is primarily about the efforts of the Mothers of East Los Angeles to organize opposition to the construction of a prison in their neighborhood, you'll also read about the savvy media tactics they used to rally the general public in support of their cause, and how they successfully lobbied their state legislature to keep the prison out. And while the lead story in Chapter 8, "Action at the Polls," recounts the pins-and-needles referendum campaign waged over handgun control in the state of Maryland, it also describes the breathtaking tactics activists on both sides of the issue used to raise money and manipulate public opinion through the media.

Sprinkled throughout each chapter are shorter examples further illustrating how local hell-raisers effectively used the tool being highlighted in that chapter. Most of the chapters also contain "Step by Step" checklists that summarize at a glance the actions you need to consider to successfully undertake the activity being discussed. "Check It Out" boxes offer immediate sources of additional information. And "Do's and Don'ts" at the end of each chapter review essential observations and recommendations.

In the appendix, "Ammunition and Allies," you'll find more resources to investigate and some organizations you may be able to turn to for help, as well as a complete relisting of the sources offered in "Check It Out" boxes.

You'll get the most out of the book by reading it all the way through, then going back to specific sections as questions arise about your own organizing efforts.

One word of caution: All of the phone numbers and addresses included in this book were verified at the time we went to press. But that doesn't mean that they haven't changed. If you can't reach an organization by using the information we've listed, don't give up right away. Check with directory assistance in the metropolitan area where the group is supposed to be located.

While I've tried to provide a diverse selection of inspiring hell-raising examples, I suspect that the best stories are yours, and they're just waiting to be told. Please, send me your stories, care of Vanguard Communications, 1835 K Street NW, Suite 805, Washington, DC 20006.

Good luck. Now go raise a little hell!

Diane MacEachern
February 1994

*"What we do for ourselves dies with us.
What we do for our community lives long after we are gone."*

—THEODORE ROOSEVELT

I

LAYING THE
FOUNDATION
An Overview of Successful
Hell-raising

Community activism in its broadest form is the policy of taking positive, direct action to improve the quality of life where you live, whether you define that place as your apartment building, the neighborhood where you're raising your kids, or the city, state—even country—in which you reside. When you start to do more than complain, you've become an activist.

Most people are community activists whether they know it or not. Voting is an activist act; so is joining the PTA, cleaning up a local park, participating in a neighborhood crime watch, or holding a yard sale to raise money for the homeless. But, although lone deeds of community activism—such as the casting of a single vote on election day—can certainly have an important impact, true social change usually occurs only after individual activists begin working together to persuade others to join with them in some common cause. In other words, they organize.

But what exactly does that mean?

Organizing for change is a little like organizing other, simpler things in your life, like Thanksgiving dinner.

First, you focus on the issue: Thanksgiving is coming and you're hosting the family dinner—but complications are involved (such as Aunt Matilda's refusal to eat turkey and the old feud between Cousin Fred and Uncle Bert). So you figure out a strategy—what you're going to serve when, how long it's going to take to prepare, and who you're going to invite. Next you choose the right tools—the ingredients you need for your entrée, vegetable, side dish, and dessert; the right pots and pans in which to prepare the meal; the dishes on which to serve it. Finally, you begin cooking.

Throughout, you build and nurture your organization—in this case, your family. You discuss what you're planning to fix and ask others for menu suggestions, such as whether you should switch from turkey to ham, beef, or even a vegetarian dish. You ask for help in inviting your guests, shopping, cleaning the house, and set-

3

ting the table. You tell stories and jokes to keep spirits up when the tasks start to get a little tedious. And when the meal is over and all your guests have gone home, you sit down together with your feet up and talk about what you're going to do or prepare differently *next* year.

Community organizing involves exactly the same planning and implementation skills. What makes it different is the fact that so many more people are involved, that these people may be motivated by many different (and conflicting) factors, and that most community organizing battles are just that: conflicts that pit one side against another in confrontations that can be extremely time-consuming, expensive, and even demoralizing. It's a lot tougher to organize fifty people to oppose the closing of a school than it is to decide whether to serve turkey or ham, or whether to invite just the immediate family or all the in-laws.

As John O'Connor, the founder of the National Toxics Campaign, once put it, "Organizing is the power to enlist and lead people in ways that are bold and dramatic. It is the ability to challenge people's limitations and help them overcome those limitations. It is an understanding of human nature and an ability to engage people in action with precise timing. It is good judgment and organizational skill, but it is also the ability to have fun as you get there."*
In other words, it's not simple—but it can be done. And for most activists, it *must* be done if their communities are to become safer, healthier, and more humane.

Most community organizing revolves around "issue campaigns"; that is, campaigns that reflect shared fundamental values and attempt to change the status quo by setting and pursuing specific goals and objectives. A group of citizens might organize themselves into a community activist association because they're worried that drug trafficking is endangering their neighborhood, then create and implement a "citizen's patrol" program to help police eradicate drug dealers and clean out crack houses. Residents of a community who oppose the placement of a new incinerator in their area because they fear its emissions could endanger their health may organize an issue campaign to ban its construction and propose alternative measures to reduce solid waste. The most effective issue campaigns offer this kind of "one-two punch," putting

---

Fighting Toxics: A Manual for Protecting Your Family, Community, & Workplace, edited by Gary Cohen and John O'Connor, Island Press, 1990.

a halt to the problem and offering a solution.

Community organizing doesn't always involve candidates for office or elected officials. But it is almost always political in the sense that there usually are two conflicting sides maneuvering behind the scenes and in public to advance their positions and achieve their goals and objectives. Decisions about whether to fund an AIDS clinic, provide school lunches, or institute a jobs training program for disadvantaged youths often are made in a political context. The stronger the disagreement over a proposal, the more political the proposal becomes.

Community activism almost always involves the law—obeying it, changing it, and even breaking it.

Yes, while this book does not necessarily advocate using illegal tactics to achieve social change, it recognizes that efforts to bring about change are not always undertaken in ways that are condoned by the American justice system—but that still have a needed impact on the public and on decision makers. When Greenpeace illegally trespasses on the property of a corporation it believes is polluting the environment in order to unfurl a banner or plug a discharge pipe, the perpetrators usually are arrested for breaking the law. But Greenpeace activists would argue—and convincingly so—that such civil disobedience is the only way they can draw attention to the problem and achieve their goals of social change.

According to the nationally renowned Midwest Academy, a Chicago-based training center for grass roots activists, the most successful campaigns are those that achieve three "principles of direct action":

1. They win real, immediate, and concrete improvements in people's lives.

2. They give people a sense of their own power.

3. They alter the relations of power, allowing grass roots organizations to become key players in the decisions that affect their communities.*

As the stories in this book can attest, people who turn to citizen activism as a way of "raising hell" help determine their own future by taking charge of it. The following chapters will tell you how.

---

*See *Organize! A Manual for Social Change,* by Kim Bobo, Jackie Kendall, and Steve Max, Seven Locks Press, 1992.

1.
GETTING STARTED

THE AMERICAN REVOLUTION
WAS NO ACCIDENT—IT WAS ORGANIZED!

EVERY GREAT SOCIAL reformation—from the American Revolution to the rise of the labor, women's, and civil rights movements; to community struggles for environmental justice, improved health care, and drug- and crime-free neighborhoods—has occurred because individuals or organizations got together to change the system.

Such movements aren't a thing of the past. Today, every day of the week, hundreds if not thousands of citizens groups across the United States determine to take steps to improve their lives and enhance the well-being of their communities.

Most organizing efforts start with two questions: "What in the world is going on?" and "What can I do to remedy it?" If you're unusually lucky, the problem you're trying to solve can be fixed in the course of a few phone calls or by writing a couple of letters. But it's more likely that you're going to need to engage the help of others. As the Mothers of East Los Angeles discovered, if you choose a good solid issue, develop a sensible strategy, use the right tactics, and build a strong organization, you can prevail.

ACTIVISTS in ACTION ➤ ➤ ➤ ➤

The Mothers of East Los Angeles

For most people, church is a place to fortify one's faith, share news and gossip, perhaps offer a sympathetic ear to an ailing friend. For the Latinos of East Los Angeles, church also plays a vital role in preserving the safety and integrity of their community: It has become a place to mobilize and plan, to paint signs bearing strong messages, and to empower one another.

In April of 1985, Father John Moretta of the community's Resurrection Catholic Church asked two hundred mothers in the congregation to stay after mass. When the surprised group gathered in the sanctuary, the concerned priest quietly unloaded some disheartening news: The state was planning to build a 1,450-bed prison in their community—the eighth lockup facility in the neighbor-

Erlinda Robles (left), Aurora Castillo and Henrietta Castillo, decked in their white scarves, pause at a Mothers of East Los Angeles demonstration at the California Department of Corrections in October of 1987.   *Friezer Photography*

hood. Almost overnight, the church became an organizing center for the determined mothers of a low-income, overcrowded barrio that, already plagued by some of the worst pollution and crime in the nation, desperately sought to shelter itself from additional degradation.

An Unwanted Distinction

Father Moretta and Frank Villalobos, a local architect, had been involved in an effort by businessmen and merchants in the community to oppose the prison since 1984—months before the Mothers got involved. In fact, the Coalition Against the Prison had lobbied against the proposal during the spring 1985 legislative session, but made little progress. One day Assemblywoman Gloria Molina, a coalition member who eventually became a Mother, surveyed the groups of businessmen who made up the coalition and asked, "Where are the women?" Villalobos and Moretta conferred and agreed. "The press wouldn't listen to us; we're suits. We thought, why don't we bring the grass roots into the picture?" Villalobos recalled.

When Father Moretta gathered the matrons of the community in April, his announcement left them in shock. The proposed prison would give East L.A. the dubious distinction of hosting the largest concentration of lockup facilities in the free world. "We didn't want our community turned into a penal colony!" exclaimed Aurora Castillo, spokesperson for the Mothers. "We were, figuratively speaking, up in arms. The ladies kept saying *Ya basta!* No way!"

The knowing priest, who had a reputation as a peaceful activist, had targeted the right group to fight the prison: all two hundred mothers immediately signed up to help. "Father knew that it was time to get the ladies involved," Castillo observed. "The mother is the soul of the Latino family, the heartbeat is the children. The head is the father, and he is looked upon with great respect and admiration, but the mother rules the roost."

They began demonstrating the next day on Olympic Boulevard Bridge, a highly visible location just a block away from the proposed prison site. Donning white scarves, torn from bed sheets, that signified peace, dignity, and respect, the defiant mothers wielded their natural positions of authority—plus a little familial pressure—to recruit husbands, children, and friends to their cause. Close to five hundred people showed up for the first march—and, remembered Villalobos, "With the scarves, it looked like twice as many people!"

The marches turned into weekly, rain-or-shine events. One Monday Father Moretta, with his eye on an approaching thunderstorm, suggested that they cancel that evening's demonstration. The Mothers responded with a determined chorus: "We have umbrellas!" The number of demonstrators grew weekly; by August, seven thousand were participating each Monday at seven P.M. sharp.

The Strategy: Make News

Villalobos, Moretta and the Mothers knew that to defeat the prison plan they would have to enlist the rest of the community—and that the media would energize the recruiting drive. They also realized that they had the potential—given the hundreds of Latino mothers demonstrating on the bridge—to attract considerable attention.

"At first the media ignored us," Castillo said, but before long, radio, TV, and local newspapers sent reporters to check out the commotion. "The press asked, 'Who are these ladies?' " chuckled Villalobos. "And Father said, 'These are the mothers of East Los Angeles!' " The name stuck.

The third weekly march became the activists' television debut. The placard-bearing women made a compelling visual for TV cameras. Young mothers in casual garb marched alongside protesters like Castillo, who, in a cardigan sweater, skirt, and dress shoes, her black hair streaked with gray, still managed to look dignified while carrying a sign reading "IS THIS A DEAL?"

The Mothers soon became media savvy, alerting reporters when celebrities indicated that they planned to march, issuing press releases, and printing up flyers, which they distributed at churches, businesses, stores, markets, and even schools, inviting every citizen in California to participate. The costly printing was donated by local businesses that were just as opposed to the prison as were the Mothers.

People from communities all around Los Angeles gathered on the bridge. United Farm Workers president Cesar Chavez, actor Robert Blake, and Cardinal Roger Mahoney, recruited by Villalobos, came to offer support; one enthusiastic woman showed up from neighboring San Pedro wearing high heels but ready to protest. Cars driving over the bridge honked their approval and gave the thumbs-up sign to the Mothers, who brandished signs reading "WE NEED SCHOOLS, NOT PRISONS" and "DON'T FENCE US IN." Eventually the marches had to be discontinued—participation had grown so large that safety became a problem. But the Mothers continued to meet two or three times a week at the Resurrection school hall or annex, plotting to beat the state and preserve what little remained of East L.A.'s integrity and self-respect as a community.

Mariachi Mexicapan leads the Mothers of East Los Angeles in their final march on Olympic Boulevard Bridge, July 20, 1992.

Changing the Law

By August 1985, Villalobos felt the Mothers were ready to approach a more influential setting: the state capitol. They needed to defeat the prison proposal at its origin. Villalobos coordinated a four-hundred-mile bus trip to Sacramento, where the Mothers planned to lobby legislators. "We were so naive when we started. We had no idea how to lobby!" remembers Castillo. But two hundred Mothers and children were willing to learn quickly. After a brief, intense discussion of legislative procedures and how to influence the lawmakers, the group traveled all night in three rented buses to the state capitol. They crowded into the state senate chamber the next day, waving their signs, just as the lawmakers were slated to discuss a bill ordering that the prison be built. More Mothers were present than there was standing room available.

"The legislators were unnerved by the sight of all those Mothers and children," chuckled Castillo. The activists energetically waved their signs and spoke to the lawmakers, urging them to vote against the bill. A press conference was hastily arranged, and the Mothers appeared on local TV to denounce the prison plans. The group traveled to Sacramento from Los Angeles an average of twice a week after that—every time the bill was slated for discussion. The first few trips were paid for with $3,000 raised at a pancake breakfast. Later, the lengthy bus rides evolved into short plane trips funded by concerned local businesses. The media covered every trip. "After a while, every time Governor George Deukmejian [a strong proponent of the prison] burped, they called us to find out our reaction," laughed Castillo.

Working with Others

The Mothers' inexperience as activists, though daunting, in some ways worked to their benefit. The women supplied the soul the movement needed, while the Coalition Against the Prison provided monetary support and a suitably professional image. "They said, 'You get the Mothers, we'll do the rest,' " Castillo remembers.

By combining their respective strengths, the two groups helped persuade the California courts to uphold challenges to two drafts of the Department of Corrections' environmental impact statement that would have allowed construction of the prison to begin. Meanwhile, a member of the state assembly, Lucille Roybal-Allard (who was later elected to Congress), was gearing up to propose legislation that would shift the East L.A. prison money to another prison that was about to open in Lancaster County. The steady grass roots opposition the zealous women mounted in the interest of their children's and grandchildren's futures finally swayed lawmakers. After a previous veto, Governor Pete Wilson, Deukmejian's successor, signed a bill killing the 1987 mandate for the East L.A. prison. The date was September 16, 1992—Mexican Independence Day. The Mothers learned about their victory from the members of the press who had covered their battles, and celebrated with them afterward at Resurrection Church.

The Mothers now meet on the last Tuesday of each month. Their current roster of members contains more than four hundred names. "If you are a mother and live in East Los Angeles, you're a Mother," Castillo says. Being concerned citizens in a largely ethnic community means no time for rest; in 1988, while the prison battle was in full swing, California Thermal Treatment Services (CTTS) of Garden Grove announced plans to build a hazardous waste incinerator in the nearby city of Vernon. This project would also make East Los Angeles a dubious trendsetter as the first metropolitan area to host a commercial hazardous waste incinerator.

The Mothers and several state politicians with whom they were working to defeat the prison proposal claimed discrimination against East L.A. As evidence, they cited a report prepared by Cerrell and Associates, a consulting firm based in L.A., and commissioned by the federal government that recommended targeting low-income communities for potentially unacceptable projects because such locales frequently lack the resources to fight back. The Mothers joined forces with the environmental group Greenpeace to march against the incinerator. They attended pubic hearings, demanding more meeting space and interpreters when necessary. "Our culture is being tampered with," Castillo said, explaining the emotion that fueled the Mothers' campaigns.

The group incorporated so they could solicit tax-deductible contributions from businesses. Following legal advice from the Natural Resources Defense Council, another environmental organization, the community sued CTTS in October 1989 for failing to file an environmental impact statement (EIS). Three years later, CTTS abandoned its incinerator plans when the California Supreme Court upheld a lower court's ruling requiring the EIS.

The Mothers of East L.A. also succeeded in blocking a second waste incinerator and an above-ground oil pipeline that had been proposed to run through their community. They attribute their successes in part to grim determination and in part to their ability to use tactics other than grass roots demonstrations to accomplish their goals. "They're now considered one of the most powerful groups in Los Angeles. They can make or break a politician," Villalobos boasts.

Working to get "their people" elected to public office has also become an important part of the Mothers' organizing strategy. First, they campaign to encourage local people to become citizens if they are immigrants, then to register, then to vote, and in some cases to run for office themselves. Two former Mothers have achieved elected positions of considerable influence: Congresswoman Roybal-Allard and Gloria Molina, who became county supervisor. Approximately 26 percent of California's population is Latino, and with the Mothers' help, says Castillo, "we now have grass roots representation. More and more Latinos are voting, and more are being elected. Sooner or later our children—because of the percentages—are going to run this state."

It is only fitting that the Mothers are headquartered in their place of worship, because, as Father Moretta put it when he called that first meeting in 1985, to win their battles would take a miracle.

WHY DID THEY WIN?

The Four Keys to Successful Organizing

The Mothers of East L.A. started with nothing, yet became a force to be reckoned with. How did they become so powerful?

1. **They had a good issue.** The Mothers felt strongly enough about blocking the prison to want to do something about it. When they sat down and figured out what it would take to win—in terms of people, money, time, and tactics—they felt they could prevail, even though victory wouldn't occur overnight.

2. **They had a good strategy.** The Mothers anticipated what was likely to happen and mapped out how they were going to achieve their goals. At the same time, their approach was flexible enough to allow for unexpected developments, and realistic in reflecting what people could and could not do. Perhaps most important, their strategy involved each member every step of the way, empowering all of the Mothers until they knew they could fight—and win.

3. **They used the right tactics.** They identified specific actions—such as the marches and other media events, their collaboration with the antiprison coalition, and their lobbying of the state legislature—that helped move their issue along while building their organization into a solid institution. The Mothers didn't do everything they might have done—but what they did made sense and worked for them.

4. **They built a solid organization.** The Mothers held regular meetings, kept their membership active through frequent opportunities for demonstrations and lobbying, allowed many of their members to exert leadership, and established themselves legally to provide more fund-raising opportunities. Each incremental win empowered their members even more.

Choosing a compelling issue is key to the success of any organizing strategy, and the remainder of this chapter offers basic guidelines to help you do so. The rest of Part I provides tips and suggestions for developing a successful strategy, picking the right tactics, and building a solid organization.

IS YOUR ISSUE A WINNER?

Are you working on a winning issue or just spinning your wheels?

The question isn't about values—whether or not the issue intrinsically merits your attention or concern. Ultimately, it's about resources. You can't tackle every issue, and you probably can't win every campaign you wage. Given the fact that most activist groups are underfunded, understaffed, and overcommitted, organizations need to pick their fights carefully. Waging one campaign successfully is a better social change strategy than struggling through several worthwhile—but losing—ones.

The following questions may help you whittle down the issues you care about into the one or two you should pursue now.

- **Do people feel strongly enough about the issue to want to work on its behalf?** Is this an issue you expect many people to rally around? Or is it going to be tough to round up even ten volunteers?

- **Will working on this issue strengthen your organization?** Will it involve people in the fight, empower your members, and build team spirit?

- **Does the issue affect a lot of people, especially those who have the potential to become members in your group?** Or will its impact be felt by only a few "special interests" who care little for your ongoing work?

- **Does the issue have a clear right and wrong side?** Or are there many possible positions that confuse potential supporters and make it more difficult to recruit volunteers and raise money?

As important as they may be, issues that are underrated, divisive, of consequence to just a few, or ambiguous shouldn't be a priority for community groups with scarce funds or resources; put these issues on your organizational back burner until they become more compelling to your members and the public.

Then return to the list of issues that now seem winnable, and prioritize again. Consider the following factors:

- **Timing.** How immediate is the issue? Must you act now because a decision is due in the legislature or by the city council? Like the Mothers of East L.A., are you faced with the impending construction of a facility you want kept out of your neighborhood or perhaps not built at all? What other time considerations are driving you to tackle this issue sooner rather than later?

- **Fund-raising.** Waging any campaign takes money, and some issues are simply much easier to fund-raise around than others. Are the donors in your community—whether they be individual citizens, foundations, corporations, churches, or civic associations—willing to support one issue campaign over another?

- **Organizational goals and objectives.** Do your priority issues effectively complement and advance the overall mission of your group?

If clarifying any of these considerations isn't obvious, you can get help from several sources. Review newspaper and magazine articles for a sense of where the issue is being placed on the public agenda. Talk with elected officials and directors and staff at governmental agencies to identify when issues will be subject to public debates and votes. Meet with the legislative directors of governmental committees to determine how much help or hindrance they'll be. Ask other organizations with whom you are working for their advice. Check with lawyers (or law students) to see if a law you think needs passing is already on the books; maybe what it really needs is to be enforced.

Make the best informed decisions you can by learning everything possible about the issues you're considering working on. Check the library, governmental agencies, and other organizations. Consult with professors and graduate students at local universities. Speak with journalists who cover the topic. The appendix of this book references other books, reports, and organizations that can also help.

Even when you've prioritized the issues, done the research, developed the strategy, and chosen the tactics, you can never be sure you'll win the battles you choose to fight. (Yogi Berra, the famed manager of both the New York Yankees and the Mets, used to say "It ain't over till it's over," and that's as true of most issue campaigns as it is of baseball games.) What you can do is improve your chances significantly by giving yourself advantages from the start—and organizing around issues that you have a chance of winning.

ISSUE
DO's and DON'Ts

• **Do** prioritize your issues. Focus on those around which you can recruit volunteers, raise money, and achieve your overall goals and objectives.

• **Don't** try to tackle every issue at once. Waging one campaign successfully is a better social change strategy than struggling through several worthwhile but losing ones.

2.
DEVELOPING YOUR STRATEGY

A STRATEGY IS a comprehensive plan, formulated in advance of the action at which it is aimed, that identifies the combination of steps needed to achieve specific goals and objectives. *Webster's New World Dictionary* compares developing a strategy to "maneuvering forces into the most advantageous position prior to actual engagement with the enemy." Basically, it's the virtual opposite of operating by the seat of your pants.

WHAT MAKES A STRATEGY EFFECTIVE?

The operative words are "advance planning." A solid strategy is thought through in advance to allow planning for resource needs and to help identify specific tactics that are needed to win. It anticipates opportunities and pitfalls, preparing organizers to take full advantage of beneficial circumstances while minimizing the fallout from potential roadblocks. Developing a strategy readies organizers to dispatch volunteers, staff, money, equipment, and other assets where they are needed when they are needed without wasting time. A good strategy keeps a campaign moving forward, creating the momentum needed to win.

An effective strategy also uses people's skills to best advantage. It remains flexible to allow for unexpected developments. And it reflects a realistic understanding of what members can and can't do.

Strategies that achieve their objectives have six characteristics in common:

1. **Long-range goal.** An effective strategy sets out to achieve a specific mission. Increasing the availability of housing, creating a new park, teaching children to read, reducing infant mortality, and improving public health are examples of the many kinds of long-range goals community activists have set for themselves over the decades.

2. **Short-term goal.** It is almost impossible to realize a long-range goal in one fell swoop. Usually, you have to pursue a series of short-term goals that help maintain the momentum needed to reach your long-term goal. For example, suppose a community's long-term goal is to build a new park. A short-term goal might be to get a majority of the community's residents to support having a park. A second short-term goal might be to get the city council to set aside the land. A third might be to raise the money for playground equipment. When all these short-term goals are achieved, you can fulfill your long-term goal of building the park.

3. **Organizational concerns.** An effective strategy will take into account the impact the campaign will have on organizational resources, strength, and morale. Strategies that work maximize the use of those resources—volunteers, money, equipment, facilities—and help secure additional resources where needed.

4. **Campaign targets.** Every campaign has a target—an entity that has the power to meet the group's demands or solve the problems it's raised. For most issue campaigns this target is either an institution, a corporation (or corporate representative), an elected official, a governmental agency, or a competing activist group. A good strategy not only pinpoints the target of the campaign but identifies its strengths and

weaknesses and suggests how the organization can exert power over the target.

5. **Tactics.** These are specific actions undertaken to move the strategy along—the "building blocks" that, positioned on the right foundation, will lead to success. (Tactics that help strengthen the foundation of your organizing campaign are discussed in greater detail in Chapter 3. Chapters 5 through 8 address tactics that pertain specifically to developing fund-raising, media, lobbying, and electoral strategies.)

6. **Time frame.** Effective strategies reflect the period of time that is available to accomplish the goals set out in the strategy. Unfortunately, unrealistic time constraints often prevent activists from taking on a campaign they just don't have enough days, weeks, months or even years to win. It is critical to identify the amount of time you have to wage a campaign and the amount of time you need to deploy specific tactics in order to achieve your short- and long-term goals. You can do this by counting backward from a specific date—a vote in the legislature, an election, a filing deadline—to the present, then taking stock of your tactical options. (See Chapter 3 for more guidelines on choosing the right tactics.)

When setting goals and developing other components of your strategic plan, keep in mind that you will frequently be subject to forces beyond your control . . . forces that could adversely affect your entire strategy. For example, your group's decision to organize a city-wide demonstration to protest unfair housing laws could be derailed by an unexpected budget crunch that forces you to lay off staff and makes it impossible to rent equipment or pay for publicity. Strategies—and the activists who develop and implement them—must be flexible to avoid being stopped in their tracks by unforeseen circumstances.

The answers to some other important queries also shape strategies and, eventually, the tactics selected to implement them: Are you trying to shift the public debate on your issue to cast your issue in a more favorable light? Can you take advantage of existing strong support for your agenda to implement your program as you have designed it? Or must you prepare to compromise with a suggestion that's already been put "on the table" in order to get the best deal available given the circumstances? Anticipate these questions at the very beginning of your strategic planning process, and revisit them often as your organizing efforts evolve.

—————————————— **Strategy Checklist** ——————————————

As you plan your strategy, keep the following considerations in mind:

✓ **Goals**

___ What are your long-range goals?

___ What are the short-term objectives that will keep your members motivated and focused on your long-range plans?

___ What specific change are you seeking?

___ How will you measure victory?

___ Are there any issues or events that are peripheral to your campaign but that adversely affect your strategy?

✓ **Organizational Concerns**

___ What impact will the campaign have on your organization? Will it strengthen it or make it weaker? In what ways?

___ Will new members join your organization as a result of the campaign?

___ Will new allies join your coalition? Will you make new enemies that could cause serious problems for your group?

___ What will the campaign cost in terms of money, resources, and credibility?

___ What skills, resources, and talent does your organization already have that will help you win the campaign?

___ What skills, resources, and talent does your organization need to acquire in order to win the campaign?

✓ **Targets**

___ Who has the power to meet your demands or solve the problems you've identified?

___ What are the target's strengths?

___ What are the target's weaknesses?

___ How can you exert power over the target?

✓ **Tactics**

___ What tactics can you use to get the target to adopt your position?

___ What will these tactics cost you, not only in organizational resources but in credibility with your members, the media, and other organizational supporters?

✓ **Time Frame**
___ When is the best time to launch the campaign?
___ How long should the campaign go on?
___ How many phases will the campaign have?

The Mothers of East L.A.'s Organizing Strategy

Here's how it worked with the Mothers of East Los Angeles:

GOALS
- Long-term: Stop the prison.
- Short-term: Build public support in the community; build general support in the legislature.

ORGANIZATIONAL CONCERNS
- Offset lack of resources and clout by joining a coalition and appealing to businesses in the community for donations.

TARGETS
- State legislators.
- Governors Deukmejian and, later, Wilson.

TACTICS
- Generate widespread publicity (through demonstrations and lobbying efforts) to put the issue on the public agenda.
- Lobby state legislators to convince them to change their plans for the prison.

TIME FRAME
- Dictated by impending votes in the legislature.

REFINE YOUR MESSAGE

The most successful campaigns reflect values that have universal appeal to people of all political persuasions. In a contentious neighborhood comprised of liberals and conservatives; Democrats, Republicans, and Independents; baby boomers and senior citizens, and a variety of ethnic groups, everyone may unite to fight a local toxic waste dump that endangers them all, or to install speed bumps to protect all of their children, or to mount a citizens' crime patrol to safeguard all of their homes.

The way you position your issue will determine whether many people will join you in your efforts—or few. It is critical to be truthful in how you present your issue to the public. Blatant lies serve no useful purpose, and even "stretching the truth" will eventually stretch your credibility with the media, your members, and other audiences.

Strive to frame the issue as an important social concern on which public action is imperative. Your campaign will succeed if you reinforce the positive aspects of your issue while reminding audiences of your opponents' negatives.

If you need help in crystallizing your campaign theme, budget permitting, focus groups could help.

Focus groups offer an excellent research tool for gaining a better understanding of complex attitudes, perceptions, and motivations. The technique is referred to as a "qualitative" research approach since it attempts to assess anecdotal information (as opposed to the more rigorously quantitative methodologies used in opinion surveys).

In a focus group, ten participants are engaged in an in-depth discussion that explores a wide range of attitudes and perceptions. Discussions generally last two hours, are led by a trained moderator, and follow a detailed discussion guide that is developed in conjunction with the group that has commissioned the research.

Participants are offered a cooperation fee ranging from $35 to $125. (The fee depends on the location, socio-economic status of the respondent group, and availability of potential participants.) Participants are screened during the recruiting process to ensure that they meet the specific characteristics called for by the research. Discussions are usually held in a focus group facility equipped with a one-way mirror and an observation room from which clients can view the proceedings, which can be audiotaped, videotaped, or both.

One of the main advantages of focus groups over other kinds of opinion research is that they allow the researcher a greater degree of give-and-take with participants, providing the opportunity for follow-up questioning and probing to unearth the reasons behind attitudes and perceptions. Focus group research can also be used to get immediate reactions to materials shown to participants. In a focus group, you can pinpoint specific words and phrases that could be incorporated into campaign slogans, posters, brochures, and bumper stickers while developing an overall campaign theme.

In addition to the participation fee, focus groups often cost at least $5,000 per group. But if the stakes are high, you're uncertain about your message, and you have the money, they're probably worth it.

WHO DEVELOPS THE STRATEGY?

People feel like they "belong" to a group when they are allowed to help develop, implement, and revise strategies for achieving the group's organizational goals and objectives. Thus, in any group, both members and leaders should help conceptualize and implement strategy. Because the process involves identifying and defining issues, it also helps anyone who participates in strategy making to learn to think in long-range terms, see the "big picture," and understand how to maximize whatever resources the group has at its disposal. Besides, involving people in strategy decisions gives them an investment in its success. Those involved in planning and developing a strategy are far more likely to want to help carry it out.

To develop your strategy, convene your leaders, staff and volunteers in a large room stocked with flip charts or blackboards. Using the Strategy Checklist on page 22 as the outline for a blueprint, attempt to answer each of the questions the checklist raises. Ask for input from the entire group, and take notes on the flip chart for all to see. (You may need to hold more than one session to get through all the questions and come up with a comprehensive blueprint for the campaign.)

Using the flip charts as a point of reference, type up the strategy into a document that you can refer to over and over again as the campaign proceeds, modifying your plans depending on how the campaign evolves. If you feel the need to keep the strategy document confidential, ask those who participated in developing the strategy not to discuss it with the media, your opponents, or the public at large. When the campaign is over, review the strategy as you originally conceptualized it. Take note of what worked and what didn't—and apply this knowledge to the next campaign you plan.

HOW WILL YOU KNOW IF YOU'VE WON?

The answer to this question isn't as obvious as it may seem. Some battles are very clear-cut—you either succeed in blocking a project or passing legislation or you don't. But even when you lose a vote, you may have won a moral victory if you've managed to educate your constituents about the issue and lay solid groundwork for round two of the campaign. Evaluate every campaign based on how many of your short- and long-term goals you've achieved, as well as on whether the campaign helped to augment organizational resources, teach members and volunteers new skills, and educate the public and decision makers about your issue.

Notes Jay Sherman, a seasoned grass roots activist who organizes for the Chesapeake Bay Foundation: "The more profound an issue is, the longer it will take to win. Understand the evolution of your success if it is not immediate."

LOCAL HELL-RAISER
Converting Rails to Trails:
The Strategy Made the Difference

During the day, Doug Cheever designs construction equipment at the John Deere Corporation in Dubuque, Iowa. But in his spare time, he helps save abandoned railroad corridors and turns them into recreational trails that thousands of people enjoy.

In 1980, Cheever was a fledgling activist who set his sights on saving a twenty-six-mile stretch of railway that was being abandoned by the Chicago and North Western Railroad outside Dubuque. Knowing he needed help, he innocently asked fifteen friends if they would "spend five hours a week for six months" to figure out a way to buy the abandoned track and convert it into a recreational trail to be enjoyed by bikers, hikers, and runners.

Cheever's cohorts formed Heritage Trail, Inc., developed an "action plan," produced a slide show and began attending public hearings. At one point, Cheever and other rail-trail supporters even pledged $12,000 of their own money to make a nonrefundable deposit on the land they were trying to buy.

At almost every step of the way, the activists encountered heavy resistance to their project from neighbors who feared the trail would entice a "criminal element" into the region. Wooden railroad trestles and bridges were burned, partly in protest, partly to intimidate the organizers.

Doug Cheever developed a winning strategy to convert an abandoned railroad corridor into a recreational path.

But Cheever and his friends pressed on. And seven years after Cheever's six-month promise was long forgotten, the activists jubilantly completed the Heritage Trail to give momentum to a rails-to-trails movement in Iowa and Illinois that has reverberated throughout the Midwest.

When asked for the secret to his success, Cheever told the Washington, DC–based Rails-to-Trails Conservancy, which has helped activists in over five hundred communities convert abandoned railways into recreational trails, "What ultimately brought us through was having an overall plan—a sequence of events. We knew what it would take to win, even if we were somewhat naive about the amount of effort required.

"We organized as if the odds were against us because they were."

STRATEGY
DO's and DON'Ts

- **Do** set specific short- and long-term goals.

- **Don't** expect to win just because you think you're "right." Without a well-conceived strategy, you'll end up wasting time and other resources and maybe never achieve your goals.

- **Do** identify the resources you need to win.

- **Don't** overestimate *or* underestimate the level of effort involved in mounting your campaign.

- **Do** focus on targets who can be swayed by your campaign—and who have the power to solve the problem you're working on.

- **Don't** aim the campaign too broadly. Focus on targets who really matter.

- **Do** use tactics that will help meet your goals and objectives.

- **Don't** try "every trick in the book." Some tactics make more sense for your effort than others; those are the ones to use.

- **Do** give yourself enough time.

- **Don't** waste time.

3.
BASIC TACTICS

IF THE STRATEGY is figuring out how to win the race, the tactics are the individual steps you take to get to the finish line.

In other words, tactics are the specific activities an organization engages in to achieve its strategic goals and objectives. Tactics create the pressure you need in the right place at the right time to win on a particular issue, strengthening your organization in the process. An effective tactic is one that impacts a lot of people, unites and involves people, is strongly felt, is simple, builds the organization, and is fun.

Some tactics defy classification but still are effective because they use creativity and ingenuity to make their point. For example, a North Carolina activist who opposed the construction of a hazardous waste disposal complex in his community bought forty-eight acres of the land that was being considered for the complex. He then subdivided his forty-eight acres into 70,000 different parcels and began selling them off at $5 a share. The income from the sales was used to offset the $70,000 the activist had borrowed to pay for the land. In order for the developer to obtain the property, it has to notify each landowner of its intent to condemn the land,

and negotiate a settlement for the purchase of each plot. If the landowners don't cooperate—and few are expected to—the developer may have to sue each one to gain possession of the land, a time-consuming process that could cost between $500 and $1,000 per landholder—and effectively terminate the project.

There are many simpler but equally creative tactics you can use to advance your campaign strategy. The Mothers of East L.A. relied upon several tactics to achieve their organizing objective—the defeat of the prison. Through their marches on the bridge and their bus trips to Sacramento, they generated statewide publicity. By lobbying in the state capitol, they swayed lawmakers' opinions. And by getting their constituents to register to vote—and elected to political office—the Mothers are garnering even more power.

As every case study in this book demonstrates, it is extremely rare for an organizing strategy to revolve around just one tactic, particularly when the focus of the strategy is to affect the law. Legislators are exceedingly susceptible to the tide of public opinion, forcing most activists to wage their campaigns in the media as forcefully as they do in their state house or city hall.

Yet most groups will not (and cannot afford to) engage in all possible tactics during the course of a campaign; nor will they need to. Instead, the most successful groups selectively choose which tactics will have a real impact on their opposition. Some options—like private meetings and letter writing—allow you to work quietly behind the scenes when diplomacy is called for and when there's less urgency to make something happen immediately. Use tactics that will help you generate a lot of publicity—like demonstrations, sit-ins, pickets, and press conferences—when you need to focus the spotlight on your target, educate the public at large, and rally significant numbers of people behind your cause. Obviously, when your goal is to pass a new law, defeat a proposed law, or amend an existing law, you'll have to lobby. When you're trying to hold a corporation accountable for its actions, you may man picket lines, organize a boycott, or launch shareholder actions. And when your only recourse is to take an issue before the courts, you'll have to engage in litigation.

Remember one important tactical rule: Let your campaign build. Start out with the easiest tactics first; graduate to more difficult but meaningful activities as your campaign advances and you need to mount greater and greater shows of strength. Don't "shoot your wad" organizing a march of five hundred people at the beginning of your campaign when a picket manned by fifty activists will do.

Save the most dramatic and impressive tactics for those times in the campaign when you really have to turn the heat up a few degrees.

• •STEP BY STEP• •

How to Choose Your Tactics

When You Want to ....**What Should You Do?**

Affect legislation ....
- lobby (visit legislators, write letters)
- make phone calls and organize telephone "trees"
- stage special events (prayer services, silent vigils,
- rallies, marches)
- attend public hearings
- demonstrate at public hearings
- testify at hearings
- hold accountability sessions
- advertise
- circulate a petition
- build coalitions

Pressure a corporation..............
- organize a strike or picket line
- mount a shareholder action
- take legal action
- negotiate
- hold a rally or stage other media events
- launch a boycott
- build coalitions

Mobilize the public ....
- stage special events
- issue reports, news releases
- distribute posters, fliers, brochures, pamphlets, public service announcements
- write letters to the editor, opinion editorials

Lobbying, generating publicity, and taking action at the polls are multifaceted tactics that require a strategy of their own to be used effectively. Tips for developing and implementing these strategies can be found in Part II. The rest of this chapter focuses on tactics that are fundamental to any organizing campaign.

LETTER WRITING

Letters are among the oldest and most effective organizing tactics. In fact, more than two hundred years ago, Committees of Correspondence during the American Revolution used letters to keep fellow patriots informed and involved. Letters can provide information to someone who matters, bolster support for a position when the person who holds it is under pressure to change his or her mind, thank people or tell them you agree with them, try to influence the way someone votes, ask for help, tell people you don't agree with them, identify a problem or propose a solution.

The National Audubon Society's Armchair Activist Program offers a terrific model for any group, local or national, that wants to use letter writing in its arsenal of activist tools. The program is aimed at Audubon chapter members who are too busy to attend meetings or other activities but who still want to make a difference.

Activists who participate in the program's Letter of the Month Club spend no more than a half hour per month writing a letter to a targeted official. Audubon provides a concise description of the problem, tells activists when to write the letter, provides names and addresses of targeted officials, and provides a sample letter to copy or rewrite. Armchair Activist organizer Jeff Lippert estimates that over 3,500 letters are generated this way each month on federal issues alone. The activists are also encouraged to write to their state or local officials. (Be aware that a letter has greater impact if it is original, rather than a copy of the same letter other activists are submitting.)

• •STEP BY STEP• •

How to Write a Letter

• **Be neat.** If you want to create a personal, nonslick effect, handwrite your letter, but make it very clear and easy to read. Otherwise, use a typewriter or computer.

• **Get the details right.** Include your return address on the letter, because envelopes tend to get thrown away. Also, include a telephone number where you can be reached during the day. If you're writing a letter to a newspaper or television station, the editors may want to call you and personally get your permission to run your letter.

• **Address the letter properly.** "Dear Editor" if sending to a newspaper; "The Honorable . . ." if sending to a legislator; and "Mr." or "Ms." if writing to anyone else.

• **Be brief, clear, and to the point.** Write in your own words. Don't be intimidated by whomever you're writing to. If it's an elected official, remember: He or she works for you!

• **Mention the bill number** if you're writing concerning a specific piece of legislation.

• **Write from the heart.** Be personal.

• **Request a reply.** Otherwise, your letter may be ignored.

• **Time your letter so that it arrives when it will do some good.** If it reaches its destination too early or too late, it won't get printed or affect the outcome of the debate. If you're writing your letter as part of a campaign, call the campaign organizer to find out when to write. You can also call the governmental committee (of the state legislature or city council, for example) overseeing the issue to find out when a vote is coming up or when a decision is in the works.

☞ **CHECK IT OUT:** For more information about letter writing, contact The Armchair Activist at 1415 Braeburn, Flossmoor, IL 60422.

BOYCOTTS

According to *National Boycott News,* which reports on boycotts under way around the world, a boycott is "the withholding of economic, social or political participation as a means of protesting or forcing the alteration of various policies or practices deemed unjust or unfair by the boycotter."

Boycotts have been practiced in this country since at least 1773, when colonists angered by the policy of taxation without representation dumped tons of British tea into Boston Harbor. But the term "boycott" didn't come into use until more than one hundred years later when, in 1880, the Irish Land League called upon the public to ostracize Captain Charles Cunningham Boycott, a scurrilous lowlife who ruthlessly collected rents for an absentee British landlord. The term "boycott" stuck.

Some boycotts single out one consumer product on which to focus their attention: Union organizers in North Carolina have admonished consumers not to "sleep with J. P. Stevens" by avoiding the purchase of the company's bed linens to force the textiles manufacturer to negotiate with employees upset over poor working conditions. The United Farm Workers still tell shoppers not to buy table grapes to gain safer and more humane working conditions for the primarily migrant fruit pickers, a message they've been disseminating since the late sixties.

Other boycotts may be broader. The San Francisco–based Rainforest Action Network has organized a boycott of all products produced by the Mitsubishi Corporation to pressure the company to suspend logging practices in endangered rain forests in Malaysia, while Earth Island Institute, also based in the Bay Area, has launched a boycott of Taiwanese products to protest that country's traffic in endangered species.

One of America's most famous boycotts was started on December 1, 1955, when Rosa Parks, an unassuming black woman living in segregated Birmingham, Alabama, refused to move to the back of a bus as local law then required all blacks to do. Her act of courage started a 381-day bus boycott in Montgomery that ultimately resulted in a Supreme Court ruling that outlawed segregation and changed the face of civil rights in this country forever.

Boycotts used to be a tactic that companies airily dismissed and activists launched just as warily. But today, most companies consider boycotts and boycotters seriously. And whereas boycotts used to take five to ten years to make their mark, today many boycotts have an impact in far less time—although you should still be prepared to devote months, if not years, to a successful boycott campaign.

Boycott organizers have learned three things: first, to succeed, it helps to get the message to as many people as quickly as possible; second, it is important to aim the boycott at the company's image, as well as at the offending product; and third, it is critical to set a clear and tangible goal so that you know when to declare victory and get out.

Boycotts work by cutting profits or demonstrating the potential for cutting profits by changing consumer purchasing habits. Boycotts also are effective in generating media coverage that could influence purchasing decisions, and in creating negative public relations for the company or product that is the target of the boycott.

In the past, groups that organized boycotts would rely on press releases, press conferences, and constituency newsletters and publications to help spread the word and mobilize consumers. Better-financed boycotts have begun using direct mail, involving celebrities, taking out full-page ads in popular newspapers and magazines, and producing videotapes documenting their cases against the companies. Some groups stage dramatic, mediagenic actions. For example, Mitsubishi boycotters have handcuffed themselves to the company's cars while they're on the assembly line. In the boycott against General Electric, organized to protest the company's substantial role as a nuclear weapons manufacturer, activists went beyond consumers to convince many hospital administrators to boycott G.E.'s hospital equipment and to persuade shop owners to carry light bulbs manufactured by firms other than G.E. In 1991, five years into its boycott campaign, the INFACT/Boycott G.E. organization got a big boost when its twenty-eight-minute film, *Deadly Deception: General Electric, Nuclear Weapons and Our Environment,* won an Academy Award for its portrayal of the mega-company as "America's #1 Polluter." Millions of consumers worldwide wrote letters, made phone calls, and signed petitions; eventually G.E. sold its aerospace division, which housed its nuclear weapons operation. (David Warshaw, manager of corporate communications at General Electric, claimed: "The decision to transfer our aerospace businesses to Martin Marietta was due to the status of the defense industry, which is undergoing consolidations as a result of government decisions to reduce defense spending." He would not comment on whether G.E. felt an economic impact from the boycott.)

Where Do You Start?

Before you launch a boycott, conduct a careful analysis of the product or company you're targeting and your ability to affect the company's profits. First, make sure you're right about the damage the product or company is doing. Remember that a product that everyone buys frequently, is easily identifiable by brand name, is nonessential, or—what's even better—for which there are competing brands or substitutes is much easier to boycott than one most people don't buy or use anyway. Boycotts against local retail businesses are more manageable than are boycotts of products made by some multinational corporation based far away.

Don't lose sight of the fact that the most effective boycotts raise a significant moral issue. To get people to participate in a boycott, you've got to make them angry about the injustice, then motivate them to wield their power as consumers to bring about change.

• •STEP BY STEP• •

How to Organize a Boycott

• **Pick your target carefully.** Is the offender the industry leader? The biggest offender in the industry? The company most sympathetic to your demands, and thus the one most likely to meet your demands and set an example? The company with the highest visibility or the products that are most easily recognized by consumers? The one that is most economically unstable or has the most vulnerable image? Most boycotters base their decision upon a combination of these criteria.

• **Write to the targeted company about your issue.** Begin with a letter from your organization expressing your concerns. Depending on the company's response, you'll probably have to write another letter stating why the company's reply is insufficient and requesting that your interests receive more serious consideration.

• **Attempt to work with a person from the company directly involved in the issue.** Try to define precisely the points of disagreement. Ask the company to develop a proposal that addresses your concerns, or devise your own proposal for how the company might address the points of disagreement.

• **Set deadlines.** If, by chance, the company agrees to your proposal, don't let it stall. Create a time line for how these points will be dealt with. If the company refuses to accept your proposal or time line, advise the company that if action is not taken by a certain date, you will consider a boycott against it. If the company fails to meet the deadline, inform it that you are researching the possibility of launching a boycott against it, but are willing to keep the lines of communication open. Include a list of your demands. If you hear nothing new from the company, notify it that you will hold a press conference announcing the boycott on a particular date. Reiterate that you are willing to keep the lines of communication open. Tell the company what a positive step it would be if it were to meet your demands.

• **Before announcing the boycott, outline your long- and short-term goals.** Decide how you will measure the boycott's effec tiveness. Select which products to target, whether just a few high-profile items or all consumer products manufactured

How to Organize a Boycott (cont.)

by the target of your boycott, and whether to include products of subsidiaries. Ask other groups to co-sponsor or endorse the boycott. Define the process to be used to reach any settlement of the boycott. Be prepared to take advantage of the visibility generated immediately after calling the boycott.

• **Ask co-sponsors and endorsers of the boycott to send letters to the company expressing their support for the future boycott action.**

• **Send updated lists of the endorsements and co-sponsors to the company and to the media.**

• **Organize a rally and press conference to formally kick off the boycott.** Efforts should include simultaneous press conferences and rallies in other cities where the boycott is being organized. The primary press conference should be held where the company maintains its headquarters—where its local image is highly important. Be upbeat, clever, colorful, and even humorous if possible.

• **When the company meets your demands, call off the boycott!** INFACT announced the April 2, 1993, end to the G.E. boycott by writing and distributing a press release, phoning local and national outlets, mailing a victory letter to its now-huge activist network, and, of course, notifying G.E. in writing that the boycott was over. Depending on the visibility and timeliness of the boycott, a press conference is also an effective way to let consumers know, not just that you won, but that you are keeping your promise and taking pressure off the company. "It's really easy to add more and more to the demands as a company responds," said Kelle Louaillier of INFACT/Boycott G.E., "but in fairness to the movement you need to respect those demands and move on."

• **Constantly evaluate the company's response and whether it is worthwhile to continue your campaign.** Don't be too proud to call off a boycott that is making no progress and thus is losing credibility. You usually can end a campaign quietly without too much hoopla. Redirect that energy to another tactic that may work better for you.

(Adapted from "Tips for Calling a Boycott," *National Boycott News,* Winter, 1992–93.)

Local Hell-Raiser
Boycotting the Big Mac

Students from Lake Region High School in Newport, VT, picket their local McDonald's in October 1987, asking patrons to boycott the restaurant's foam packaging.

Who'd ever think a bunch of kids could bring a mega-corporation to its knees? Certainly not the McDonald's fast-food chain. But it didn't take McDonald's long to figure out it was wrong.

In 1990 groups of New Jersey elementary and high school students had become increasingly concerned with the environmental impact of the company's nonbiodegradable plastic foam packaging. The groups, Kids Against Pollution, from Tenakill Elementary School, and the Environmental Club of West Milford High School, feared that the gases used to produce the company's plastic foam packaging were damaging the Earth's protective ozone layer. The students weren't too wild about the amount of garbage McDonald's was creating, either. At the time, McDonald's was the nation's largest user of polystyrene food packaging, producing enough "styro-trash" each day to fill a football stadium forty feet deep.

The kids asked McDonald's to stop using foam packaging and to switch to paper or reusable products. They gave the company six months to consider their requests, asking for a response by Earth Day 1990.

When McDonald's failed to reply, the two groups issued a call for an international boycott and began picketing McDonald's restaurants in their area. The art department at the high school printed "BOYCOTT MCDONALD'S" bumper stickers that were mailed to concerned youths across the country.

Kurtiz Schneid, a senior at West Milford High School and president of the Environmental Club, attracted enormous attention at demonstrations by dressing as a clown named "Ronald McToxic." In fact, when Schneid was invited to speak before the United Nations General Assembly as part of the UN's 1990 Environmental Education Day, he used the occasion to issue an appeal to the world to boycott McDonald's dressed in full "McToxic" attire. Borrowing from a well-known McDonald's jingle, the clown concluded: "The planet deserves a break today."

Meanwhile, another grass roots group, Vermonters Organized for Clean-up (VOC) had been urging friends and neighbors to stop buying Big Macs—and all foam packaging—since 1987. For their ban of Styrofoam (foamed polystyrene), Vermont boycotters had also chosen McDonald's as a target, and capitalized on holidays by organizing rallies at local McDonald's outlets. The slogan for yearly April Fool's Day activities was "You Can't McFool Us!" On Valentine's Day, huge hearts were constructed reading "Dear Earth, I'm Yours Forever—Love, Styrofoam." Vermont students got into the act, too, plastering a McDonald's front window with boycott bumper stickers.

VOC saw the national potential for the boycott and contacted the Citizen's Clearinghouse for Hazardous Wastes (CCHW), which plugged the boycott into its grass roots network of activists around the country concerned about the manufacture and disposal of toxic substances. "It became a people's campaign," said Theresa Freeman of VOC.

To attract people from all over the country to participate in the movement, CCHW developed the newly dubbed McToxics Campaign to "fit the people in our movement," wrote Penny Newman, a seasoned CCHW organizer, in a postmortem of the campaign that appeared in *Everyone's Backyard* (February 1991, vol. 9, no. 1, published by the Citizen's Clearinghouse for Hazardous Wastes, Inc.). Those people were "hardworking, everyday people scattered throughout the country," she said. "We looked for a campaign that was nationwide—people taking action on their local McDonald's and tying it to their local problem."

According to Newman, as groups took on McDonald's in their communities, they tailored their protest to their own circumstances. Groups concerned about shrinking landfill space talked in terms of McDonald's' contribution to the garbage crisis. Organizations fighting incinerators argued that burning packaging material like the McDonald's clamshells would release toxic gases into the atmosphere.

Groups tailored their tactics to their own resources as well. In addition to the boycott, some groups raucously picketed their local McDonald's; others tied up order lines demanding alternative packaging.

In November 1990, McDonald's announced an end to its use of plastic foam packaging and a switch to a paper-based wrap.

On the day McDonald's announced its new policy, the Environmental Defense Fund, with whom the company worked closely to develop new ways to reduce the amount of garbage created at its 8,500 restaurants, noted that the "end of the polystyrene clamshell represents the turning of the tide on the throwaway society." (Today, the company reports that the packaging switch has reduced the volume of its sandwich container waste by 90 percent. The switch has also prompted other fast-food chains to adopt more environmentally friendly practices.)

In the end, the campaign to get McDonald's to stop using Styrofoam was won, said Newman, "in our towns, by our own people, involving more and more people from all walks of life."

(Excerpted from "McVictory: Kids Mount Successful McBoycott," *National Boycott News,* Winter, 1992–93.)

☞ **CHECK IT OUT:** *National Boycott News* provides updates on many of the boycotts under way. The Institute for Consumer Responsibility publishes this newspaper irregularly; 8 issues have been put out since 1984. For a subscription cost of $10 for individuals; $20 for schools, libraries, and nonprofits; and $40 for corporations, you'll receive a major issue of 80 to 120 pages and three smaller updates. Write NBN at 6506 28th Avenue, N.E., Seattle, WA 98115.

Co-op America helps its members, both individuals and groups, invest in a socially and environmentally responsible manner. Its newsletter, *Co-op America Quarterly,* includes "Boycott Action News," a regular update on active boycotts. To join Co-op America, send $20 to 1850 M Street, N.W., Suite 700, Washington, DC 20036.

SLAPP Suits

Companies that are feeling pressured by a boycott sometimes threaten to sue boycotters for libel, slander, or defamation. These suits are largely intended to intimidate the activists and drive them to drop their campaign. Dubbed SLAPP suits—for Strategic Lawsuit Against Political Participation—these actions usually fail to recover damages. Todd Putnam of *National Boycott News* said that of the hundreds of boycotts the publication has tracked, he knows of only three SLAPP suits that have been carried to closure. Still, suits prey upon activists' fears and pocketbooks, since hiring a lawyer, even as a precaution, can get expensive.

Sometimes SLAPP suits can backfire, even for powerful corporations. During INFACT's seven-year boycott of Nestlé infant formula, INFACT was sued by Nestlé. Many newspapers criticized the corporate giant for trying to squeeze money out of a small grass roots organization, and INFACT, flush with the free publicity generated by the suit, won its demands from the company shortly after the lawsuit fizzled.

Boycotters frequently seek legal counsel early in their campaigns to identify potential areas where they may be vulnerable legally and to avoid creating a situation in which they might libel or slander the target of the boycott. Some groups have a lawyer review all their materials before they release them, while others incorporate to prevent individuals within their ranks from being sued. Still other hell-raisers file SLAPPback suits, which sue the suers for suing.

If SLAPPed, says Will Collette of the Citizen's Clearinghouse for Hazardous Wastes, publicize. Once you are absolutely sure that the issues raised in the suit are groundless, complain to the media and the public that free speech is being threatened by the company that is bringing suit, and keep the heat on until the suit is dropped or thrown out of court.

PUBLIC HEARINGS

A public hearing is a forum that allows both sides of a controversy to air their concerns and raise questions about an issue on the public agenda. Though intended to be a civilized debate, it may turn into a downright brouhaha if tempers flair and passions rise.

In general, public hearings are organized by public officials—for example, representatives of the federal government, city or county administrators or council members, school board members, PTA officials—in order to receive varying viewpoints on contentious topics. The hearing may be held because a vote on a city or town ordinance is approaching, or because an agency is considering taking an action that affects many people in different and significant ways. Though votes favoring or opposing the proposal being debated are rarely taken at a public hearing, the event may serve as an important barometer of public opinion that influences the vote when it eventually occurs. If you know going into the hearing whether the event is perfunctory, or whether the issue will in fact be determined largely by what transpires at the hearing, you can develop your strategy accordingly.

Because public hearings can be so volatile, they are usually scheduled at a neutral location, such as a school or city hall. The convener of the hearing may alert the public by putting a notice containing the hearing date, time, and location in the local paper or by posting announcements on the local cable television channel and on fliers at city hall, often publishing the meeting's agenda as an invitation for all residents to attend. (The federal government posts notices of the hearings it holds in the *Federal Register.*) Some agencies will mail notices of impending hearings to lists of citizens who have requested that they be kept informed. Activists who worry they might somehow miss the hearing are free to contact the city council or agency themselves to find out when the hearing will be held, then publicize the meeting to their members or other volunteers by posting flyers, running their own advertisements, and activating their telephone trees (for a description of telephone trees, see page 51).

Public hearings provide activists with an important opportunity to demonstrate their own strength on the position they've taken and give those convening the hearing a sense of how concerned people are about the issue. Hearings also let activists observe their opponents "up close and personal" and get a sense of the arguments they can expect to be made against them during the campaign. Finally, through their testimony at the hearing (and the written comments usually submitted afterward), activists can place additional facts and comments about their position on public record.

Local hell-raisers frequently use public hearings to generate favorable media coverage about their position and to establish credibility for their organization in the public eye (Chapter 6, "Communications," offers many more tips for using the media effectively). For example, community activists in Lorton, Virginia, who were opposed to the poorly planned siting of an incinerator in their community, attended a public hearing wearing surgical masks to demonstrate their concerns for the negative air quality impact the incinerator would have on the region. The next day's newspaper carried a small story about the hearing but a large photo of the activists bedecked in their white surgical masks. When Lois Gibbs, catalyst of the now-famous victory over the Hooker Chemical Company in Love Canal, New York, saw that officials were not listening to her testimony at a public hearing in Maryland, she stopped speaking and stood silently at the microphone. Eventually, the officials noticed and asked if she was finished. "No, sir," she said. "I was simply waiting for you to start listening. When you're ready, I'll continue."*

Potential speakers at a public hearing often have to sign up at the beginning of the meeting; the first five, for example, from the "pro" side as well as from the "con" side may be given time at the podium. Sometimes interested speakers can call the sponsoring council member's office in advance and sign up there. Either way, slots should be set aside for an equal number of speakers on each side of the issue. Some hearings limit the time per speaker and the time allotted for the entire proceeding; some offer a chance for audience or council members to respond to or question speakers.

---

*As reported in *How to Win in Public Hearings*, Citizen's Clearinghouse for Hazardous Wastes, Inc., 1990.

• •STEP BY STEP• •

How to Successfully Participate in a Public Hearing

• **Make the most of your time at the microphone.** You probably won't have a chance to address every issue you'd like. So strategize with others in your organization or coalition and divvy up the topics that need to be addressed among several speakers. That way all of your issues get aired.

• **Even though your oral remarks may be brief, you can submit lengthy supporting documents to the entity convening the hearing for the public record.** You can also distribute news releases, fact sheets, and charts at the hearing to help educate the public and the media along with those holding the hearing.

• **During the hearing, make notes of what others say on the issues you feel need clarification or rebuttal.** You can respond either during your statement or in written comments you submit after the hearing.

• **Make your group's presence at the hearing felt with signs, posters, and other visual effects.** These will not only help pressure the officials who have convened the hearing, but will increase your chances of securing media coverage.

• **Be polite.** Emotions tend to escalate at public hearings as charges fly and people begin to vent their anger and frustration. Try to keep your feelings in check, even as you hear accusations or assertions you believe to be untrue. You will have your chance to respond either at the podium or in writing; by doing so in a controlled, mature manner, you will earn the respect of those who are making the decisions and will not be portrayed as unreasonable or out of control by the media.

☞ **CHECK IT OUT:** *How to Win in Public Hearings,* published by the Citizen's Clearinghouse for Hazardous Wastes, Inc., P.O. Box 6806, Falls Church, VA 22040, $6.95.

Anatomy of a Hearing, League of Women Voters, 1730 M Street, N.W., Washington, DC 20036, $.35.

LITIGATION

Litigation—filing lawsuits and going to court—is a way to enforce existing laws and to protect the rights that are inherent in them. According to the League of Women Voters (LWV), litigation is "both a practical method for effecting change and a powerful means of working within the system for the best interests of the public."

Filing lawsuits can be extremely time-consuming and expensive, so it's not a tactic you want to employ on a regular basis. There will be occasions, however, when your adversary leaves you no choice but to use the law to force a specific action, delay a decision, punish wrongdoers, or give your group the leverage it needs to win compensation for people who have suffered personal injuries or whose property has been damaged. LWV writes that public officials sometimes take groups that have brought suit more seriously in future negotiations, and often the mere announcement of plans to sue can cause officials to change course.

How do you know when a lawsuit is the right plan of action? In its publication *Contemplating a Lawsuit,* the Northern Rockies Action Group (NRAG) has compiled a list of questions organizations should ask themselves when considering legal action.

1. **Does the lawsuit aim at an activity that is squarely within the established purposes of the organization?** The cause for the suit should be very near to the group's core issue or issues. A suit is not a good means for broadening a group's horizons.

2. **Will the suit be a diversion and a drain?** "A lawsuit is a

luxury," NRAG author Scott W. Reed writes. If it doesn't further one or more of the goals of the group, choose another tactic.

3. **Is the suit likely to boost membership or at least not alienate present members?** In other words, will the lawsuit maintain the organization's current health and foster, rather than hamper, future growth?

4. **Are there alternatives to litigation that might achieve comparable or better results?** Don't forget about the effect political pressure—leveraged through letters, phone calls, lobbying, public hearings, boycotts, media coverage, and other tactics—can have on a public official, governmental agency, or corporation.

5. **Can you afford a lawyer?** Check your resources and commitment to a potential suit before taking on the expense that accompanies legal counsel.

6. **Have you been sued?** Do you need to defend yourself against a lawsuit that has been filed against you or your organization? Lawyers may be extremely useful in helping you unearth important information about your opponent in the event you are sued. In a process known as "discovery," both you and your opponent will have to respond to written questions (interrogatories); make sworn, pretrial oral statements in response to verbal questions (depositions); and file certain motions such as "motions to produce" documents, records, or files, or "motions to inspect" buildings or facilities.

Your lawyer will also represent you before the judge in the case, and argue your case before a jury if necessary.

While you can read the laws and regulations applicable to your business, a lawyer may be able to help you find ways to avoid legal pitfalls, intercede with the government, and represent you in the event you are sued. A lawyer can also help you review the details of any agreements or contracts you enter into, read proposed legislation to determine if the language a bill contains is constitutional or reflects your intent, and help you determine when to sue others.

The following example highlights one situation in which hellraisers had no option other than litigation to change a company's discriminatory business practices.

Local Hell-Raiser
Suing to Shatter the "Glass Ceiling"

In any workplace, individual cases of discriminatory hiring and firing are pretty disturbing. But when an entire group of employees finds itself victimized, a chilling sense of alarm can begin to infect everyone. Frequently, the most successful—and sometimes the only—way to fight job discrimination is in court.

In the 1970s, Muriel Kraszewski, Wilda Tipton, and Daisy Jackson worked as administrative assistants and office managers in various State Farm Insurance Company agencies in California. Each was interested in becoming a sales agent trainee. Yet every time they asked their agency managers about the possibility of advancement, they were told that they could not possibly be considered. The reasons the management gave were weak and kept changing: The women did not have college degrees; they didn't have enough sales experience; the company did not encourage transfers from operations to sales divisions. For each excuse they encountered, however, all three employees could think of several male State Farm agents who lacked the very qualifications that were supposedly barring the women from sales. The women were highly efficient workers who juggled heavy loads of responsibilities, including discussing complicated policies and premiums with clients over the phone; why couldn't they, too, advance?

When Muriel and Daisy quickly found successful positions as sales agents with one of State Farm's competitors, it became clear that not only were they qualified to sell insurance but that State Farm was actively discriminating against women.

The three women contacted Guy Saperstein, an Oakland attorney known for class action suits and a member of Trial Lawyers for Public Justice. Together they filed suit against State Farm for sex discrimination in its hiring and promotion policies. Then Saperstein embarked on a complicated, twofold mission: to review the company's recruiting and personnel materials, which he believed would prove the company's bias against recruiting women; and to locate the hundreds of plaintiffs who had experiences similar to the plaintiffs'. By acting on behalf of more women, he not only could win them their deserved compensation but also could build a stronger class action suit against State Farm.

State Farm dug in its heels, claiming that it did not keep information on job applicants who took aptitude tests, which would have provided the plaintiffs with a crucial body of evidence. But even without the proof, when the trial began in 1982, the lawyer painted a picture of blatant sex discrimination by State Farm's agency managers. Many witnesses backed up the three women's assertions, recounting statements that had sent the subtle but unmistakable message to women that they need not apply for promotions.

Saperstein continued to build his case against State Farm, showing that the company recruited mainly within social circles and among personal contacts. None of its recruitment literature pictured or even mentioned women. And then a bombshell hit: Slipping up, one witness, a State Farm research employee, referred to the number of women who had taken the company's aptitude test and to the data base from which he got the number. Here was the critical evidence Saperstein had been looking for all along, and the company had blatantly withheld it! The judge was as enraged as Saperstein, and ordered State Farm to produce the records immediately.

Besides slapping State Farm with $430,000 worth of sanctions for withholding discoverable evidence, the judge wrote a 175-page opinion finding State Farm liable for intentional sex discrimination in its recruitment, selection, and hiring of sales agent trainees.

But Saperstein's work was far from done. He demanded individual trials for each woman who had been discriminated against. State Farm agreed to help track down these women, who numbered an estimated 60,000 to 70,000.

After individual hearings for ninety victims of discrimination resulted in seventy-five settlements, State

Farm realized that it could be drowning in attorney's fees for years. The insurance company opted to end the case once and for all with a $157 million settlement—at that time, the largest employment discrimination payment in U.S. history. Meanwhile, the judge mandated that all of the company's California branches fill 50 percent of their sales positions with women for the next ten years.

(Condensed from the Trial Lawyers for Public Justice publication *Trial Lawyers Doing Public Justice*, 1992.)

How Do You Find a Lawyer?

Hire a "hometown" lawyer if possible, preferably someone in your community who cares about the issues your group is working on and may even be a member. Interview several candidates in an effort to find an experienced lawyer who has carried suits similar to the one you're contemplating. Get references from other organizations you trust. And check with your local bar association to determine whether any complaints have been filed against the attorney you're considering hiring.

When you do hire a lawyer, draw up a clear contract detailing what the lawyer is expected to do and by when. You should remain in control of key decisions, but use your lawyer for advice. Don't let your lawyer decide your organizing strategy; most lawyers are not experts on organizing, and they may shy away from tactics you need to engage in for organizing purposes.

Everybody knows that legal services are not cheap. Still, you may be able to find a lawyer or a group of lawyers willing to take your case on a *pro bono* ("for the public good") basis, which would allow you to get their services for free and pay only their expenses.

Most lawyers charge either on a simple fee basis or on a contingency fee basis. A simple fee basis, charged by the hour, may be expensive but also gives you more control over how and when legal bills are incurred.

A contingency fee will be paid only when and if the lawyer wins your case in court. But the disadvantage of this approach is that it may encourage your lawyer to settle too quickly and under conditions that you find unsatisfactory just so the attorney can get paid. Talk through both approaches with other experienced advisers in your organization before choosing an option.

CHECK IT OUT: Trial Lawyers for Public Justice maintains a bank of 1,200 lawyers who "work to bring about social change and vindicate individual rights by making wrongdoers pay for their misconduct," according to a TLPJ publication. For more information about obtaining public interest legal aid, call Trial Lawyers for Public Justice, 1625 Massachusetts Avenue, N.W., Suite 100, Washington, DC 20036 (202-797-8600).

Also see:
Contemplating a Lawsuit? A Practical Guide for Citizen Organizing, available for $7.50 from the Northern Rockies Action Group, 9 Placer Street, Helena, MT 59601.

A User's Guide to Lawyers, available for $5.75 from The Citizen's Clearinghouse for Hazardous Wastes, Inc., P.O. Box 6806, Falls Church, VA 22040.

Going to Court in the Public Interest: A Guide for Community Groups, from the League of Women Voters, $.85, 1730 M Street, N.W., Washington, DC 20036.

 ACTIVIST'S NOTEBOOK

ORGANIZING TELEPHONE TREES AND COMPUTER NETWORKS

In the heat of a campaign, you may need to ask many members of your group to participate in a demonstration, write letters, send telegrams, or take other important "last-minute" actions.

The fastest way to mobilize people may be by calling them on the telephone.

In smaller organizations, telephone "trees" work by circulating a list of names and phone numbers to all members and asking each person to call the next name on the list. If someone can't be reached, the caller skips to the next person who can be contacted.

In larger groups, your tree might consist of a coordinator to keep the tree organized; activators, who oversee groups of phoners; and phoners, who will actually contact the members.

Using either a card file or a computer, the coordinator establishes and maintains the names, addresses, and phone numbers of members who participate in the telephone tree. The coordinator also helps recruit activators and phoners, making sure that each activator who is recruited has the names and phone numbers of the members he or she is to call.

Each activator helps recruit and maintain a team of five to seven phoners. These are the people the activator will contact when an emergency arises and the callers need to begin telephoning.

When the alarm goes out, each phoner contacts five to ten more members and urges them to act.

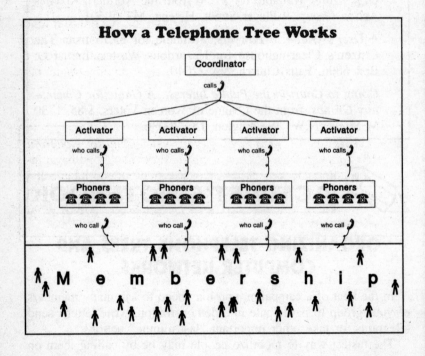

You can also post your message on a computer bulletin board system (BBS), just one service available in the vast world of electronic networking. You need only a computer, a phone line, a modem, and bulletin board software to read and transmit announcements throughout your activist network. America Online, CompuServe, Prodigy, and Internet are networks popular among general users; EcoNet, HandsNet, PeaceNet, and the WELL are

networks specially designed for nonprofits and social interest organizations. The modem works with the phone system to connect one
computer to another; you talk and listen through your keyboard
and monitor. The software and membership in a network allows
you to take advantage of services such as message boards, electronic mail, file exchange and storage, and data bases.

"Many nonprofits have set up BBSs to provide their members
and related agencies with information and the opportunity to learn
from each other," writes Tom Sherman in the Benton Foundation's
excellent primer, *Electronic Networking for Nonprofit Groups: A
Guide for Getting Started.*

Through their computers, groups can send out a request to others on the network urging them to write a letter, make a phone call,
or participate in a demonstration. (They can also announce a fund-
raising event, request research on a certain topic, or share their
own information about the issue they're working on. Chapter 6,
"Communications," offers many more tips on electronic networking.)

☞ CHECK IT OUT: *Electronic Networking for Nonprofit
Groups: A Guide to Getting Started* provides good basic in-
formation about electronic networks and bulletin boards and
lists major networks, services, and training. Available from
The Benton Foundation, 1634 I Street, N.W., Twelfth floor,
Washington, DC 20036, $7.00.

NEGOTIATING

Negotiate means "to confer with another so as to arrive at the settlement of some matter." It also means "to get through, around or over successfully." Both definitions are instructive in the way community groups can use negotiations to achieve their goals.

You may not like talking to your "enemies" much, but there are times when it makes sense to negotiate with them. Why? Perhaps you've directed successful actions against them but you still don't have written guarantees to secure their commitments; negotiations will formalize or solidify on paper the gains you've made through more direct action. You may want to negotiate if doing so will strengthen your position vis-à-vis your opponent. You may also agree to negotiate in order to gain information about your target that you can't get any other way.

On the other hand, do not negotiate when you suspect your negotiating team can be outmaneuvered, or if you must cease direct action while the negotiations are under way. And certainly don't negotiate a bargain that benefits individuals within the organization over the membership at large.

• •STEP BY STEP• •

How to Negotiate

• **Appoint a negotiating team for your group.** Give those representing you explicit instructions about how far they can go in the session and what kinds of issues must be brought back for a group decision.

• **Do not surrender your power.** Don't give up the right to picket, sue, organize, or do whatever else is needed just because a company is willing to talk with you.

• **Plan your strategy carefully.** Decide who from your group will chair the meeting, what will be said and not be said during the session. Be clear about what kinds of disclosures you want to make and avoid. Don't disclose your "bottom line" or reveal any hidden weaknesses in your group's power or strategy.

• **Be prepared for give-and-take.** Set initial demands high enough to give you room to compromise on some points. Be

How to Negotiate (cont.)

clear about your most essential interests, as well as the difference between your interests and your negotiating posture.

• **Agree on an overall timetable.** Establish a specific schedule with dates and times and a bargaining location most advantageous to your organization.

• **Be sure you're talking to the right people.** Verify that the people sent to negotiate with you have the authority to make the decisions needed.

• **Caucus during and between negotiating sessions.** If any member of your team is unsure about what to say during a session, take a break out of earshot from the company officials and other outsiders and consider your strategy. Don't hesitate to walk out of the session if the other side is being uncooperative. Between negotiating sessions, get together with other members of your group and discuss strategy, progress, and issues raised in your discussion.

• **Get agreements in writing.** Look for loopholes that could enable those with whom you're negotiating to renege on their commitments.

• **Determine in advance what you want the outcome to be.** Without realistic expectations, you won't know where to start negotiating, you'll have trouble figuring out what steps to take along the way, and you won't know where to stop.

Remember the words of James Freund, author of *Smart Negotiating: How to Make Good Deals in the Real World* (Simon and Schuster, 1992): "It's winning him over that counts; not winning over him."

Freund compares negotiations to sports: You need to approach each negotiation with a well-conceived game plan the same way a football team prepares for each week's contest. But, he warns, you can't follow the plan blindly; you must remain alert to unexpected revelations, new developments, emerging risks, and opportunities. As the activists in the following story have learned, if you start out with a sense of where you're heading and a well-conceived strategy to get there, you'll make out much better over time.

LocaL Hell-RaiseR
Everybody Wins

Rikki Spears is making a difference. As a community organizer for the Maryland Low-Income Housing Information Service, she travels throughout the state in an effort to help disadvantaged renters establish effective tenant groups that can negotiate for protection against crime, control of rodent infestations, and improved housing services.

"I try to get [people] a voice in the community so they can see themselves as part of the decision-making process," she told the *Baltimore Sun.* "They have to realize that they are powerful as a unit."

One way they can increase their power, Spears believes, is by negotiating with landlords about their demands over maintenance and crime, laundry and recreational facilities, and other important daily issues.

In one case, she worked closely with the Madison Park North Tenants Association in Baltimore to teach its members how to hold constructive meetings and to negotiate with the management company that oversees the building. "Rikki has given us a lot of guidance about how to deal with management in a very positive way," Brenda Stokes, the association's vice president, told the *Sun.*

"Negotiation with management had always been the weakness of our organization. . . . We'd end up in a huff and nothing would be resolved," Stokes said. "Now, before I go to a meeting I'm looking for the leverage points that we have and thinking about what it is we want at the end. . . . You have to be able to negotiate and come to that middle ground."

☞ **CHECK IT OUT:** *Negotiations for Public Interest Groups,* The NRAG Papers, vol. 3, no. 4, 1980, Northern Rockies Action Group, 9 Placer Street, Helena, MT 59601, $7.50, 45 pages.

SHAREHOLDER ACTIONS

According to federal law, shareholders in any corporation have the right to obtain information and to vote on major policy issues involving that corporation. The U.S. Securities and Exchange Commission (SEC) requires "public-issue" corporations—those registered with the SEC—to disclose lawsuits or agency proceedings that might require significant spending to represent or defend the corporation. (For example, environmental proceedings must be reported if it's obvious that the typical investor would want to know about them, or if the resulting costs could add up to more than 10 percent of the firm's assets, or more than $100,000 if the government became a party to the case.) The SEC also allows shareholders to vote on social and political questions relevant to the corporation. Any holder of voting stock can pose questions for all shareholders to vote on. Thus, activists can use shareholder meetings to raise issues and place pressure on corporate officials to change the company's policies.

What Can You Do?

- **Become a shareholder.** Buy shares of stock in a company whose policies you want to change.

- **Formulate a proposal for action by shareholders.** Recommend action by the firm's management or suggest amendment of the corporate bylaws. Address your proposal to overall company policy rather than day-to-day decisions. The SEC has particular guidelines that will help you prepare your proposal; contact them at 450 Fifth Street, N.W., Washington, DC 20549.

- **Begin a divestment campaign.** Encourage shareholders, institutions, unions, and public leaders to sell their stock and reinvest in company stocks and money market funds unsullied by objectionable policies or practices.

If butting heads with a corporation seems like a daunting task, take heart from the activists profiled in the following story, who have begun a series of shareholder actions to save their jobs.

Local Hell-Raiser

Influencing Shareholders:
Hitting a Company Where It Counts

Members of the Alternative Use Committee sported these buttons at the 1992 Unisys shareholder's meeting in Philadelphia.

Declining defense contracts have made for tough times at Unisys Corporation's Paramax division, maker of computer and electronic parts for the military. Approximately 50,000 workers around the country have been laid off since 1989, 7,000 of whom lived in the Minneapolis and Saint Paul area. Although workers formed an Alternative Use Committee and identified twelve viable "hot" products that could be produced at Unisys's existing plants—like "smart" irrigation systems and traffic lights that change in response to traffic conditions—more layoffs loomed, and the company turned a deaf ear to the committee's suggestions. Angered by Unisys's lack of interest in their efforts to save jobs and bolster profits, the committee members wrote a shareholder resolution to be presented to Unisys and voted on by shareholders in April of 1992.

The resolution, written in consultation with the New York–based Interfaith Center for Corporate Responsibility, asked that the Unisys board of directors involve management, labor, and community representatives in alternative economic planning—called "economic conversion"—and make a progress report available to shareholders. Employees who owned Unisys stock presented the document to the company for approval for the ballot, but Unisys succeeded in blocking it. According to Janet Groat of Minneapolis's Jobs With Peace Campaign, who was active on the Alternative Use Committee and in writing the resolution, Unisys claimed that alternative planning is "ordinary business" and therefore exempt from shareholder intervention.

Meanwhile, Unisys workers in Flemington, New Jersey, had also been rebuffed in their attempts to buy out their closing plant. Disgruntled Flemington workers boarded a bus to attend the company's 1992 national shareholders meeting in Philadelphia, and the Minneapolis committee decided to make the trip too, resolution or none.

It turned out to be a wise move. The two busloads of activists leafletted the shareholders as they entered the meeting, asking attendees to "Support Employee Efforts to Save Jobs." They carried signs declaring "THE U IN UNISYS MEANS UNEMPLOYED!" and wore buttons stating "NEW PRODUCTS SAVE JOBS."

"We were very polite," Janet said, "but we were very noticeable." They seized opportunities during the meeting to ask strategic questions: Claudette Munson, chair of the Alternative Use Committee, jumped in during a discussion of company plans to retain top-level executives. "We also have a retention plan for employees," she told shareholders and directors. "It's called the Employee Stock Ownership Plan [the worker buy out] in Flemington and economic conversion in Minnesota." She handed CEO James Unruh the committee's packet of information and asked him to study the alternative use proposals. Shareholders in attendance, who had themselves suffered three years of consecutive losses,

UNISYS

WHEREAS   the sudden and unplanned shift away from weapons procurement has resulted in defense industry plant closings nationwide as well as thousands of worker layoffs.

WHEREAS   civilian manufacturing is the cornerstone of the economic security of our nation.

WHEREAS   economic conversion in a company or at a plant dependent upon defense contracts requires the planned transfer of productive resources to more stable, diversified operations.

WHEREAS   job loss in the U.S. is a broad social and economic issue of major national significance. Anxieties over declining defense spending lead the sponsors of this resolution to emphasize the need for federal policy providing incentives for corporate conversion which includes alternative planning for production facilities, job creation and retraining for employees.

WHEREAS   in light of Unisys' dependency on military and other government contracts, increasing numbers of lost and relocated jobs, the growing citizen concerns and national policy shifts, we believe it in the best interests of our employees and shareholders to plan for alternative products.

THEREFORE BE IT RESOLVED: The shareholders request the Board of Directors to review Unisys plans for workers and facilities engaged in defense contracts and provide a summary report of the findings of the review and recommendations for changes in policy and plans in light of this report. The report should be available to shareholders on request, may omit proprietary information and be prepared at reasonable cost.

STATEMENT OF SUPPORT

We believe the combination of projected decline in arms procurement and weapons development and the relocation of work to nations with lower wages and less stringent enforcement of labor and environmental laws is affecting all companies, their employees and the communities involved. The review requested will demonstrate our Company's concern for shareholders, employees and the environment. We propose the review and report include information on the following.

1.   Mandatory alternate-use planning committee at teach Unisys plant site.
2.   Process for cooperation with unions or labor representatives at each Unisys plant site.
3.   Strategic planning for environmentally sustainable commercial business and non-defense government contracts, including worker retraining, retooling of production facilities, identification of new products and markets.
4.   Working relationship with community and/or state agencies specializing in planning for alternative products.

Just one year ago, the employees of Unisys -- Fleming, New Jersey and St. Paul, Minnesota -- brought their alternative-use plans and efforts to shareholder attention in an attempt to prod management action.

As shareholders, we hope to avoid further financial devastation and further suffering by workers. We encourage Unisys to be an innovative planner and responsible employer by demonstrating leadership in planning for alternative products. Unisys can be #1 in creating vital local economies by utilizing its highly trained professionals and employees to produce needed commercial and non-defense government products that meet today's public needs such as transportation improvements, infra-structure improvements and environmental conservation for Unisys markets at home and world wide. If you agree, please support this resolution by voting YES.   11-19-92

Fifteen percent of Unisys's shareholders voted "yes" to this resolution in April 1993.

joined her in voicing their own discontent with the company's lack of planning.

The turbulent meeting received good coverage in the *Philadelphia Inquirer* and the *Trenton Times.*

After the meeting, Unruh contacted Munson to say that he had sent a new business representative to Minneapolis, who proved surprisingly receptive and helpful. "We believe her responsiveness was due to our presence at the meeting. There is now a more open door policy," Groat said.

The Alternative Use Committee wrote a second resolution for the 1993 ballot that Unisys was unable to block. It spoke more specifically about economic conversion, individual communities and their workers. Though committee members were out of resources and unable to attend the annual Philadelphia meeting, the document still received a 15 percent affirmative vote.

The next step, Groat said, is to try to get a sponsor for economic conversion from within that 15 percent block of shareholders. "The resolution is a good foot-in-the-door tool," she said. They will be targeting one of the supporting shareholders—who just happens to be a Minneapolis native.

Challenging a large company from such a close vantage point is not easy; thus, groups don't hope for immediate policy change when submitting a resolution. According to Janet Groat, companies naturally do whatever they can to keep resolutions off their proxy ballots. Unisys's cry of "ordinary business" is an argument many companies use to prevent shareholders from interfering in business decisions.

"However," she said, "it should be ordinary business to involve workers in discussions, and they were not."

Once the resolution lands safely on the ballot, a 3 percent positive vote is considered a victory; to gather a 15 percent vote is almost unheard of, according to Valerie Heinonen of the Interfaith Center on Corporate Responsibility. "It was astounding," Heinonen said. "1992 was the first time we've had such high response on a military-related issue." The goal of a resolution is to alert shareholders to your issue, secure their sympathy, and, ideally, get them to back your efforts. The company realizes that you mean business, and faced with potential shareholder pressure, may consider changing the policies or practices that you are protesting.

☞ **CHECK IT OUT:** The United Shareholder's Association (USA) publishes *The Shareholder Proposal Process: A Step-by-Step Guide to Shareholder Advisory for Individuals and Institutions* for those interested in submitting shareholder resolutions. You have to join USA to receive it; membership rates vary and the price includes the 16-page booklet, USA's bi-monthly newsletter, and various studies ranking companies' performance versus compensation and their responsiveness to shareholders. For membership information, call USA at 202-393-4600, or write to 1667 K Street N.W., Suite 770, Washington, DC 20006.

ACCOUNTABILITY SESSIONS

Accountability sessions are meetings you hold with targeted officials to gain their support for a specific project or proposal and to remind them about the promises they've made in the past. They offer a way to make public officials responsible for their votes, policies, written and verbal statements, and other actions by letting them experience the displeasure or concern of their constituents firsthand and warning them that they will be held accountable for their actions through some kind of reprisal. Those reprisals may take several forms, depending on who the target of the accountability session is. Withholding votes, boycotting products, organizing pickets and sit-ins, and generating negative publicity are among the tactics groups use to hold officials accountable.

Accountability sessions are usually held toward the end of an issue campaign after you've built up a great deal of strength, when your group has the ability to turn out hundreds of people, and when your leaders can run an effective meeting that really puts heat on the target.

Invite the target of the session to attend in a polite letter in which you propose as many dates for a meeting as possible so your target can't refuse. Follow up with a phone call to the target's office. If the target refuses to meet with you, publicize the situation through the media, putting the target under more pressure to comply with your request.

Choose the "right" people to speak: In addition to a panel of people who can most powerfully represent your position, include other elected officials who support you (and who want visibility), and general community members. Limit the session to two hours.

During the session, make specific demands on the target. These demands could include such actions as passing legislation, funding a specific program, enforcing regulations, or providing community services. Whatever they are, they should be demands you can win: You should go into the session with a substantive list of at least three main demands, as well as a fallback position that still helps you achieve your goals.

After a substantial amount of discussion between the audience (your members) and the "guest speaker" (the target of the session), ask the target to agree to your demands—or at least to your fallback compromise. If you get agreement, ask for a time frame in which you can expect action to occur. Decide in advance what steps you will take if no agreement is reached.

Before holding an accountability session, analyze your target to determine how susceptible he or she is to pressure from your group. Are your members likely to vote for or against this official? The more you represent the politician's constituency, the more clout you'll have. A novice politician who barely won the last election, has a tough time raising campaign money, and is already being pursued by potential opponents for the next election may feel compelled to make more concessions to your group—for the purposes of getting your votes—than a seasoned pol who doesn't need your support to win reelection or swell his campaign war chest.

For example, legislators who accept large campaign contributions from the National Rifle Association—and who by and large are elected by opponents of gun control—aren't going to be distressed if gun control advocates try to hold them accountable for pro-handgun votes. On the other hand, candidates who are elected by gun-control advocates have more to fear if they oppose legislation to restrict weapons access.

Not all targets of accountability sessions are elected officials. Appointed officials of government agencies or employees representing a corporation may also be the subject of an accountability session if they have failed to uphold commitments they made to help improve the community. Though you can't "punish" these targets by switching votes, you can use other tactics that can convince the target to take you seriously. Urging the public to buy products from a competing company or staging protest demonstrations outside the official's office or the company's headquarters are just two of the accountability tactics at your disposal.

Here's just one example of how local activists incorporated accountability sessions into their organizing strategy.

Local hell-Raiser
SWOP

Residents of the Sawmill neighborhood in Albuquerque, New Mexico, complained frequently about the sawdust falling on their homes from a nearby particle board factory and the odd taste of their well water, but to no avail. Then the Southwest Organizing Project (SWOP),* a community-based nonprofit group formed in 1980 to work with Chicanos, Native Americans, and African Americans, intervened.

SWOP helped organize neighbors in the area into the Sawmill Advisory Council, then held an accountability session at which more than two hundred residents confronted local elected officials and agency administrators about the air and water quality problems they were experiencing. Organizers asked the state legislature to intervene, circulated petitions, contacted the media, and saw to it that the New Mexico Environmental Improvement Division tested the air and the water, which was found to be contaminated with formaldehyde, nitrates, toluene, and other toxins.

After a six-year fight, state officials finally demanded that the mill comply with hazardous waste management

*A more complete description of SWOP's organizing efforts can be found in *Our Earth, Ourselves* by Ruth Caplan, Bantam, 1990.

regulations and begin a $2 million cleanup program to contain the sawdust, reduce noise pollution, prevent the discharge of hazardous substances by lining the pits that stored the plant's waste, and purifying the aquifer that fed most of the area's wells.

Said Richard Moore, SWOP's co-founder and executive director, "There had been numerous complaints for years, but until this community organized, nothing was done."

PETITIONS

A petition is a document that outlines your position on an issue and on which you accumulate the signatures of people who agree with your views.

A petition is a useful organizing tool on a number of levels. It provides a way to demonstrate the strength of your group by offering concrete proof that a substantial number of people (at least, those whose names appear on your petition) concur with you and support the action you are proposing. A petition drive may generate positive publicity that you can use to pressure the target of your petition while educating the public about the focus of the petition. Asking people to sign a petition may even help you raise money; generous donors may be willing to make a large contribution to support the petition drive itself, and those who sign the petition are frequently willing to contribute a few dollars along with their signature.

The petition can be printed on either 8½-by-11 or 8½-by-14-inch paper. It usually has a title, and clearly identifies the group circulating the petition and the person intended to receive the petition.

A statement at the top of the petition describes the reason you're circulating the document in the first place: Is there a construction project you oppose, a new anticrime bill you support, a candidate whose name you're trying to get on the ballot? Whatever the reasons for action, they should be explained straightforwardly so that they create no confusion about your position or the action you're recommending.

Below this statement of purpose should be blank lines so that people can print and sign their names and provide their addresses

OPERATION VALLEY SHIELD
PETITION

STOP THE CONSTRUCTION OF THE PROPOSED CONTAMINATED SOIL BURNING
PLANT TO BE LOCATED IN WAYNE TOWNSHIP NEAR THE BORDER WITH SOUTH
MANHEIM TOWNSHIP ALONG ROUTE 895 IN SCHUYLKILL COUNTY

NAME(S)	FULL ADDRESS	TELEPHONE NUMBER
1.		
2.		
3.		
4.		
5.		
6.		
7.		
8.		
9.		
10.		
11.		
12.		
13.		
14.		
15.		
16.		
17.		
18.		
19.		
20.		
21.		
22.		
23.		
24.		
25.		
26.		
27.		
28.		
29.		
30.		

A petition put together by Citizens for a Healthy Environment in Schuylkill County, Pennsylvania. Its almost 6,000 signatures helped defeat a soil-burning incinerator in December 1992.

To bolster its petition drive, CHE created this arresting flier based on a design from Organizations United
for the Environment, another Pennsylvania group. The
fluorescent orange flier glares from the windows of
many residences in the county, and many activists
wear buttons of the same design.

(street, city, state, and zip code) and telephone numbers. (Number
the signature lines so you can keep track of how many signatures
you've collected.)

When you're finished collecting signatures, photocopy each
page of the petition that's been signed and keep the copies in a safe
place to maintain proof that the signatures were collected in the
event the original petition is lost or accidentally destroyed. Present
your petition to an individual or governmental body with the
power to act on your ideas.

Petitions can be used to launch statewide campaigns or to make
change at the community level. Read how a Girl Scout troop
joined a community group in circulating petitions to build support
for their local library.

Local Hell-Raiser
The Boyertown Community Library

Boyertown, Pennsylvania, known for its winning American Legion baseball teams, hosted visiting opponents in a shiny new $600,000 stadium. But on rainy days, residents of the little town had to settle for TV: the Pennsylvania Dutch community didn't have a library!

In 1986, a group of book-deprived residents, including representatives from the Berks County Public Library System, local teachers and school librarians, the superintendent of schools and his wife, and the local Girl Scout troop, coalesced into the Boyertown Community Library Committee. Was there any way to generate public support for a library, they wondered?

One scout, Becky Brown, took on the role of project chairman for the Girl Scouts troop, which volunteered to whet readers' interests by setting up a "washbasket library." They collected old books from their friends and neighbors and lugged them in washbaskets to barbershops, beauty parlors, and Laundromats around town. Customers waiting to get their hair cut or for their clothes to dry could choose a book, read it there or take it home, and return it the next time they stopped in.

The book exchange got people reading, and got the community excited about having a true library that would offer many more books. But that was just the first step. To gather more concrete evidence that the community endorsed the idea, the committee drew up petitions declaring "I am in favor of a public library in the Boyertown area."

The scouts circulated the petitions at school, asking adults there to sign. They also set up brightly colored signature boxes in supermarkets and other stores in town, with coupons that passerby could tear off, sign, and drop in. The *Boyertown Times* ran and article entitled "Do YOU want a library in Boyertown? Return coupon with comments." The accompanying clip-out coupon read, "YES, I would be in favor of a library in

The Scouts and their washbaskets, with troop leader Marian Borneman.

Burt Swayze

A LIBRARY IN BOYERTOWN?

CLIP THIS COUPON

YES, I (we) Would Be In Favor Of A Library In Boyertown!

NAME _____

ADDRESS _____

_____ PHONE _____

PLEASE DEPOSIT THIS COUPON AT ANY ONE OF THE FOLLOWING LOCATIONS:
National Bank of Boyertown (Boyertown, Bally and New Hanover Branches), Bause's Drug Store, Schaeffer's Family Restaurant, The Acme Market, Library Stand at Fun Day and the Sidewalk Sale. For More Information Call 369-0805.

☐ CHECK HERE IF YOU WOULD ALSO BE WILLING TO VOLUNTEER YOUR HELP IN THE LIBRARY PROGRAM.

Readers of the *Boyertown Times* added their names to the petition drive in support of the library by clipping, signing and dropping this coupon into collection boxes at stores and banks around town.

Reprinted courtesy the Boyertown Times

Boyertown!" Coupon names were added to the petition signatures.

The petition drive amassed 3,000 signatures, an impressive number considering that the library's readership base totaled only 6,000. The activists then submitted the petitions to officials in Boyertown and neighboring Colebrookdale Township, which the library would also serve.

When town officials saw that fully half the community desired a local library enough to stop and sign their name or mail in the coupon, they eventually approved and helped fund the library.

The little building opened in 1988 amid great celebration. Today state and local funds fuel its modest monetary needs. For example, officials in Colebrookdale Township contribute $500 a year as a result of the petition drive orchestrated by the committee. The effort seems to be worth it: The Boyertown Community Library now proudly claims the distinction of being the third most used library (per capita) in the county.

NONVIOLENT ACTION

Throughout history, people have been effecting change through nonviolent direct action—peaceful demonstrations that protest objectionable laws and institutions. Some activist groups organize an occasional nonviolent action such as a boycott or sit-in as part of a larger political or public relations campaign. But many activists adopt nonviolence as a philosophy by which to live. Martin Luther King, Jr., considered nonviolence the only social change tool that was both socially effective and morally acceptable, writing in 1958 that it is "one of the most potent weapons available to oppressed people in their quest for social justice" *(Stride Toward Freedom,* Harper and Row). Nonviolent activists resist with love instead of hate and answer even violence with peaceful activity. Their willingness to take personal risk without threatening others makes the tactic a powerful one.

A well-known form of nonviolence is civil disobedience, wherein activists consciously break a law or oppose a policy they regard as unjust. By this means they raise awareness and spotlight their dissatisfaction, ideally to induce lawmakers to change the law or policy they oppose. Civil disobedients may also violate a law that is integral to a system or institution they protest—such as trespassing on the grounds of a nuclear plant to protest that plant's manufacture of weapons.

Henry David Thoreau, one of the first to discuss civil disobedience in his writings in 1849, was jailed for refusing to pay his poll tax—his way of protesting the government for supporting slavery and warring against Mexico. In the 1920s and 1930s in India and earlier in South Africa, Mohandas Gandhi engaged in both individual and mass acts of civil disobedience to protest oppressive British rule and to win independence for India. Rosa Parks committed civil disobedience when she sat in the front of a bus, then legally reserved for whites, in 1955 in Montgomery, Alabama. Today, citizen activists turn to civil disobedience to protest war, abortion rights, and environmental injustice; to support civil, voting, and labor rights; and to counter dictatorial governments around the world.

King identified three important characteristics of nonviolent action:

1. It does not seek to defeat the opponent but to win his or her friendship and understanding.

2. It is directed against the objectionable policy itself rather than the people enacting the policy.

3. It involves a willingness to suffer. Gandhi found that when opponents won't listen to reason, suffering is a powerful tool that opens their eyes and ears.

Nonviolence is not for cowards. When activists engage in civil disobedience by knowingly breaking the law, arrest must be an acceptable consequence. The tactic's goal is an assertive one: to ensure that law- and policy-making remain fluid and adapt to changing public opinion.

According to Gene Sharp, a leading figure in nonviolence theory and founder of The Albert Einstein Institution, a nonprofit organization advancing the study and use of strategic nonviolence, nonviolent action can take three forms: *Nonviolent intervention,* in which activists challenge an opponent directly, can be the most dangerous and likely to result in arrest and therefore requires the most discipline among participants. Examples include sit-ins, fasts, or nonviolent obstruction (such as lying down in a roadway to prevent trucks hauling waste from getting through). Civil disobedience most often falls under this heading. Strikes, boycotts, and political noncooperation are categorized as *nonviolent noncooperation.* *Nonviolent protests,* such as marches, picketing, leafletting, vigils, and public meetings, are largely symbolic activities whose goal is not to break the law (although they may involve trespassing) but to build public awareness and show your opponent that you are serious.

Why Demonstrate Peacefully?

Why choose this "peacenik" tactic and philosophy in a world where violence seems to be the rule rather than the exception? King explained that "We adopt the means of nonviolence because our end is a community at peace with itself." Most people find violence detrimental to their cause as well as morally reprehensible.

In his essay "On Civil Disobedience" Thoreau maintained that people have the right to question and resist government when its "tyranny or inefficiency are great and unendurable." He personally could not consider himself ruled by the same governing body that allowed slavery. His reasoning was simple and sensible: If a law is so unjust that it "requires you to be the agent of injustice to another, then, I say, break the law."

Dr. William Coffin, a debater in the book *Civil Disobedience: Aid or Hindrance to Justice?* (American Enterprise Institute for Public Policy Research, 1972), held that one should engage in civil disobedience only as a last resort—"having written your congressmen, having visited senators time and time again" with no results.

Nonviolence Training

Groups planning to engage in nonviolent action benefit from some form of nonviolence training. According to the War Resisters League's *Handbook for Nonviolent Action,* training helps participants understand nonviolence both as a historic and enduring philosophy and to prepare for particular actions. Activists can share fears, build one another's confidence, and contribute toward group unity. A typical training session, which lasts eight hours and involves two trainers and ten to twenty-five people, asks participants to role play various scenarios that will prepare them for the personal and legal confrontations their action could entail.

In addition to the traditional antiwar and antinuclear campaigns, nonviolence training is being used as part of the nationwide movement against violent crime. Dorie Wilsnack of the War Resisters League said that groups such as Alternatives to Violence in New York City are teaching nonviolence philosophy and tactics to prison inmates and high school kids to help them learn how to resolve conflicts peacefully.

Is Civil Disobedience for You?

LeRoy Moore of the Rocky Mountain Peace Center, a grass roots group committed to nonviolence as a way of life that organized, among other things, to shut down the Rocky Flats nuclear weapons plant in Colorado, runs the Center's Nonviolence Education Collective in Boulder, Colorado. To determine whether civil disobedience is the appropriate tactic for your particular fight, he recommends the following four steps, designed by King.

●●STEP BY STEP●●

How to Engage in a Nonviolent Action

• **Investigation.** Collect information about your particular issue and the laws or policies you feel are unjust.

• **Negotiation.** Communicate with the people who have the power to change the law or policy, and request change. Publicize the issue, meet with legislators and policy makers, take some of the other actions suggested in this book before resorting to breaking the law.

• **Self-purification.** If negotiations break down and other tools fail, do some soul searching and spiritual preparation for the action. Determine whether your membership has the sense of peace and mental discipline to demonstrate without lashing out at opponents who may themselves resort to violence.

(Some may worry that arrests will generate negative publicity that will set back their cause, or that donors might discontinue their support. For activists working under the philosophy of nonviolence, however, Moore says, economic considerations are not among their top concerns.)

How to Engage in a Nonviolent Action (cont.)

• **Direct action.** Organize your action carefully, planning for both positive and negative outcomes. Prepare those who plan to participate with a training session that suggests the various scenarios they could encounter. Communicate with fellow demonstrators throughout the action to provide guidance and reassurance. If you have chosen not to participate personally, offer constant moral support.

☞ **CHECK IT OUT:** The War Resisters League publishes the *Handbook for Nonviolent Action* (1989, 34 pages) which provides both detailed instructions on how to engage in civil disobedience and discussions of the various issues nonviolent action may be useful in promoting. To purchase, send $3 to the League at 339 Lafayette Street, New York, NY 10012.

Gene Sharp's three-volume book *The Politics of Nonviolent Action* (1973) offers a good introduction to nonviolence principles and methods. The set is available from Porter Sargent Publishers, 11 Beacon Street, Suite 1400, Boston, MA 02108 for $14.85 or at your local library.

USING THE FREEDOM OF INFORMATION ACT (FOIA)

FOIA provides a way for citizens to gain access to tens of thousands of executive branch documents that would otherwise remain confidential. Correspondence between government officials and corporate executives, and documents that may provide hints about who lobbied whom and how laws and policies took shape, are all available through FOIA.

Under the terms of the FOIA, an agency has ten days to respond to an FOIA request. An FOIA "response" can be a simple form letter saying your request for information has been received. You may receive notice that files on the information you want have been found and are being reviewed. More months can go by during this "review"—especially if the information is at the Federal Bureau of Investigation. Upon occasion, you may have to sue an agency to comply with your request and provide the documentation you are seeking.

••STEP BY STEP••

How to File an FOIA Request

• **Contact the agency that has the documents you want.** Ask who at the agency should receive the specific request.

• **Put your request in writing.** Make the request as clear as possible (see page 76). Send it to the agency's FOIA officer, and write "Freedom of Information Request" on the envelope.

• **Mail the request to the agency** via registered mail if possible, return receipt requested. Keep a copy of the letter and the receipt.

• **Don't be surprised if the agency turns down your request.** Oftentimes, a group is forced to sue a government agency to get the documents released.

• **If the agency will release the documents, ask it to waive the photocopying or search fees.** It's up to the agency to decide if your research is "in the public interest," in which case it may waive the fees.

Sample FOIA Letter

Date
Freedom of Information Unit
(name of person; name and address of government agency)
Concerning: Freedom of Information Request

Dear (Name):

Under the Freedom of Information Act, 5 U.S.C. 552, I wish to request access to (or a copy of) the following document(s): (fill in document numbers).

In accordance with 5 U.S.C. 552 (a)(4)(A), I request that you waive any fees or expenses incurred during the course of fulfilling my request, as disclosure of the records I seek is in the public interest (with one more paragraph explaining how the public interest will be served).

If you do not grant my request within 10 working days, I will assume that you have denied my request. Should that be the case, please inform me in writing why you have chosen to deny my request.

Thank you.

Sincerely,

(Your name, name of organization, address, telephone number)

Many states also have freedom of information laws that grant citizens the right to obtain state government documents. Inquire at your public library.

SPECIAL EVENTS

A special event could be a demonstration, fair, festival, celebration, march, walk, run, picket, protest, civil disobedience, or any other activity you organize to apply political pressure, generate media coverage, demonstrate support for legislation, or mobilize members. Though you can have a successful impromptu event, the best events are well thought out and organized in advance.

The following guidelines were developed with Peter Harnik of the Rails-to-Trails Conservancy in Washington, DC, to help rail-trail activists participate in the National Rail-Trail Celebration, which commemorated the conversion of the five-hundredth abandoned railroad corridor into a multiuse recreational trail. But the tips it contains are equally relevant to a local sit-in or a positive picketing demonstration.

The organizing schedule is based on a two-month scenario. It is possible to organize an event in less time, but it won't work as well if you try to hurry it along. Several activities must be completed during the first month so that a more intense effort can be undertaken during the second month.

Organizing a Special Event

KEY ACTIVITIES IN THE FIRST MONTH

- Organize planning, logistics, and media committees.
- Recruit other organizations to participate.
- Hire or appoint a coordinator.
- Secure the site for the event, choosing one that is accessible, symbolically meaningful, and provides good background visuals for media coverage.
- Begin publicity.

KEY ACTIVITIES IN THE SECOND MONTH

- Finalize plan for event.
- Publicize heavily.
- Recruit volunteers.

———————————— Special Event Checklist ————————————

✓ **Month 1**

___ Hold initial organizing meeting.

___ Establish planning committee.

___ Secure commitments from participating organizations.

___ Hire or appoint someone to coordinate all of the elements of the event.

___ Select the date, time, and duration of the event.

___ Obtain permission to use the event site.

___ Design the overall event plan. (What time will it begin and end? How many different activities will there be? Who will speak? etc.)

___ Develop a detailed budget.

___ Begin fund-raising for the event.

___ Begin publicity.

___ Begin scheduling speakers and entertainment if there are to be any.

___ Acquire necessary permits and insurance.

___ Establish logistics committee.

___ Establish media committee.

___ Determine whether or not you need liability insurance.

___ Determine whether you need to pay tax on any merchandise you plan to sell at the event. Check with your local city government or a tax attorney familiar with your city's laws.

✓ **Month 2**

___ Write first news release announcing event.

___ Design posters and flyers.

___ Distribute posters and flyers.

___ Send out promotional mailing to potential participants (including organizations and individuals).

___ Develop press list.

___ Recruit volunteers.

___ Contact press for interviews, feature stories, and talk show appearances.

___ Follow up with press efforts to publicize celebration.

___ Send out PSAs and announcements to community calendars.

✓ **3 Weeks Before**

___ Schedule volunteer crews (security, cleanup, stage).

___ Confirm date with all the speakers.

──────────── **Special Event Checklist (cont.)** ────────────

___ Reserve stage risers, podium, and sound equipment.

___ Arrange for banners, background music, and other attractions at site.

___ Send out complete event update with speakers list, celebrities, activities, etc.

✓ **2 Weeks Before**

___ Keep up press work. (Secure commitments for coverage, meet with editorial boards, do talk show interviews.)

✓ **1 Week Before**

___ Double-check with everyone—media and logistics committees, volunteers, vendors.

___ Double-check publicity progress and redistribute leaflets and posters as necessary.

✓ **Day of Event**

___ Anticipate problems.

___ Double-check instructions with everyone.

___ Have fun!

───

SUPPLIES LIST FOR THE DAY OF THE EVENT

- Sign-in lists (for the media), pens, paper, pencils.
- Name tags for organizers and volunteers.
- Cash for emergencies, including change for the pay phone; cash boxes that lock; receipts if you are selling merchandise.
- A first-aid kit.
- Telephone numbers for police and fire departments and ambulance services.
- Name of doctor, nurse, or paramedic who is attending the event.
- Copy of your permit to hold the event.
- Literature for your sign-in table: Include background information on your organization, membership applications, etc.
- Posterboard, black felt-tip markers, masking tape.
- Fire extinguisher.
- A special media sign-in table to hold press releases, supporting documentation, and other information.

- Microphone and "mult box" at the podium. A "mult box" allows several radio and television station crews to plug into the same sound system for the best sound.

OTHER CONSIDERATIONS

- **Set deadlines.** Put clear deadlines down on a day-by-day time line and circulate it to all volunteers and workers. External deadlines, such as a printer's schedule or the need to secure a permit by a certain time, can help keep your group on track.

- **Plan for disasters.** Something will surely go wrong—it may rain, your speakers may not show up, your equipment might not work—so anticipate all possible disruptions of the event. Add "disaster contingency plans" to the task list (for example, know where you can get a tent in the event of rain).

- **Consider your audience.** The design, location, date, time, and duration of your event will determine the audience you attract, as will the types of promotion and media outlets you use. Will people travel in order to attend your event, or just wander in because they are already in the area? If they are just wandering by, how can they be drawn in? Banners, balloons, and music attract people on the street. Station leaflet distributors on nearby corners on the day of the event to invite visitors to attend.

- **How long will people stay?** If your event happens during the week, the crowd will probably peak between noon and 2:00 P.M., when people are on their lunch hour. Plan accordingly.

- **What age will your audience be?** Children, teenagers, adults, and seniors will be attracted to very different kinds of events and services. And they can be reached by different media sources. (Not many teenagers read the daily paper and not many elderly people listen to rock stations.) Decide what kind of audience you want to attract, then organize activities those participants will find appealing.

- **Should you hold a press conference?** You may want to have a press conference prior to the commencement of the event to give reporters a chance to ask your speakers questions and to give your speakers the opportunity to present their statements exactly as they would like to be repeated by the media. (Look for more tips on dealing with reporters in Chapter 6, "Communications.")

- **Consider recruiting a radio sponsor.** Radio stations are willing to co-sponsor many events, especially fund-raisers, public education campaigns, and celebrations. Should you decide to pursue radio sponsorship for an activity you are organizing, pick a station you feel will have the most appeal to your target audience, whether it be country, rock, rap, or all talk. Then call the station and ask to speak with either the director of marketing or promotions, or the disc jockey you want to approach. Convince the station that the event will generate as much goodwill for the station as it will for the event.

Local Hell-Raiser
National Night Out Against Crime

Like so many neighborhoods in this country, the New York City community of southeast Queens has a drug problem. Most dealing is done on the streets among the area's youth, and relations between kids and police can get pretty tense. The South East Queens Community Partnership (SEQCP) has made every attempt to curb potential aggression and improve relationships between cops and kids by setting up civil patrols and training some cops as CPOPs—Community Policing Officer Patrolmen—to be more sensitive to kids, minorities, and the cultural challenges that have developed in the community, which is 75 to 80 percent African American and Caribbean American. Despite growing hostilities toward police around the country, SEQCP also helps bridge the gap between civilians and police with its annual Night Out Against Crime.

As an umbrella organization of block and civic associations in southeast Queens that focuses on drug prevention and community organizing, SEQCP wanted to have an event some summer evening to encourage kids

A reporter from WNBC-TV in New York gives viewers a taste of the double Dutch contest, a highlight of SEQCP's National Night Out Against Crime.   *Nat Valentine*

and cops to get to know each other and have fun doing so. Other communities around the country were holding a Night Out Against Crime the first Tuesday night in August, staking out a central area or park, turning on the floodlights, and offering contests, music, and food for all the kids and cops in town to enjoy together. Why not them?

In early summer 1991, Partnership staff members and volunteers from member associations began planning for that August 6. Through a little persuasion, they got a local supermarket owner in Cambria Heights to give up half his parking lot for the event.

Security and permitting were by far the group's biggest hurdles. By working closely with the police force, who backed the event wholeheartedly, they secured permits to close the street and allow the ruckus that would undoubtedly be raised by the four bands they had invited to perform. SEQCP and the police collaborated on rerouting traffic to avoid interfering with bus routes. A local housing development donated maintenance people and trucks, the City Parks Department provided the stage apparatus, and the police brought in the floodlight unit.

"One of the hallmarks of our group is our ability to coordinate among different agencies," Robert Sinclair, Jr., a spokesperson for the Partnership, said. When organizing large events, he advised groups: "Get the police on your side; everything else is gravy." SEQCP's civil patrols worked with police to provide on-site security during the event. Sinclair said that SEQCP's own law enforcement coordinator efficiently delegated the numerous tasks among planners; often during the event, he said, "heads of civilian patrols were directing the police!"

The group publicized the event by sending press releases and public service announcements to the local media, concentrating on the African American press. There are four sizable African American newspapers in New York City and many smaller papers published on the community level. SEQCP and its civic and block association members spread the word via their newsletters. Sinclair, who is a former radio reporter, landed an interview on a morning radio show, "Drive-Time Dialogue," aired by a gospel station. He discussed substance abuse and the need for a community-bonding event like the Night Out. SEQCP received some promotional material from the National Association of Town Watch, which sponsors the National Night Out, and were able to reproduce a poster and purchase bumper stickers to give away.

The event was self-funded; money was raised by renting space for $25 to $75, depending on the space. Street vendors offering every kind of food, drink, and souvenir set up shop in the parking lot. Kids got their own operations going, too: the Positive Youth Connection, one of the Partnership's programs, bought soda in bulk, and the kids sold sodas for $1 each and kept the profits. "We have some enterprising youngsters here," Sinclair laughed. A letter to the New York Mets organization produced a carload of donated paraphernalia—such as posters, packets of team photos from 1961 to 1992, and figurines of former players—that was either given away or sold throughout the night.

There was lots for the youths to do under the lights: They bounced in a jumping bubble; played softball against the police; joined in dance, rap (the favorite), and poster essay contests that focused on drug abuse;

listened to a local church choir perform and a Baptist preacher preach. Many local officials showed up, including New York City Mayor David Dinkins, and forty agencies disseminated information on health, drugs, and teenage pregnancy. The City Health Department gave away boxes and boxes of condoms.

Due to its success in building positive relationships between the police and the community, the Night Out Against Crime was dubbed the "first annual," laying expectations for subsequent events. "The biggest thing is the opportunity to meet and greet the police officers," Sinclair said. Kids had a great time, and parents felt secure letting their children attend an event where police were both on staff and on the guest list!

One of SEQCP's civilian patrollers hangs out with a child at the Night Out.   *Robert Sinclair, Jr.*

TACTICS
DO's and DON'Ts

▲

•**Do** familiarize yourself with the many tactical options you have at your disposal.

•**Do** choose tactics that will affect a lot of people, unite and involve people, be strongly felt, are simple, build your organization, and are fun.

•**Don't** limit yourself to one or two actions. In most cases, you'll have to employ a combination of tactics to make your strategy work.

4.
BUILDING YOUR ORGANIZATION

BEING A SUCCESSFUL hell-raiser is no easy task. What sustains many groups facing enormous challenges, occasional demoralizing defeats, and the constant stress of insufficient capital is a solid organization that has learned how to command a troupe of loyal members and stretch resources by working in coalition with others. Organizations with lesser attributes may be able to win a few battles, but if they want to win the war, they need to build a solid organization that, ultimately, gives people power over their own lives.

INCORPORATING

Your organization doesn't have to be a legal entity to meet as a group, use a name, fund-raise, and accomplish great things. In fact, activists who wage just an occasional campaign often find it easier to remain an ad hoc operation that can disband when the campaign ends, than to go through the legal hassle and paperwork of incorporating into an institution that will require some level of resources to maintain.

On the other hand, groups that have a long-term agenda often decide to incorporate, a legal act that legitimizes the group as a business and allows a group to apply to the Internal Revenue Service for nonprofit, tax-exempt status, a qualification that significantly enhances fund-raising opportunities.

Most of the information presented in this book is relevant whether you incorporate or not. But should you decide to do so, these tips will help you out.

How Do You Incorporate?

You'll need to come up with a name for your group, a statement of purpose, and the legal documents for incorporation used in your state. These will probably include your Articles of Incorporation, which describe the nature of your business and list the officers of the group; and your bylaws, which describe how you plan to operate (vacation policy, sick leave, employee review process, availability of health benefits, and so on).

Though you may be able to take care of this paperwork yourself, many groups feel more comfortable if a lawyer handles incorporation for them. Eventually, you will probably want to design an emblem or logo to represent your group visually, and have stationery printed, but this is not necessary for incorporation. While you're at it, determine whether you need a business license with the city or county in which you'll be operating. Contact the business licensing division, tell them what you plan to do, and they'll direct you to the right department.

A business can be incorporated as either a for-profit or nonprofit concern. Activists who band together into an organization usually opt for nonprofit status, primarily to take advantage of the fund-raising benefits that status allows them to pursue.

Key among these is that a nonprofit can become recognized by the Internal Revenue Service (IRS) as tax-exempt and thereby receive tax-deductible contributions, an attribute that makes the group very appealing to potential donors who wish to make contributions but want a tax advantage for doing so.

In addition, the assets of a nonprofit cannot accrue to a private individual. Whatever assets a nonprofit organization possesses have got to be used for the express purpose around which the organization has been established. Other benefits nonprofits enjoy include the opportunity to obtain a bulk mail permit, which essentially deeply discounts postage costs, and to participate in a simpler tax filing process.

There are several nonprofit categories an organization can consider, but most activist groups will opt to become either a 501(c)(3) or a 501(c)(4) organization.

A 501(c)(3) organization is one that is classified by the Internal Revenue Service as both nonprofit and capable of receiving tax-deductible contributions.

According to *Nonprofit Organizations, Public Policy, and the Political Process: A Guide to the Internal Revenue Code and Federal Election Campaign Act,* prepared by the Washington, DC, law firm of Perkins Coie, a 501(c)(3) organization is a nonprofit corporation, unincorporated association, or trust that engages in educational, religious, scientific, or other charitable activities and is exempt from federal income tax under the Internal Revenue Code. Contributions to a 501(c)(3) organization may be accepted from individuals and corporations, and are deductible from the donor's federal income tax. In addition, a 501(c)(3) organization may be funded by foundation grants.

As a condition of its tax-exempt status, a 501(c)(3) organization is subject to restrictions on its lobbying activity, by which the organization seeks to influence action by the Congress or a state or local legislature. Tax-exempt 501(c)(3) organizations are also prohibited from engaging in any partisan political campaign activity. (A 501(c)(3) is permitted to conduct some lobbying and voter education activities, as outlined in Chapter 8, "Action at the Polls.")

An organization must apply to the Internal Revenue Service for recognition of tax-exempt status under section 501(c)(3). The Application for Recognition of Exemption, IRS Form 1023, requests a detailed financial statement and information about the organization's present and proposed activities, purpose for formation, sources of funding, and governing structure. In addition, copies of the organizing documents, such as Articles of Incorporation and bylaws, must be provided.

Forms are available through the national and regional offices of the IRS. The IRS will review the application and either deny exemption, request additional information, or issue a determination letter recognizing 501(c)(3) status. In order to avoid an adverse decision, the application should be filled out with care; it is advisable to consult someone familiar with the procedure and knowledgeable about tax-exempt organizations.

In addition, many states have laws, including tax laws, governing nonprofit organizations. Activity in any given state requires that you determine the relevance of that state's laws to your operations.

A nonprofit corporation or association designed to develop and implement programs for the promotion of "social welfare" may qualify for tax-exempt status under Internal Revenue Code section 501(c)(4). A broad spectrum of public policy issues has been recognized by the Internal Revenue Service as promoting "social welfare," including environmental protection, arms control, women's rights, poverty, housing, civil and minority rights, and handgun control. While 501(c)(4) organizations are exempt from paying federal income tax, contributions and membership dues to a 501(c)(4) organization are not tax deductible.

Generally 501(c)(4) organizations are more politically active than 501(c)(3) organizations and are permitted to engage in a broader range of activities. A 501(c)(4) is not limited in the amount of lobbying it may conduct and may advocate a particular point of view even on controversial issues, as long as the issues are related to the primary purpose of the organization. The Application of Recognition of Exemption for 501(c)(4) organizations, IRS Form 1024, requests a detailed financial statement and information about the organization's present and proposed activities, purpose for formation, sources of funding, and governing structure. In addition, copies of the organizing documents, such as the Articles of Incorporation and bylaws, must be provided. See Chapter 8, "Action at the Polls," for more insights into what activities you can engage in as either a 501(c)(3) or 501(c)(4) organization.

If you are uncertain as to the category of tax status you should pursue, consult with a lawyer who specializes in federal tax law. Some groups opt to form both a 501(c)(3) and a 501(c)(4) organization to have maximum flexibility in developing strategies to help them achieve their long-term goals.

☞ **CHECK IT OUT:** For more information, see *Should Your Group Incorporate? Practical Advice On How to Decide On Your Group's Structure,* published by the Citizen's Clearinghouse for Hazardous Wastes, P.O. Box 6806, Falls Church, VA 22040, $6.50.

Nonprofit Organizations, Public Policy, and the Political Process: A Guide to the Internal Revenue Code and Federal Election Campaign Act, Perkins Coie, 607 Fourteenth Street, N.W., Suite 800, Washington, DC 20005, free.

RECRUITING MEMBERS

Saul Alinsky, the legendary community organizer from Chicago who wrote *Rules for Radicals* (Random House, 1971), said there are two sources of power: people and money. Most citizens groups rarely have much money (see Chapter 5, "Fund-raising," to learn how to raise more). What they do have is people. And the more they build up their people capital, the more powerful they become.

Still, the size of your membership alone does not automatically translate into political clout. The level and nature of membership involvement in the group's issues have much greater political significance. You may have a lot of members on paper, but unless they march, write letters, attend meetings, and testify, they won't make their presence felt—and the target of your campaign will have little to fear.

Who should you recruit? Obviously you can't afford to recruit everybody. (Which is fine, because not everybody is going to want to join you anyway.)

In most cases, and especially if you are just beginning your campaign, your resources are going to be pretty limited. So you need to think carefully: What kind of organization do you want to have? A neighborhood alliance composed of dues-paying members who live near the problem? A coalition of disparate groups who share a common concern? An association with an elected leadership and formal committees to help organize, raise money, and promote your cause?

To answer these questions, review your strategic goals and objectives. Identify the constituencies in your community that will most likely want to help you achieve your objectives. And figure out how much help you're going to need to win—and from whom.

Remember that people join organizations for very different reasons. Some are altruists: They have a genuine concern for the public good, and even though the issue at hand may not affect them at all, they'll join the organization or contribute money because it is the "right" thing to do.

Others have a direct self-interest in the work the organization is doing. Parents who joined the Mothers of East L.A. did so because they were personally concerned about their children's future. Low-income citizens in Connecticut who join the Connecticut Citizens Action Group do so because they know CCAG will fight to protect them from utility rate increases and loan discrimination by banks against their disadvantaged neighborhoods.

Sometimes, people join a group just for the information it publishes or the service it provides: They like the group's magazine or newsletter, or they want lower-priced fuel (so they'll "join" an oil-buying co-op) or organic food (for which they'll "belong" to a food co-op). But these "members" see themselves more as subscribers or consumers and are unlikely to participate actively in the organization. Citizen groups committed to building involved memberships should focus on recruiting people who have a direct and personal interest in the outcome of the campaigns that the organization plans to wage.

• •STEP BY STEP• •

How to Get People to Join Your Organization

• **Get to know the community you're trying to organize.** Does the neighborhood have a name? Washington, DC, is a big city of 600,000 residents. But people organize there by neighborhoods: Cleveland Park, Dupont Circle, Shaw, Capitol Hill— all are communities with distinct racial, ethnic, and economic characteristics that provide a point of departure for organizing campaigns.

• **Identify the key institutions that you may be organizing with or around.** These could include religious institutions, employers, hospitals, schools, senior citizens organizations, and so forth.

• **Target community "celebrities,"** the people of influence who can make a difference, especially if they're on your side. These include members of the clergy, union leaders, politicians, and perhaps even sports figures.

• **Decide who else needs to be contacted.** Start small and close to home. Initially, recruit people you know well rather than those you haven't met yet.

• **Hold a meeting.** Decide in advance what you would like the response to be, then organize accordingly. Are you asking your neighbors to get a stop sign installed on your street? Then invite people over to your home for coffee and a strategy ses-

How to Get People to Join Your Organization (cont.)

sion some afternoon. Do you need to let hundreds of people know about the hazardous waste facility that's been proposed for a nearby lot? Hold the meeting at a local church, the library, or the community center, then post fliers, ask radio stations to mention the meeting as a public service announcement, and urge newspapers to list it in the community calendar.

• **Provide information.** Educate people about the nature of the problem and the power they have to do something about it. At organizing meetings, present your suggestions, offer success stories from other groups, and solicit ideas from others.

• **Identify activities that will move people into some form of mass action.** Establish committees to get people to work and recruit new members. Set up information tables at shopping malls, county fairs, farmers' markets, political forums, town meetings, other community events. Hand out literature, buttons, bumper stickers. Talk to people. Circulate petitions. Ask people to sign their names, addresses, and phone numbers so you can contact them later. (See Chapter 3 for more information about organizing a petition drive.) Post brochures, fliers, leaflets, or other material on bulletin boards at churches and temples, supermarkets, on college campuses, and in libraries.

• **Canvass.** Cesar Chavez, the famous community and labor organizer who was president of the United Farm Workers Union, was once asked, "How do you organize?" He said, "First, you talk to one person, face-to-face, then you talk to another. . . ." This is the essence of canvassing (also called doorknocking): making a personal appeal to someone else to join your organization, take part in the campaign, and contribute to the cause.

To get started, identify a neighborhood; start with areas most likely to be concerned about the problem. Take along a clipboard that includes petitions or contact sheets with spaces for names, addresses, and phone numbers. You should also have membership information to allow people to sign up on the spot, a letter from someone else in the community endorsing your or-

How to Get People to Join
Your Organization (cont.)

ganization, fact sheets about the problem and about your organization, and other supporting materials.

Rehearse your pitch with others in your organization to prepare mentally for going door-to-door. Try different scenarios involving various responses—friendly, indifferent, hostile—so you won't be taken aback by the reactions you get from people who open their door to you.

Knock. Introduce yourself and your organization. Make it clear why your issue is legitimate. Have a conversation. Ask the person who answers the door about the issues that concern him or her most. Describe solutions you feel would address the issue at hand; ask your neighbor's opinion. Ask if she or he will become a member and invite him or her to the next event or meeting. Regardless of whether the person becomes a member, request a contribution (or the price of membership) to help support the campaign.

Some groups also engage in direct mail to recruit new members. Advocates of this tactic believe it gives you the opportunity to make a careful, complete and thoughtful pitch to potential recruits about the important work your organization is doing and what you have to offer. It also allows you to reach very large numbers of potential members with very little staff time.

But the down side is significant. Usually, the return rate—the number of people who respond to the mailing by joining your group—will be quite low, less than 1 percent. The mail package itself, which includes a letter, envelope, return envelope, membership card, and perhaps a window decal—is expensive, and you need funds upfront to put the package together, print it, and mail it. Sometimes these expenses are never recovered.

Furthermore, getting members through the mail may create a passive, uninvolved membership that doesn't respond to your calls for direct action.

If you decide to go the direct mail route, you can identify lists of potential recruits by borrowing or exchanging lists with similar organizations, signing up people at events, and by obtaining state voter lists. (For more tips on direct mail, see Chapter 5, "Fundraising.")

BUILDING COALITIONS

Sometimes you just can't get a job done by yourself (the bigger the issue, the more this is true!) You need more resources, clout, or credibility than your organization can provide alone. That's when you need to build a coalition.

The coalition doesn't need to be a formal, legal entity. It could consist of an ad hoc arrangement among representatives of various groups united by a common position on a single issue. Or it could be an organization itself, set up as an ongoing network of groups that share a similar approach on several issues. As a more formal entity, it could have its own letterhead and staff and require members to pay dues.

When should you form or join a coalition? Bill Robinson, executive director of the Lawyers' Committee for Civil Rights Under Law, believes: "You get massive coalition efforts when the stakes are truly high and there is an ultimate goal that everybody can subscribe to, and understand why they must submerge their individual and organizational interests."

Answering these questions will help you decide whether to participate in a coalition:

- **What power would other groups bring to the coalition?** Do other groups have greater access to elected officials, the media, or other institutions whose support you need? Do they have more resources—in terms of money, people, or equipment?

- **What problems would other groups bring?** Have they received bad publicity that would somehow reflect negatively on you? Have they alienated organizations that you hoped would also join your coalition? Would you have to support them financially if they joined your effort?

- **What does your group stand to gain or lose by working in a coalition?** Will you get less credit than you deserve because you have to share the limelight with others? Will the coalition cost you more because you have to support the work of others? Or will you profit from being associated with organizations and individuals who have more credibility or resources than you do?

- **What issues will you need to avoid if you work in coalition with others?** Are these issues at the core of your organization's identity, and therefore impossible to subsume, even for the sake of the coalition? Or can these issues be put aside until the coalition effort succeeds?

- **Under what sort of structure will the coalition function?** Will it incorporate, in which case you may be asked to become a board member or officer of the new group? Will it be a formal entity that may not have legal standing but still prints up stationery and has members who develop and implement strategy? Or will it be an informal coalition that only kicks into gear on occasion to testify at a public hearing or write a letter to the editor?

- **What groups are potential coalition members?** What other organizations share your values completely and can operate as full partners in the coalition? What organizations don't you agree with on every issue but can put differences aside for the time being to coalesce behind the issue at hand?

- **How will the coalition function?** Will members consist only of those who pay dues? Or provide some in-kind service? Will officers be elected from the members? Who makes decisions? Officers, members, or both?

Sometimes you may cooperate closely with other groups without joining their coalition. You can share information, attend meetings, participate when appropriate in activities, like signing coalition letters and appearing at news conferences, and offer organizational resources when doing so doesn't compromise your positions. Working in this way allows you to add your strengths to those of other organizations while maintaining your independence, thereby avoiding association with any positions you have not taken.

Coalitions should be as broad as possible in order to demonstrate wide support for an issue and make it difficult if not impossible for your opponents to ignore your demands. (Coalitions of *unlikely* partners, such as labor and industry, urban and rural communities, agricultural interests, and environmentalists, are almost unbeatable.) To repeat an important point, build coalitions with natural allies, but look for new allies, too. An issue that can unite adversarial partners makes a powerful statement about that issue, proving it is of widespread public concern.

••STEP BY STEP••

How to Build a Coalition

- **Identify organizations and individuals who want to be involved in a joint effort on the issue.** Compile a list of those who are most likely to share your concerns, then add to it names of people and groups who may have a peripheral interest in the issue but are still worth contacting.

- **Familiarize these activists with each other, the substance of the issue, and the goals of the campaign.** Hold a meeting of all interested parties. Prepare a written summary of the issue, augmenting it with newspaper clippings, research reports, testimony, or other documentation. Discuss the goals of the campaign and ask for input from participants in refining those goals.

- **Develop specific approaches, strategies, and time lines that are action-oriented.** Work either in the whole group or within a smaller strategy committee to refine your plans and identify specific tactics to use in pursuing your goals. Develop a realistic time line that sets deadlines for services to be rendered by coalition members and that anticipates deadlines that may be imposed on you, such as a specific vote or a construction date.

- **Share the work.** In the most effective coalitions, everyone pitches in. Some groups lobby, others maintain mailing lists, others write press releases and testimony, and others organize events. Give every member of the coalition a job. Each organization should do what it does best. The whole should be greater than the sum of its parts.

- **Take one issue at a time.** Don't try to create a new coalition around several controversial issues all at the same time. Begin with one issue, and as trust and understanding develop among coalition members, explore opportunities to expand the interests of the coalition into other arenas.

- **Spread the credit and praise for those who complete coalition tasks.** A little public recognition goes a long way. Recognizing the efforts of those who work with you will make them feel good about collaborating with you on another campaign.

Local Hell-Raiser
The Right Time for a Coalition

For such a tiny state, Rhode Island has had more than its share of political corruption. The long list of scandalous elected officials included the former mayor of Pawtucket, jailed for accepting $250,000 worth of kickbacks, and two corrupt judges accused of accepting money from lawyers who appeared in their courts. Common Cause of Rhode Island had been unsuccessfully promoting campaign finance and ethics reform for several years when, in January 1991, an event occurred that drove public interest groups into action: The private fund that insured forty-five banks and credit unions around Rhode Island collapsed, freezing the savings of one-third of the state's residents.

By the fall of 1991, Common Cause had recruited a variety of community and state interest groups to join a coalition and help lobby for ethics and campaign reform legislation. Their diverse roster represented almost every type of Rhode Islander. "There was so much anger in the state about the collapse!" explained Phil West, executive director of Common Cause and eventual vice chair of the coalition.

The coalition was built in part by inviting specific groups to join and partially in response to organizations

Phil West of RIght Now!
Anne Grant

that contacted Common Cause to obtain information about the effort, such as the Environment Council, the State Council of Churches, and eight Chambers of Commerce. West said that he sometimes spoke to four groups per day about the need for political reform and the coalition's ambitious but realistic legislative plan.

The activists chose a name for the coalition that told lawmakers that they meant business—RIght Now! Members wore buttons that boldly gave voice to their organization's objectives and to the activists' own determination.

West and others appeared on several radio talk shows around the state to discuss the issue and publicize the coalition; many groups that called in during these programs eventually added their names to the RIght Now! roster. An article in the *Providence Sunday Journal* opened the floodgates of support even farther. Groups that West had never heard of before, such as the Executives Association of Rhode Island, signed up as a result of the feature story.

By early spring 1992, the tally of coalition member groups numbered over two hundred. The American Association of Retired Persons, the United Way, the League of Women Voters, the Sierra Club, the Girl Scouts, the State Council of Churches, Heal the Bay, the Urban League, the Jewish Federation, and the Rhode Island Public Expenditure Council were among those who had joined RIght Now! "We sought from the beginning to make it broad," West said, believing that greater diversity among the coalition's members would have greater impact on the statehouse.

Common Cause provided organizing and lobbying guidance, while other members of RIght Now! provided the grass roots manpower. Behind their three-part agenda: sweeping campaign reform, ethics reform, and four-year term limitations for bank officers. Common Cause had researched and written much of the legislation before the coalition formed, but everyone worked to perfect it for passage. RIght Now! mailed out fund-raising letters to supporters and wrote letters to all 150 legislators asking their opinion on six different reform bills. Responses were printed in local papers, enabling lawmakers to send a direct message to their constituents.

RIght Now! logo

West described a spectacular showing of grass roots power in the form of a rally on January 5, 1992, the day before the General Assembly opened its session. With church bells tolling around Providence, activists marched to the statehouse and linked arms to form a roiling human ring around the building. Representatives from the Council of Churches, the Rhode Island Board of Rabbis, and Native American tribes—all coalition members—led prayers and passionately addressed the urgent need for political reform, calling for repentance within government walls. All major religious denominations were present.

"We used the traditions of each group," West remembered, including Baptist hymns, Jewish ritual horns, and blunt activist placards. "DON'T YOU DARE MAKE AN ILLEGAL DEAL, YOU QUAHOG, YOU!" one demonstrator's sign warned, referring to a large clam found in New England waters.

"It was nip and tuck for a while," said West, but the group eventually succeeded in passing twenty-seven out of twenty-eight bills, including a ballot referendum that increased the governor's term of office from two years to four, which won by a 60-to-40 margin. RIght Now! had made nepotism illegal, imposed campaign contribution ceilings, won four-year terms for general officers, and established a year-long waiting period that officials must honor before taking other state jobs after they leave public office.

Remember Common Cause's admonition: "No permanent allies and no permanent enemies." Do not hesitate to contact organizations you have opposed in the past on other issues but which may share a common interest in your current battle. Make it a priority to approach groups who you know have influence in your community or with a legislator you've targeted.

If an organization will not join your coalition, perhaps it will communicate on its own with the legislator, or individuals in the group may take some personal action on their own. Identify a contact person at each organization who can be invited to meetings. Clear joint actions of the coalition with the organization if at all possible.

For coalitions to be successful, participating groups need to establish a common understanding of the issue and goals they share. Yet they must also have the flexibility to accommodate political and organizational differences. Clearly defining the purpose and goal of the coalition at the onset of the campaign will help avoid problems down the road. No one who is not authorized to do so should ever speak for the coalition on issues not previously agreed upon.

 CHECK IT OUT: For more information on organizing coalitions, see *Elements of Successful Public Interest Advocacy Campaigns,* by the Advocacy Institute, 1730 Rhode Island Avenue, N.W., Suite 600, Washington, DC 20036, 202-659-8475, $7.50.

THE BOTTOM LINE: GIVE PEOPLE POWER

Don't recruit people simply to boost membership figures or raise money. Recruit them because it will empower the organization as well as give them a say in determining their own future.

Once people become members, strive to make them feel like they're really part of the organization.

- **Ask them to do a specific job:** answer the phone, stuff envelopes, man the telephone tree, hand out leaflets, collect signatures, and so forth. Make sure everyone understands how their function fits into the "grand scheme" of things. They'll work harder if they know that others are depending on them.

- **Make volunteers and members feel important.** They need to know that their help is essential—and appreciated. Extend heartfelt thanks when they complete one task and prepare for the next.

- **Get to know people who join and volunteer.** Why did they join? What other issues concern them? Let people know this is *their* organization, too.

BUILDING YOUR ORGANIZATION
DO's and DON'Ts

• **Do** identify individuals who have something to gain by joining your effort or group.

• **Do** be creative in how you recruit new members and volunteers. Knock on doors, distribute literature, circulate petitions, hold meetings.

• **Do** build coalitions. Identify organizations that share your concerns and are willing to work with you.

• **Do** incorporate into a 501(c)(3) or 501(c)(4) if it will help you raise money or achieve your political objectives.

• **Do** give members and volunteers meaningful tasks.

• **Do** say thank you.

• **Don't** think you don't need help. You do.

• **Don't** rely upon just one approach. Use many tactics to broaden your membership base.

• **Don't** expect supporters to come to you. Actively seek them out.

• **Don't** be afraid to work with groups you may have opposed in the past. If you both agree on the issue at hand, put your differences temporarily aside.

• **Don't** waste people's time.

5.
FUND-RAISING

EVEN THE MOST clever organization will be short-circuited before long if it doesn't have the funds it needs to pay basic bills. Though many groups operate on a shoestring, it still takes money to keep that shoestring from unraveling.

Actually, for most groups, money is there for the asking—if they'll only figure out how, who, and when to ask. At least that's what a small band of determined women learned when they launched a year-long, heart-pounding drive to raise an enormous sum of money to save a portion of their community's soul.

ACTIVISTS in ACTION > > > >

Pleas for Peaslee

Once a full-scale elementary school, the century-old Peaslee Primary School was located in Over-the-Rhine, a low-income, racially integrated neighborhood just north of Cincinnati's downtown. Like many inner-city neighborhoods, Over-the-Rhine experienced decline during the seventies and eighties. Its population dwindled, and many of its buildings were redeveloped or boarded up; six schools were shut down between 1972 and 1982, greatly weakening the community's educational resources. As the *Cincinnati Post* put it, the community knew "hardship and deprivation firsthand."

As the neighborhood deteriorated, the Peaslee elementary school's student body shrank, and its original, century-old building was torn down in 1974, leaving only a three-story, twelve-classroom annex that served approximately 280 children from kindergarten to the third grade.

Reduced in size though it was, the school earned the fervent devotion of the parents whose young children attended classes there. "Peaslee was a beautiful school," one of the parents told *Cincinnati* magazine. "It was designed for small children," with large, airy windows and teachers who gave their pupils lots of individual attention. The parents considered Peaslee's small size one of its greatest assets, but given decreasing state and federal funds for education, the Ohio State Board of Education had recommended that smaller schools would be the first to shut down. Rumors began circulating in 1979 that the Cincinnati Board of Education had targeted Peaslee for closure.

A small band of Peaslee mothers stormily protested, beginning a long, frustrating, and ultimately unsuccessful campaign to save the school. Noting that Peaslee was ranked sixteenth academically among Cincinnati's seventy-five public schools, they managed to persuade the school board to keep the institution open for the next two years, but at the end of the 1982 school year the board decided to pull the plug. Soon after the decision was made, six community mothers—three Appalachian whites and three blacks—tried suing in federal court to prevent the school from closing permanently, charging discrimination against black and poor Appalachian youngsters. The suit was dismissed and the mothers disheartened, but they were far from defeated.

A School Dies But a Dream Is Born

Among the leaders of the Peaslee mothers' group were Bonnie Neumeier of the Over-the-Rhine Community Council; Kathleen Prudence, a white mother of three daughters, two of whom had attended Peaslee; and Everleen Leary, a black mother of seven who had a son attending the school at the time it closed. Prudence and Leary, who had met on Peaslee's playground, became co-chairs of the Peaslee for the People Project, which was determined to assume control of the now-shuttered school and maintain it as a community center, a focal point of neighborhood pride that could offer programs designed to address the community's various problems.

Rival Suitors Pursue Peaslee

The Peaslee for the People Project organizers were not the only ones with their eye on the school building. The Young Lawyers section of the Cincinnati Bar Association wanted to rent the facility from the school board for use as a day-care center for their children, given its proximity to downtown law firms. The affluent young lawyers were willing to pay $6,000 a year in day-care tuition for each of an anticipated sixty children to be placed in the old school. The city's Community Action Commission (CAC) also wanted to rent Peaslee for offices, a Head Start Program, and day-care facilities.

At a public hearing in June 1983, the young lawyers tried to sweeten their offer by pledging to guarantee at least six places in the proposed day-care center for low-income youngsters from Over-the-Rhine; they also suggested that an endowment might be created to fully or partially subsidize the tuition costs for poor children. Over-the-Rhine residents at the hearing jeered and laughed at the inadequate proposal and kept their sights focused on acquiring the building themselves.

Responding to the community's pleas, the school board postponed consideration of the young lawyers' request and gave the CAC and the Over-the-Rhine community two months to develop proposals for Peaslee. A separate, nonprofit neighborhood development corporation, Owning-the-Realty, Inc. (O-T-R), joined forces with the Over-the-Rhine Community Council and offered to purchase the school building from the Board of Education, then reopen it as a center housing not only the Head Start program but also programs for remedial tutoring for adults, job training, and other social services.

Meanwhile, the neighborhood activists gained the support of Sylvester Murray, then Cincinnati's city manager (whose role is similar to that of a mayor in other cities), who wrote to the school board urging it to sell the Peaslee school to O-T-R so that it would be used "with the best community interests at heart."

In September 1983, as Kathleen Prudence hugged herself and nervously sat on the edge of her seat, the Board of Education voted to sell the Peaslee School to the area's residents. Prudence and some eighty-eight other Over-the-Rhine residents at the meeting cheered, but they knew the price of their victory would not be cheap: In order to get the city to approve the deal, the Over-the-Rhine group would have only two months to come up with $15,000 as nonrefundable "earnest money," representing the rent the city would be losing while the Peaslee people spent the next year raising the remaining $240,000 for the property.

Off and Running

A six-person fund-raising committee was organized, meeting once a week with Kathleen Prudence as its chairperson; a six-person development committee, headed by Bonnie Neumeier, met every other week to decide what new activities would take place inside the school and how best to convince the "big guys" of Cincinnati—the representatives of large corporations and foundations—to contribute to the cause.

A Peaslee for the People Project fund-raising committee meeting (Kathleen Prudence is seated at right side of table). Note that the activists were close to reaching their goal; the thermometer is filled in nearly to the top!

The Peaslee for the People Project managed to raise much of the first $15,000 in donations from concerned citizens who lived in Over-the-Rhine, and by November 1983 were able to pay the city the first installment of the school's purchase price. It was an astonishing show of neighborhood solidarity in a section of Cincinnati where the average family income was just $6,000.

Keys to Opening the City's Heart— and Coffers

Key components of the fund-raising drive involved face-to-face presentations to dozens of civic, religious, labor, and educational groups throughout the metropolitan area; solicitations of endorsements from similar, but larger, umbrella organizations; and approaches to foundations and charitable trusts located in Cincinnati and elsewhere.

No community newsletter was too small to be sent information about the fund-raising drive, while major newspapers, magazines, television, and radio stations in the area received regular updates on each development.

By obtaining the mailing lists of such organizations as the Metropolitan Area Religious Coalition of Cincinnati, the Chamber of Commerce, and the Cincinnati Women's Network, and by receiving in-kind donations of stationery, address labels, envelopes, and the services of local printing firms, copying centers, and stationery stores, the Peaslee for the People Project succeeded in sending out at least twenty thousand pieces of mail on its own. Groups such as the Urban Appalachian Council and the AFL-CIO's three hundred union locals organized additional mailings on behalf of the Peaslee parents that brought in many individual donations.

To raise the bulk of the money needed to buy the school building, the Peaslee activists determined to capture the attention of the greater Cincinnati community. The tactics they used were both traditional and innovative.

For example, they began a Buy-a-Brick Campaign, charging contributors $10 to "buy" a symbolic brick in the school. Silk-screened posters were designed and sold at a "Rally for Peace and Justice in Central America." An endorsement was solicited from famed folk singer Pete Seeger, who invited the Peaslee parents to attend a concert he was giving in Cincinnati and allowed them to pass out brochures there. Bake sales were held by block clubs, church groups, and student organizations. One high school with rules regarding student attire permitted the youngsters there to contribute fifty cents each for a "pass" that gave them the right to wear jeans to school on a particular day instead of the school's regular uniform.

Students at the College of Mount St. Joseph held a fast and contributed their lunch money; the MUSE Cincinnati Women's Choir performed a benefit concert at Xavier University, while the First Unitarian Church in another community held a spaghetti dinner and contributed its proceeds to Peaslee. One woman shared a year's worth of her bingo winnings; another Cincinnatian left Peaslee a $2,000 bequest in her will. And an insurance firm, John J. and Thomas R. Schiff & Company, donated $5,000 to cover the insurance the parents would need to have once they obtained control of the school.

On the anniversary of Martin Luther King, Jr.'s,

birthday in January 1984, the Peaslee organizers held a "cleanup" event at the school, inviting neighbors to mop down and straighten up the structure and its surroundings. King's "I Have a Dream" speech was read aloud as the local newspapers and radio and television stations covered the activities.

"Self-Help at Its Best"

"What Over-the-Rhine residents are attempting is self-help at its best," declared the *Cincinnati Enquirer* in an editorial. The *Cincinnati Post* agreed. Over-the-Rhine had demonstrated "a cohesiveness of community interest and a core of concerned residential leaders," the paper editorialized. "We believe Over-the-Rhine can also depend on help from the wider Cincinnati community who have been heartened by the determination of the Peaslee parents to save their school."

The newspaper's view proved prophetic. Small contributions still came in from hardware stores, health clinics, and theater troupes, but as the Peaslee parents undertook the tedious yet essential paperwork necessary to obtain grants, larger sums arrived from such institutions as the Greater Cincinnati Foundation, which initially contributed $25,000, and several local charitable trusts, which gave $10,000 each. Cincinnati's city council itself awarded the Over-the-Rhine activists a $50,000 Community Development Block Grant (CDBG) in February 1984, while the Greater Cincinnati Foundation upped its contribution to $50,000 that May.

Nevertheless, the Peaslee for the People Project was $60,294 shy of its $240,000 goal as the November 1984 deadline approached. Even though a few local and national foundations had pledged between $5,000 and $10,000, and contributions were continuing to come in, time was running out.

The Impossible Dream

In October 1984, with a month to go before their deadline for payment arrived, the Peaslee activists went before the Board of Education and asked it to reduce the price of the school by $40,000. The parents said they had a grand total of $179,706.51 already in hand or pledged and were certain they could raise $200,000 by the deadline, as well as the nearly $10,000 annual maintenance cost for the building. What they needed now was a break. The board made no immediate decision but agreed to consider the group's request.

The clock continued ticking. On November 10, four days before the deadline for final payment, the school board agreed to lower the price for the Peaslee School to $200,000. The Peaslee leaders then approached the city's Methodist Union, the controlling body of the local Methodist churches, and asked for a $10,000 contribution. It would be the amount needed to put them over the top. They met with the church leaders with only a few days to spare before the November 14 deadline. Prior to their presentation, Prudence, Neumeier, and Leary sat in a waiting room for what seemed to them forever. On the wall was a picture of Don Quixote, the idealistic, befuddled hero of Miguel de Cervantes's seventeenth-century novel—and a twentieth-century Broadway musical—who tries to help the oppressed.

"After making the presentation and saying how much the money would mean to us, Bonnie mentioned seeing the picture. Then she broke into the musical's theme song, 'Impossible Dream,' a cappella," Prudence recalled. "It was so moving, the whole room was in tears! The next day we found out they voted us the money as a loan which would be forgiven if we kept the building for five years."

On December 14, 1984, a month to the day after the official "closing" deadline, members of the Board of Education handed over the keys to the school in return for a check from Peaslee for the People for $209,239.13, which covered the price of the property and a year's worth of maintenance costs.

The Dream Tested—and Triumphant

The Peaslee for the People Project made good on its promises for the former school building. A federally funded Head Start program serving at least thirty children between the ages of three and five began operating there, along with an after-school Homework Room Program offering tutoring and arts and music programs,

Kids show their support for Peaslee in front of the school and the "Keep Peaslee Open" banner.

with three rooms left over for general community use. Maintenance costs for the break-even community center were met by charging modest rents to the agencies and groups using the facility. Additional grants were obtained to support programs of value to the community. And in May 1989, the Peaslee Neighborhood Center incorporated on its own. The center "has been going strong ever since," says Prudence.

The various tenants and programs in the Peaslee Neighborhood Center today include the Clarke Academy Infant Day Care program, providing care for the infants of public high school dropouts who have returned to school; the Intercommunity Peace and Justice Center, a coalition of eight religious groups; a sixteen-piece steel drum band; and the Peaslee Sign Singers, a group of children between the ages of five and twelve who perform "signing" concerts throughout the city for deaf audiences.

The goal of the center is to maintain the building as a place that serves the needs of low-income people, despite the now spreading "gentrification" of the Over-the-Rhine neighborhood, Prudence explains. For example, in August 1992, a neighborhood health center for poor people opened in one section of the building. "The health center came about because some of the young health professionals who helped us stuff envelopes and gave church presentations during the fundraising period started dreaming about running their own clinic," according to Prudence. Today they do.

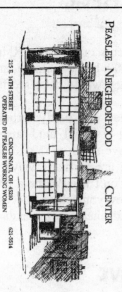

YES! I WANT TO HELP

NAME

ADDRESS

CITY/STATE/ZIP

PHONE

All contributions are tax deductible

☐ I want to support the efforts of Peaslee Working Women.

☐ Enclosed is my contribution to Peaslee Neighborhood Center.

Make checks payable to:
Peaslee Neighborhood Center
215 E. 14th Street
Cincinnati, OH 45210
621-5514

Peaslee Neighborhood Center
215 E. 14th Street
Cincinnati, OH 45210
621-5514

PEASLEE NEIGHBORHOOD CENTER

215 E. 14TH STREET
OPERATED BY PEASLEE WORKING WOMEN

CINCINNATI, OH 45210
621-5514

"NEVER DOUBT THAT A SMALL GROUP OF THOUGHTFUL COMMITTED CITIZENS CAN CHANGE THE WORLD" — Margaret Mead

Peaslee Neighborhood Center (once a public primary school) is a story of inner city women's relentless effort since 1981 to save our school as an educational resource for Over-The-Rhine. It is a story of Black and White women challenging insurmountable odds, losing battles, winning steps but having new injustices placed in our way, and finally persevering and obtaining our own deed, in December, 1988.

HISTORY

In 1981, parents organized to save Peaslee, a racially integrated school. A massive outcry tried blocking the School Board's shutdown effort. We couldn't stop them with pickets, marches, a lawsuit, and public pressure. Peaslee's doors closed in June, 1982.

We were determined Peaslee remain a center for education in the hands of our community, serving our people. After a massive fundraising drive in December, 1984, we turned over $209,000 for the keys to Peaslee. A Neighborhood Development Corp. held the deed in trust for the Peaslee Women because the women were not yet officially incorporated. From 1985-87 neighbors and volunteers repaired and decorated. By the end of 1987, we were nearly fully occupied; a day care center with a head start program and homework room.

But then tragedy reoccurred in Spring of '88. Daycare & Headstart withdrew and problems emerged with the NDC holding our deed. Summer of 1988 Peaslee was up for sale to outside developers.

We are now rebuilding.

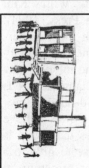

PROGRAMS INCLUDE:

• Homework Room — Bright, cheerful and safe, this room offers an environment for our children (K-6) to build skills in math, reading and writing. Volunteer tutors are essential to the success of this program M-Th, 3:00-4:30 P.M.

• Over-The-Rhine Steel Drum Band — Our children living in Over-The-Rhine, ages 10-18, have the opportunity to learn music and perform at events.

• Art Works Room — is maintained as a place for community art work — banners, brochures, silk screening, signs, etc.

• Peaslee Women's Educational Programs provide opportunity to gather together and deepen our awareness of justice concerns, both locally and globally.

HOW YOU CAN HELP

• We need FINANCIAL ASSISTANCE for ever-present operating costs. Donations from individuals, churches, unions, businesses in the form of gifts, grants or monthly pledges are most appreciated.

• VOLUNTEERS, both individuals and groups, to assist with fundraising tasks, contacts, mailings, etc.

• TUTORS are needed for the Homework Room and the Steel Drum Band.

• SPACE RENTAL: If your group or program is interested in renting permanent space or space for single events at Peaslee, call 621-5514.

• WISH LIST to meet ongoing programs and building needs, e.g. office and school supplies, folding chairs, children's books, etc.

Activists developed this flier to help raise money for the Peaslee Neighborhood Center.

"HE WHO PAYS THE PIPER, PICKS THE TUNE . . ."

If you had told the inexperienced residents of Over-the-Rhine that someday they'd raise more than $200,000, buy a building, and establish their own neighborhood center, they would have scoffed in disbelief.

Yet raise the money is exactly what they did, partly through sheer grit, and in larger part by acquiring essential fund-raising skills that continue to help their center thrive.

As the Peaslee campaign showed, there are as many ways to raise money as there are to spend it. You don't need to be a professional fund-raiser or have a lot of money yourself to successfully generate funds for your group. You do need to understand why people give.

WHY PEOPLE GIVE

"People give because it serves their self-interest," writes Kim Klein in *Fundraising for Social Change* (Chardon Press, 1994). The driving force that compels people to part with their money, whether through a purchase or a donation, is the need to satisfy their own desires for comfort or peace of mind. The interest may be purely materialistic: Some donors like to receive tangible goods in exchange for their contribution, which is why fund-raising "premiums" like calendars and books are so popular. Others believe strongly in an organization's mission and the issues the group is trying to address. Then there are donors who choose to give to a particular group because its work relates, however remotely, to their own lives.

People also donate because they feel guilty, they feel proud, or because they're angry. They may write a check because they can't write a letter. They may call five friends on the phone and ask them to give because they would feel less comfortable calling five legislators and asking for their votes. They may give money because they are loyal to the organization. Maybe they just need a tax write-off.

Whatever the reasons, giving enables people to "reinforce their image of themselves" as, for example, feminists, environmentalists, conservatives, or patriots, Klein writes. Donors thus get a boost in self-esteem by giving; they can tell themselves that they helped someone else today—someone like them.

It is important for you as a fund-raiser to remember not only that people give for themselves but specifically that they do *not* give simply because it will benefit your group. According to Klein, many nonprofits get on the wrong track by relying on singular appeals like "We need the money" and "Your gift is tax-deductible." These points do not set you apart; all groups can claim to need money and most offer tax deductions. "Imagine your donors as customers," she advises fund-raisers. "Move away from *your* need to *their* need."

Believe that people do want to help you out, and that most people will give what they can. It's your job to convey to them what an opportunity it is to support "the most valuable program in the community." They'll dig into their pockets, if only you'll ask.

Here's a terrific example of an activist who did ask—in a variety of ways—and raised the money his group needed to defeat the construction of three hazardous incinerators that were proposed for his town.

LocaL ␣ell-Raiser
A Gopher for Funds

A carpenter by trade, Bob Forney has a second occupation in the "building" field, though it has more to do with dollars and cents than hammers and nails: He raises money for Citizens for a Healthy Environment (CHE), a grass roots group located in Summit Station, Pennsylvania, that uses every dollar Forney scrapes up to build and strengthen the organization.

One of Forney's biggest challenges was to raise money to support CHE's fight against the petroleum-burning incinerator proposed for the town. The self-proclaimed "gopher for funds" claims he had great success asking for donations from local businesses. Appealing to the merchants' financial self-interests, as well as their appreciation for the facts, Forney helped them weigh the pros and cons of the incinerator proposal. "I told them that the hazards accompanying the incinerator would override the number of jobs gained," he said. He also argued that customers would move away if the facility were built.

The merchants responded with donations averaging $25 to $50 each, and dozens of businesses contributed more than once. (A used car lot eventually allowed the group unlimited use of its copy and fax machines.) People, he found, will give again if a reasonable case for additional support is presented, and especially if a deadline is approaching.

Fired up by Forney's success, CHE took advantage of public events to sell merchandise and raise even more money. They held bake sales at public meetings about the incinerator, at community fairs, and at block parties. (Forney claims that probably every one of Summit Station's 3,900 residents baked at least once for CHE!) At a public auction, CHE members sold soup they had made, as well as barbecue, hamburgers, and drinks donated by a local restaurant.

Not ones to pass up any opportunity, the CHE organizers set out donation jars at their meetings, adding the change and occasional dollars people contributed to

Members of Citizens for a Healthy Environment and Citizens Against Incineration work at their sales pavilion at the Schuylkill County Fair. Kate Potter and Bob Forney are second and third from the left. *Chris Lutchen*

their slowly growing coffers. They also sold buttons declaring "BAN THE BURNERS—SCHUYLKILL COUNTY GREEN FOREVER" and bumper stickers that announced "I'M PRO-ENVIRONMENT AND I VOTE."

In a region where the average family income is only $23,000, Bob was able to raise over $4,000 in ten months to help cover CHE's expenses.

To earn more money, Forney built a pavilion at the county fair so that the group could sell its buttons and bumper stickers, as well as plants and more baked goods, while they distributed their educational materials.

I'm Pro-Environment and I VOTE!

CHE sold these bumper stickers to members of the community for $1.

"We reached thousands of people that way," Forney said. "We had to open their eyes." For not only was CHE battling the owners of the incinerator, it was also fighting an uphill battle against residents who supported the facility because of the jobs they believed it would bring to the community.

To help CHE gather information and pinch pennies at the same time, local engineers and scientists offered their technical expertise *pro bono,* helping to analyze the implications of the proposed facility on the entire county.

The group raised enough money to keep going, producing information sheets and news releases that, in 1992, helped persuade legislators to defeat the original incinerator proposal, two additional incinerator proposals, and a new landfill.

In the meantime, several concerned lawmakers finally noticed the struggle for resources that communities like Schuylkill County must go through to assert their rights to a clean environment. One Pennsylvania state legislator, Representative Bob Allen, in 1993 proposed legislation to create a $2 million state fund to provide communities with resources to research and investigate land-use proposals. If the law passes, it will be a little easier for groups like Citizens for a Healthy Environment to make a difference.

DEVELOP A FUND-RAISING STRATEGY

Fund-raising, like the other organizing tools discussed in this book, can help build and strengthen your group—as long as you have put in place a carefully crafted strategy that meshes with your short- and long-term organizational goals and objectives. Where do you start?

- **Clarify your tax status.** Decide whether to apply for 501(c)(3) tax status, which will classify your group as a nonprofit organization. Though being classified by the IRS as such is not legally necessary for you to receive money, donors can only make tax-deductible donations if you are a 501(c)(3). Remember that in order to obtain an IRS tax status, you must incorporate. (For tips on how to incorporate and file the proper documents, see Chapter 4, "Building Your Organization.")

- **Develop a fund-raising committee.** The fund-raising coordinator keeps track of how much money is coming in, where it's coming from, what specific fund-raising events you have planned, and the fund-raising calendar. Others on the committee actually get out and raise the money, whether that means selling products, collecting donations door-to-door, or staging fund-raising events. All of those on the committee should be willing to ask for money—and collect it.

- **Set goals.** You will gain self-sufficiency from a well-planned, long-range fund-raising campaign that enables you to pay staff (if necessary), buy supplies, maintain your programs, and generally sustain the organization. Having enough money will give you not only independence but also peace of mind. Obviously, it will also help you meet program goals and build your membership.

- **Figure out how much money you need.** This is the million-dollar question. The amount of money you need will change as your organization grows in size, matures in outlook, and develops in ambition. Don't try to raise $100,000 to hire several staff, buy computers, and rent an office when you're organizing a temporary campaign and all you really need is $500 to pay the phone bill and print up fliers. On the other hand, don't continually try to scrape by on $1,000 a year when you're finally ready to turn your volunteer operation into a staff-paid organization that requires higher personnel and overhead costs.

- **Prepare a complete calendar with specific fund-raising deadlines scheduled throughout the year.** Hold major fund-raising events at the same time every year so people get used to giving to you annually. Consider whether certain holidays lend themselves to special fund-raising appeals. Try to develop discreet programs—the annual release of an assessment of the problem you're trying to correct, a day or week in which you mobilize volunteers for a community-improvement activity—around which you can raise money.

- **Ask frequently.** You have to raise money for your organization year-round if it is going to survive and grow. Keep in mind that people who go to church are used to giving every week. Many grass roots organizations solicit a donation from their members at least once a month. Others ask twice a month, while some ask every two months. Reminds expert fund-raiser Amy Leveen, "If you don't ask, you won't get. The more often you ask, the more often you'll get. And the more you ask for, the more you'll get." Adds Jenny Thompson, a direct mail specialist with Craver, Mathews, Smith & Co., "People will give again if they understand what happened with the money they gave and why additional funds are needed. The calendar doesn't determine when you need money. Need does."

- **Keep accurate records.** Follow the standard rules of accounting to keep track of your finances. If you are paying a staff, know when to file the proper state and federal taxes for employees, as well as income taxes for your group. Though you can probably read a book on how it's supposed to be done, you're better off hiring a part-time accountant who can help you keep track of it all. Perhaps you can even find an accounting firm that is willing to take you on as a *pro bono* client. Make sure you get a professional audit annually—most foundations and corporate grantmakers will want an audit as proof that you're operating in a fiscally responsible manner.

- **Find the right bank.** When looking for a bank, choose one with a history of being responsive to your community. Does the bank "red line" neighborhoods, refusing to give loans in low-income or minority communities? If so, you may want to take your business elsewhere. You may also want to explore working with a credit union or community co-op.

  Once you choose a bank, get advice on the different kinds of savings and checking accounts available. At the right time, you may also want to talk with the bank about establishing a line of credit to help you through periods when cash flow is weak.

  Also, don't forget to turn to your friends for things like interest-free loans to buy postage or cover printing.

- **Avoid debt.** As much as possible, avoid incurring debt. Nobody likes to help somebody else pay for items that have already been bought or an activity that has already taken place. Keep track of your expenses and how much they're going to cost; if you don't have the money to cover them, scale back until you do.

The Over-the-Rhine Fund-raising Strategy

Here's how the Peaslee activists managed to raise over $200,000 to save their community center:

GOALS
- Long-term: Raise $240,000 to buy the property.
- Short-term: Raise $15,000 in "earnest money."

ORGANIZATIONAL CONCERNS
- Fund-raising in low-income community.
- Lack of organizing and fund-raising experience.
- Virtually no organizational resources.

TARGETS
- Immediate neighbors.
- Residents of the greater Cincinnati community.

- Foundations.
- Businesses, civic associations, churches.

TACTICS
- Publicize campaign.
- Make face-to-face presentations to potential donor groups.
- Seek and receive in-kind contributions and services.
- Organize "buy a brick" campaign; sell posters, organize bake sales, and stage various fund-raising events.

TIME FRAME
- Two months to raise the first $15,000.
- One year to raise the remaining $240,000.

Answering the Million-Dollar Question

To set your fund-raising budget, first anticipate what your expenses for the year are going to be. Calculate the following:

- **How many paid staff members (if any) does your organization employ?** In addition to their salary, you'll have to pay federal, state, and perhaps city taxes, as well as FICA and Social Security.

- **What benefits do you offer?** Most organizations provide at least a modest health insurance plan. Is your group also contemplating child care, reimbursement for transportation or parking, or tuition rebates for continuing education courses? Do you have vacation and sick leave? Figure all of these costs into your budget.

- **How much rent do you pay?** Also calculate in related office expenses, such as telephone, heating and cooling, stationery, subscriptions, transportation, and office supplies.

- **What special publications are you planning that will require additional funds?** These could include the design and printing of a newsletter, annual report, press kit, or background brochure on your group.

- **What special events are you planning?** You should develop a separate budget for these "big ticket" events.

Try to find an accountant who will work with you, at least initially, on a *pro bono* or reduced-fee basis. Develop a line-item budget that will help show you where you need to spend money and where you can trim expenses and try to get the greatest dollar return for each member's hour of work. It's a lot easier to sell one $100 ad than twenty $5 raffle tickets.

 ACTIVIST'S NOTEBOOK

Fund-raising and Computers

People may only give to people, but computers can certainly help keep track of the transactions. Thorough fund-raising can involve a vast amount of paperwork, especially when a group is trying to maintain personal and business information on individual and company donors and tailor its appeals to those characteristics. Once your mailing list reaches five hundred names, you should consider storing it on a computer.

Chapter 6 offers some tips on how to get a computer donated. But whether you buy a computer or get one for free, make sure it has enough memory to store all your data and to run the software programs you want to use. The two most important programs are a word processing program and a mailing list program. The word processing program should be able to merge the mailing list and a form letter so that a series of letters is printed that differs only in address and salutation (this is called "mail merge"). The program should also be able to move paragraphs around and change document margins and length easily. The mailing list program should be able to sort through thousands of names and addresses in several different ways—by zip code, membership type, and so forth.

Fund-raising software packages may seem like an extravagant outlay of money, but they end up saving incalculable money and manpower—while increasing fund-raising profits. Cathy Folkes of Real Good Software, who has done extensive research on fund-raising software for grass roots groups, shakes her head in wonderment when small organizations balk at package price tags, then assign an employee or volunteer to hours and hours of paperwork recording and updating files—a job a software package could manage in minutes.

Ten to fifteen different major packages are available. Among the very cheapest functional programs is Campagne Associates'

(Nashua, New Hampshire; telephone: 800-582-3489) $975 entry-level package, which comes in either an IBM-compatible or Macintosh version. This software tracks donors, gifts, events, and phone numbers. It links to any word processing system for mailings and can run a variety of reports. More sophisticated packages may cost $4,000 or more, but some companies will lease packages so organizations can pay on a monthly basis.

ASKING FOR MONEY

Many people are intimidated by asking for money. Veteran grass roots fund-raising consultant Joan Flanagan says: "Asking for money is like going out to beat up a bear. The larger the amount, the more frightening it becomes, because you have to beat up a bigger bear."

The less fearful you are of the task, the smaller the bear will become. Review the reasons why people give, then ask yourself what they want in return. As much as possible, put yourself in your donor's shoes. Why do you support the group you're raising money for? If you have no qualms about giving the group money, chances are other donors won't be nervous about it either.

Ask yourself, what's the worst thing that can happen to you when you ask donors for money? They might say no. That's it. End of consequences.

On the other hand, they might say maybe, and ask you to send more information. And they might say yes, in which case you walk away happy and a bit richer. Thought of in these terms, asking for money doesn't seem so painful, does it?

Personalize your appeals for money as much as you can. Notes or individual attention of any kind make donors feel important. Remember the old fund-raising adage: People give to people. In *The Art of Asking* (The Taft Group, 1985), Paul H. Schneiter writes: "In-person asking consistently produces more and larger contributions than any other solicitation technique" by imposing urgency and pressure to respond on the spot. If you can't make a face-to-face request for money, use the telephone. Create fund-raising letters or brochures that are as intimate and heart-wrenching as possible. Add personal notes when you are mailing to someone you know. Hit home with a story so human that readers can easily identify with the characters, or so tragic that even the most callous reader will feel compassion for them (and feel darn glad that they are not in their shoes).

• •STEP BY STEP• •

How to Ask for Money

Just Do It!

• **Just ask.** Don't think for a minute that simply by doing a great job on an important issue, people will want to contribute. You've got to ask. The worst thing that can happen is that the donor will say no.

• **Ask family and friends.** Aunts, uncles, cousins, grandparents, neighbors, co-workers . . . the same people who buy your daughter's holiday greeting cards or son's magazine subscriptions.

• **Ask for a specific amount.** If you need $25 from all one hundred people you're asking, say so. You can also give potential donors a range to choose from, say $10, $25, $35, and $50.

• **Be direct.** Don't be coy. Explain your needs straightforwardly so that people will clearly understand how much money you need and what you need the money for.

• **Zip your lips.** Once you actually ask, "Will you give us the money?" don't say another word. Avoid the instinct to rush in and fill the pregnant pause that usually follows every request. If you wait for the prospective donor to speak next, it forces the donor to respond and prevents you from succumbing to the impulse to take back what you've just said.

• **Contribute yourself.** How can you expect others to give if you haven't anted up yourself? And don't forget to canvass your own membership and board.

• **Ask for enough money.** Don't set your expectations too low because you can't believe donors would give you as much money as you really need. This is especially true for foundations and banks. (I once nervously went to the bank where my organization had a checking account to pursue a $15,000 loan. The bank officer interviewed me extensively and said "Hmmmm" a lot; when the loan was approved, it was for $25,000. "You need more than you're asking for," the wise banker told me. He was right.)

• **Don't give up.** Just because someone says no once or twice doesn't mean he or she won't give the third time you ask. Keep trying,

How to Ask for Money (cont.)

• **Give people options.** Some donors prefer paying membership dues. Others like to get merchandise in exchange for their donation. And still others want to fund a service, like the planting of a tree or the provision of meals in a shelter. Let people "buy" something with their contribution.

• **Ask face-to-face.** It's much easier for people to turn you down if they don't have to look you in the eye and say no.

• **Ask again.** The most likely donors are those who have already given you money. Most churches ask for money at least once a week. You can probably ask a few times a year.

• **Say thank you.** Send a note, make a quick phone call, list donors' names in your newsletter, or hand out a certificate. Particularly large contributors should be recognized with a plaque or an award.

FUND-RAISING EVENTS

Why hold an event? In addition to boosting finances, fund-raising activities help build the organization as they expand your bank account, making people work together as a team to make the event a success.

What kind of event can you hold? Here's a list of the kinds of activities fund-raisers often turn to:

Bake sale	Ice cream social
Auction	House and garden tour
Yard sale	Theater party
Raffle	Carnival
Dinner dance	Concert
Benefit	Las Vegas night
Sock hop	Movie premier
Walk-a-thon or other	Telethon
marathon	Bingo
Fun run	"Celebrity" tennis
Pancake supper	tournament

If none of these appeal to you, consider selling services—baby-sitting or gardening for working parents, car washes, housecleaning, even dog walking. Here are some innovative examples:

- *The First Reformed Church of Schenectady* in New York raised money for its ecological programs by offering to "Eco-mug a House." For $30, older youths in the church removed the parishioner's name from unwanted mailing lists, installed a toilet dam to save water, checked the air in car tires for proper inflation, installed a water-saving faucet aerator, provided a handmade cloth "draft dodger" to keep cold air from getting in under the door, presented a hand-painted "Eco Mug" so the user could avoid throwaway coffee cups, supplied note cards printed on recycled paper, and took unwanted clothing to an appropriate donation center.

- Local affiliates of *Women's Action for Nuclear Disarmament* organized "Mums for Mom" fund-raising programs, particularly around Mother's Day. The Buffalo affiliate, which initially sold live flowers, turned to blooms made out of silk, accompanied by a peace message and information about their group. Using colleges and churches as their primary "storefront," the group profited approximately $5,000 per sales campaign.

- Richmond, Virginia's, *Goodwill Industries* issues vouchers to businesses, churches, schools, and other charities that collect clothes, shoes, toys, furniture, and other household items for Goodwill's outlets. Every half-truck load of donated items is "worth" ten vouchers; each voucher can be redeemed for ten pieces of clothing at any Goodwill store. The vouchers are given to disadvantaged people in the community.

- Parents of the *Spuytenduyvil Pre-School* in the Bronx organize theater nights for the parents of their students and other school supporters. Fund-raisers locate theater troupes that are willing to sell them blocks of tickets at a discount, then resell the tickets at their face value or more. One group of performers that's cooperated with the school is the Paperbag Players in New York City, a troupe whose performances are geared to children and whose scenery and props are made from cardboard and brown paper (hence the name). School fund-raisers buy $14 tickets for $9. Though they'll resell the tickets for $14, they ask for $20, guaranteeing them a profit of between $5 and $11 per ticket. The only hitch is that they have to buy at least one hundred tickets up front in order to get the discount.

- San Francisco's *Project Open Hand*, which provides meals to AIDS sufferers, raised more than $13,000 through an unusual insert it ran in its newsletter. To promote "Hand to Hand," a holiday food festival, the insert urged donors to send back a drawing of their own hand, along with a minimum $1 AIDS contribution. The heartwarming pictures were displayed at the festival, which also offered gourmet snacks and entertainment. Though the average donation amounted to only $9, several checks were significantly higher—like the one for $1,000 that accompanied a package of hand drawings that arrived from every member of the Oakland A's baseball team. It didn't hurt that the *San Francisco Chronicle* put a copy of the insert in the paper for four days, free of charge.

Project Open Hand developed this clever fund-raising idea, a flier that ran, at no charge, in the *San Francisco Chronicle* for four days and in the organization's newsletter as well. San Franciscans were encouraged to trace their hand on the flier, adorn it as they liked, and sign and date the palm. Finished art was mailed to Project Open Hand. The decorated hands, which generated over $13,000 for the organization, were arranged on a wall at Open Hand's 1992 holiday food festival and luncheon. The collage now travels as an exhibit to office buildings.

Creative fund-raising ideas have an uncanny ability to snowball into bigger and better money-raising events. Here's an example of a fund-raiser that began as a way for one teenager to repay society for his misdeeds and spawned a highly successful statewide fund-raising drive.

LOCAL HERO
Kids Helping Kids

He was sixteen years old and a juvenile delinquent. A victim of child abuse, "Smitty" had been sentenced to Oregon's MacLaren School, a juvenile detention center twenty-five miles outside Portland, for attempting to murder his foster parents.

His life could have been one long downhill slide. But tenacious counseling helped this young man realize that it was the abuse, and not some inherent evil trait, that was primarily responsible for his destructive behavior.

Vowing to raise funds to help other abused children, Smitty asked employees at the MacLaren facility to pledge money for every mile he ran in the fortified compound where he lived. One month and a hundred miles later, he had raised close to $100.

School officials were startled at the positive impact Smitty's fund-raising activities had among other incarcerated youths at the facility, many of whom had also been abused. They were pleased as well at the difference the project had made to Smitty, whose self-esteem bloomed as he began to "pay back" society for his previous misdeeds.

The school wondered if Smitty's run could become a model for involving youths at other correction facilities in special projects to benefit their communities and raise money for child abuse prevention programs.

Thus was born "Mission: Possible," an annual effort to raise money for the Children's Trust Fund of Oregon, which provides money for child abuse prevention programs throughout the state.

As part of national Child Abuse Prevention Month, one day is set aside in April each year to involve kids in physical and community challenges. Corporations, small businesses, and individual citizens sponsor activities through donations, which local banks accept and forward to the trust fund.

Mission: Possible kids dig in while helping to landscape a children's advocacy center in Medford, Oregon.

In 1989, the first year of the program, more than five hundred delinquent Oregon boys and girls participated in Mission: Possible, raising more than $10,000 in contributions for child abuse prevention programs. Today, the program involves hundreds more youths, who clean fish hatcheries, craft stuffed animals for abuse victims, chop wood for senior citizens, and even produce cassette tapes to answer parents' questions about sex abuse that are distributed by Parents Anonymous.

Because it has generated so much publicity, the program has had the added benefit of educating the public about the links between child abuse and juvenile delinquency, while helping increase revenues for the Children's Trust Fund. The theme for 1993's Mission: Possible campaign was simple: "Kids Helping Kids."

Thanks, Smitty.

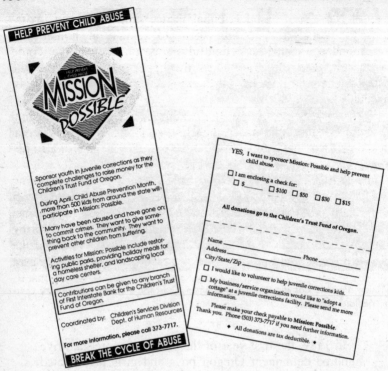

This fund-raising package was mailed to large corporations throughout Oregon and distributed with a poster to local businesses. The versatile brochure and envelope were also handed out at Portland Trailblazers basketball games and even stuffed in with the paychecks of Oregon government employees.

To organize a successful fund-raising event, develop a specific strategy, timetable, and deadline. Calculate how much it will cost you to hold the event and how much money you need to raise to make the effort worthwhile (ask the same questions to prepare an event budget as you do to budget your annual organizational expenses; see page 121). Identify a list of invitees and keep track of who attends the event this year so they can be invited again next year. Set prices for tickets and other items, organize refreshments, and publicize to the degree appropriate for the event (see Chapter 6, "Communications," for publicity tips).

On the other hand, you can ask people to stay home. Several groups have organized a "no-benefit benefit," asking for money to spend on services instead of a party. In one case, the Fortune Society, a New York–based criminal justice group, sent out invitations to supporters—inviting them to remain at home. The correspondence painted a delightful picture of an event that could have been, then offered readers the opportunity to stay home instead—and send Fortune "all the money you will save by not going out."

Keep thinking creatively. In addition to its successful annual Mission: Possible fund-raiser, the Children's Trust Fund of Oregon kicked off two new campaigns in 1993 that they hope will also become yearly events.

The Trust Fund "sold" Mother's and Father's Day cards during its Share the Love campaign in exchange for a donation ($10 per single card or $15 for a set of two) made in the parent's name. Kids submitted their colored drawings in an art contest during Child Abuse Prevention Month in Oregon in April 1992, and the two cheerful winning designs graced the covers of the 1993 cards. The Trust Fund got the public excited about its campaign by insert-

The Children's Trust Fund's 1993 Father's Day greeting card.

ing flyers in its newsletter and mailing them to children's services and abuse prevention programs around the state and handing them out at businesses. Press releases and ads ran in local media outlets. The 1993 return—a 3,000 profit—fueled plans for future campaigns. Other West Coast groups have heard about the fund-raiser and have asked Kim Walker, president of the fund, to help them set up their own projects.

"Your Buck Stops Fear" was the theme of an intriguing "point of purchase" campaign launched to coincide with the September 1993 school season. The Trust Fund, Walker said, arranged with a local department store to place at its checkout counters promotional cards that explained the campaign and contained special bar codes. Parents purchasing back-to-school gear picked up the card, which read: "Just include this card with your purchases and one dollar will be added to your total. . . . You will support over twenty abuse prevention programs statewide." The clerk scanned the mock department store tag printed on the card along with other clothing tags, collected one extra dollar from the purchaser, and then turned over all contributions to the Trust Fund. The back of the card—which the parent could keep—contained shopping tips such as "Never let your child use an escalator unattended."

According to fund-raiser Jenny Thompson, whose firm consults with many grassroots and national nonprofit organizations, the most successful fund-raising events are those that relate directly to the work of the organization. In other words, it makes more sense for a group concerned about substance abuse to host a premier about a film dealing with drugs and alcoholism than one about dinosaurs or rainforests.

 CHECK IT OUT: *The Grass Roots Fundraising Book,* by Joan Flanagan, Contemporary Books, Chicago, IL, 1992, $14.95.

**ONE DOLLAR. A SIMPLE WAY TO STOP
CHILD ABUSE AND NEGLECT IN OREGON.**
Just include this card with your purchases and
one dollar will be added to your total.
Your donation will be passed on to Children's
Trust Fund of Oregon by Meier & Frank.
You will support over 20 abuse prevention programs statewide. These programs provide outreach, education
and support to Oregon's
"high-risk" families.

MEIER & FRANK
092 385 36
892 1 P10

**EVERY DOLLAR DONATED
TO CHILDREN'S TRUST
FUND OF OREGON GOES
DIRECTLY TO PROGRAMS.**

*Thank you for your
tax-deductible donation.*

$1.00

The donation form the Trust Fund
made available for shoppers at
Meier & Frank.

MEMBERSHIP AND DUES

Are you contemplating charging people to become members of
your organization? Remember that what drives people to join is essentially the same as what motivates them to give, ratcheted up an
emotional notch or two.

Richard P. Trenbeth, author of *The Membership Mystique*
(Fundraising Institute, 1986), says many people become paying
members of a group because they need to "belong" to something
larger than themselves and crave the "elevated societal level" to
which they believe they rise when they join. Others, he believes,
hope to gain knowledge from the group that will help them solve
personal problems. Then there are those who just want to fill their
empty mailbox with membership mail. All of these motives may
be summed up as self-interest.

Thus, before you can set the price of membership dues you have to pinpoint what a person wants to get out of joining your ranks, then decide what the person is willing to pay to reap those benefits. For $10,000 to $15,000, you can conduct a scientifically based, statistically meaningful telephone or mail survey to ask new members, old members, and potential members (selected randomly to insure that they are not biased) questions that probe clues as to how much money people are willing to pay to join your group or support your cause.

One way to set the membership price is to discuss what the group wants to do in the coming year. Estimate what those activities will cost, how much money you can raise from nonmembership sources like fund-raising events and foundation grants, and set dues to recover the rest.

You may also need to create alternatives to allow all members to "pay" their dues even if they don't have money. Students, senior citizens, and the unemployed may be able to earn their membership with "sweat equity," volunteering a certain number of hours to clean the office, file documents, answer the telephone, or program computers.

While many national organizations charge up to $35 for an annual membership, local groups may start off at $5, $10, or $15. Because any potential donor base consists of people with vastly different interests and income levels, many organizations give the check writer a number of donation levels to choose from.

Hopefully, your membership will grow along with your plans. As you begin to count on a stable membership, you can set your membership price at one level, so that it doesn't change year after year.

Always be on the lookout for new members, who will soon become a source of revenue and support. Set up a membership table at every event you organize, and at those of other, supportive organizations as well. Send out membership information with the reply to every request for information, after you've given a speech or attended a public hearing, even when talking to others in your neighborhood or community whom you believe should be interested in your issue. (For more tips on membership recruitment, see Chapter 4, "Building Your Organization.")

What Members Expect in Return

"Joiners" are more likely than "givers" to expect a tangible as well as an emotional return on their dollar. To appease those who need to feel that they're getting more for their dues than the simple satisfaction of joining a worthwhile organization, consider offering organizational bumper stickers, buttons, T-shirts, tote bags, window decals, or membership cards. Many direct marketing packages automatically include the cheaper items, like stickers or membership cards, for free, compelling recipients to write that initial check to alleviate the guilt they feel at already having been given a gift. The more expensive treats, like T-shirts and tote bags, will be offered as "premiums" to members who give more than the basic dues. You can also offer free admission to events or discounts on tickets, goods, or services.

In their handbook *Dollars and Sense,* fund-raisers for the San Francisco–based Western States Shelter Network advise groups not to offer potential donors too much initially. "A quarterly newsletter is fine," they write. "A membership card which entitles the bearer to nothing, but looks nice" also provides an adequate incentive to get people to join. As a general rule, they caution: "Don't spend more than 20 percent of the membership dues on the membership benefits."

GRANTS

Your organization may be eligible to receive a grant either from a philanthropic foundation, a corporation, or an individual. You can also seek grants from city or county officials who may have access to funds through arts councils, city beautification committees, and other agency operations. Upon occasion, grants from the federal government may also be available to support your work.

It should come as no surprise that any institution that grants you money will be doing so for a specific purpose. Though foundations and corporations occasionally dole out contributions to support a group's "general operating fund," most grantmakers prefer to contribute to a specific project that has well-defined goals, revolves around a practical budget, produces a useful product, and can be accomplished in a reasonable time frame.

To get a grant, locate grantmakers who are interested in your issue. For help, contact the nearest regional office of the Foundation Center, an independent, nonprofit organization that provides free access to all of the materials necessary to research likely funding sources and develop a good proposal.

Identify a specific project around which to structure your grant request. Be precise in the purpose of the project, who is going to execute it, and how much it is going to cost.

Then, write a proposal. Many foundations offer specific guidelines for proposals they will consider, and not all guidelines are alike. Foundations will send you a copy of their guidelines if you ask (see page 137 for proposal writing tips.)

Meet deadlines. Most funding entities make grants only during certain times of the year. The guidelines they send should alert you to their "funding cycles."

Finally, do what your proposal says you're going to do. Many foundations will give to a group again and again if the group proves reliable and accomplishes the mission set out in its proposal. On the other hand, the quickest way to insure that you'll never get additional money out of a funder is by squandering the donation on something other than the project the donor was funding.

Make sure you meet the foundation's guidelines for reporting on the status of your grant. You will probably have to file a detailed final report on how you spent the money that includes some sort of evaluation on the success or failure of your project. But don't wait until you've spent that last foundation dollar to get back in touch with those who've given you money. Put them on your mailing list to receive newsletters, annual reports, white papers, and press releases. Send them copies of favorable news clippings about your organization or issue and letters of commendation from constituents and public officials. Cultivate and maintain a solid relationship that will result in additional gifts to your group in the future.

• •STEP BY STEP• •

How to Write a Grant Proposal

Here is a list of basic elements that should be included in any proposal.

• **Summary.** Briefly summarize the project for which you are seeking funding and the amount of money you are requesting.

• **Introduction.** Provide a brief history of your inception, a list of your goals and objectives, and a summary of your most important accomplishments to date.

• **Problem or Needs Assessment.** Outline the problem you intend to solve or the needs you will meet by pursuing the project described in your funding request.

• **Objectives.** Define your goals, providing explicit examples of how the project you're proposing will solve the problem you've identified. State the tangible results you hope to achieve, such as the number of people you will reach or the product that will result from your effort.

• **Methods.** Describe how you intend to achieve your goals and objectives, including the techniques you will use, the staff who will be involved, and the amount of time you expect the work to take.

• **Evaluation.** Include a plan for determining your success or failure in achieving the goals and objectives of the program for which funding is being sought. Can you quantify the number of people you reached or analyze the impact of your program on your target audience's behavior?

• **Budget.** Provide a detailed budget for the project that includes all expenses you anticipate, including rent, telephone, travel, printing, salaries and benefits, legal fees, supplies, and other expenses. Include any anticipated income from other sources, such as conference fees, membership dues, and merchandise sales. If your request will only cover a portion of your expenses, explain where you expect to raise the remainder of your budget.

How to Write a Grant Proposal (cont.)

• **Appendices.** Attach materials bolstering your arguments that the work you propose to undertake is critical and that your group is the most appropriate organization to undertake the effort. Press clippings, an annual report, letters of endorsement from respected members of the community will all help. Don't forget to include a copy of your tax-exempt classification letter—foundations will not fund you unless you are tax-exempt or have a tax-exempt sponsor.

Once your proposal is complete, develop a convincing cover letter to accompany the document. The letter should present a captivating summary of your proposal and stress its relationship to the foundation's interest.

☞ **CHECK IT OUT:** For an excellent primer on developing proposals, see *Program Planning & Proposal Writing,* by Norton J. Kiritz, published by the Grantsmanship Center, $4.00 plus shipping and handling; 1125 West Sixth Street, Fifth Floor, Los Angeles, CA 90017.

The Foundation Center operates national collections in Washington, DC, and New York City and several "Cooperating Collections" at libraries, community foundations, and other nonprofit agencies that provide a core collection of fund-raising how-to publications and a variety of supplementary materials and services in areas useful to grant seekers. To locate the Cooperating Collection nearest you, call toll-free 800-424-9836, or write The Foundation Center, Eighth Floor, 79 Fifth Avenue, New York, NY 10003.

Here's an example of how a consortium of community groups creatively combined grant writing with other fund-raising techniques to generate the money they needed to open an after-school program for sixth, seventh, and eighth graders.

LOCAL HELL-RAISER
Where There's a Will There's a Way

Day care isn't just for babies and toddlers. Older kids need after-school direction and supervision, too, yet satisfactory programs for young teenagers can be hard to come by.

At least this is what the League of Women Voters (LWV) of Metro Columbus, Ohio, found when they began to look into after-school opportunities for sixth, seventh, and eighth graders. In fact, there were virtually no opportunities at all in the area, which is 70 percent nonwhite and in which 90 percent of middle schoolers are on reduced-rate lunches.

Ten-to-thirteen-year-olds are difficult to manage, said Suzanne Fahey, then president of the League chapter, because they are old enough to require skilled instructors but still have short attention spans. League volunteers decided to create a school-age-child-care program for these youths, but had few financial resources to throw behind the effort. Traditional fund-raising techniques—such as selling fruit to the community during Christmas—simply didn't work; residents just didn't have extra money to spend on fund-raising gimmicks.

The League also found it very difficult to get parents to pay for their kids' participation. When they announced that the after-school program would cost only $25 for the year, no one signed up.

Back at square one, the League set up a "community round table" to assess what sort of resources the area would likely invest in the program. The activists found that their fellow community groups—the YMCA and the YWCA, the United Way, 4-H, the Girl Scouts—not only supported the idea but offered to help in any way possible. Each group donated either money or instructors for after-school activities, whatever it could afford.

Most groups found that by donating manpower they would not strain their limited budgets and would still provide the program with a valuable resource. The

YWCA began by sending one person one day per week to run an activity, as did the Columbus Recreation Department, despite its own lack of funds. The Salesian Boys' Club offered to transport kids from school to their facilities if need be, and offered tutoring services. The YMCA was not having any luck attracting middle school kids to its own facility, so its Black Achievers Program sent a volunteer once a week to work with the League's more successful effort.

The program made good use of the middle school's gym, equipment, and space. "To me, the beautiful thing is that the school is open. The staff has gone home; that facility has become a community resource to help the boys and girls. These kids love that school!" exclaimed the League's Fahey. The League provided supervised sports, computer instruction, cooking classes, and African dance demonstration. The kids even produced a puppet show on drug prevention, which proved both instructional and effective in building self-esteem. "Giving them an activity is the same thing as prevention," Fahey said.

League members worked nonstop to solicit grants to support their program's start-up. Fahey applied for and received a $2,500 grant from the Columbus Human Services Department that paid for two staff members who manned activities every day. The group also secured grants from the City of Columbus, the Columbus Foundation, the state's Department of Education, the Junior League, the Ohio Commission on Minority Health, and COMP DRUG, a drug prevention organization—in all, raising $17,000 to fund the program's kickoff year.

The League reduced the fee it charged kids to $1 per afternoon, and when this still proved excessive, to $.25. Though nominal, the charge proved sufficient to pay for the kids' daily snacks, which were purchased extremely cheaply from the Mid-Ohio Food Bank, a business that distributes food inexpensively to nonprofits.

Early in its first year, around twelve kids showed up. "With this age group," noted Fahey, "they'll only come if it's the cool thing to do." Often parents were working, but for the most part, kids came because they had nothing better to do after school. In 1992, the number grew to thirty kids per afternoon, which was plenty for the instructors to handle. "We have the facilities to handle more kids, but not the manpower," Fahey rues.

Since the League of Women Voters was not set up to be a fiscal agent for this type of operation, its goal from the beginning was to secure a stable sponsor to run the program and handle its funding. Eventually, the Columbus YWCA took charge, allowing the League to channel its efforts into seeding similar projects in two other Columbus middle schools.

Matching Grants

Matching grants require an organization to "match" or raise all or part of the grant with money secured in other ways. Some grants will give you a dollar for every dollar you raise; others match two-to-one or three-to-one. Creating a matching grant program is a way for a foundation to hedge its bets, to protect itself when funding a new organization that doesn't have a track record, or to prevent the organization from becoming too dependent on foundation support. Plus, grass roots groups can give their members and donors the psychological benefit of knowing that every dollar they give is at least doubled.

According to the Ms. Foundation, "Wily old Ben Franklin was the first person in America to use the matching grant. In 1750, he lobbied the Pennsylvania Assembly to create the first government matching grant—for his favorite project, the Pennsylvania Hospital. Assemblymen agreed to appropriate 2,000 pounds if Franklin's volunteers could raise 2,000 pounds."

As Franklin wrote in his autobiography: ". . . the members [of the Assembly] now conceived they might have the credit of being charitable without the expense . . . and then, soliciting subscriptions among the people, urg'd the conditional promise of the law as an additional motive to give, since every man's donation would be doubled; thus the clause work'd both ways."

CHECK IT OUT: Many excellent resources can help you identify sources of corporate, government, and foundation grants. They include:

Government Grants: The Catalogue of Federal Domestic Assistance, $46.00 for a one-year subscription, make the check out to the Superintendent of Documents at U.S. Government Printing Office, 8660 Cherry Lane, Laurel, MD 20707.

Taft Corporate Giving Directory: Comprehensive Profiles of America's Major Corporate Foundations and Corporate Charitable Giving Programs. Check your local library.

The Foundation Directory. This source contains crucial information on the nation's 8,700 largest, most influential foundations. Check your library or order from the Foundation Center for $150. Call toll-free: 800-424-9836.

DIRECT MAIL

Direct mail involves sending a letter and some supporting materials directly to members, potential members, or donors to ask them to join or to make a contribution. Because of the printing, assembling, mailing list rentals, and postage costs involved, the process is an expensive one that is not usually affordable for start-up organizations or for those that depend on low dues or few or no dues-paying members for support. A direct mail "test"—an initial mailing conducted to determine how much money you can expect to raise if you mount a complete direct mail program—could cost tens of thousands of dollars, depending on how complicated the "package" is (see pages 146 and 147) and how many people you mail it to.

Even if you can afford it, direct mail may not be an appropriate fund-raising technique for your group if you have little name recognition, are trying to raise money for an issue few people care about, have no track record solving the problems you claim in your letter to be trying to fix, or are competing with other, better known groups that are providing essentially the same services or benefits as you are. More to the point, ask Kay Partney Lautman and Henry Goldstein in their thorough direct mail manual *Dear Friend: Mastering the Art of Direct Mail Fundraising* (The Taft Group, 1991), can your organization financially survive a loss of 40 percent or more of its investment should the test mailing fail to recoup what it cost? And can you wait two to four years before realizing "spendable net income" from the mailing? If not, direct mail is probably not an option for you.

Financial Considerations

Several factors contribute to the cost of direct mail. Designing and printing the direct mail package, renting mailing lists, and postage will run into the thousands of dollars even for a modest mailing.

If you do business with a direct mail consulting firm, the firm may be willing to advance you the expenses of producing and mailing your direct mail package in anticipation of the money the package will generate in return. If you're trying to get a direct mail program off the ground yourself, you'll have to bear the burden of those expenses "up front."

Either way, the venture will cost you money. The trick to direct mail is to develop a compelling package and mail it to a responsive list of people, who will eventually, as they give subsequent gifts to your organization, generate more income than it cost you to send out the package in the first place.

Before you launch a direct mail program consult with other organizations of similar size who have either foregone or opted for this fund-raising approach. Talk with experienced direct mail consultants who will give you good advice whether they get your business or not.

Realize that you can increase the amount of money you receive from first-time donors by carefully picking the people to whom you mail, and by mailing to fewer people at one time.

Mailing to donors around the same time each year, mailing more than once, and following up with phone calls or letters to thank donors for their contributions will also help boost the size of the gift. But even in a successful campaign, don't expect more than .75 percent of those you mailed to in the first mailing to respond. Of those, 40 to 50 percent will give again, and of those, 75 percent will give when asked three times or more. But if you can't afford to get that first mailing off the ground, the rest of these numbers will be meaningless.

The Direct Mail Package

Once you are comfortable with the budget projections, you can turn your attention to developing the kind of direct mail package that will generate the most income for your group. No one knows for sure exactly what people read; most direct mail packages contain the following elements:

- **An envelope.** Many professional fund-raisers believe that the envelope is the most important part of the direct mail package. With the onslaught of mail people receive every day, your package must be interesting enough at first glance to make the recipient not only open it but read on. A catchy headline or slogan, the promise of a gift inside, the opportunity to fill out a survey, complete a questionnaire, or sign a petition are all gimmicks that fund-raisers have used successfully to get readers to open their mail.

  While the most persuasive envelopes are those that are hand-addressed to make you feel like you're getting a letter from a friend, most organizations don't have the time to address their letters by hand, especially if the mail is being sent to thousands or tens of thousands of people. More commonly, letters will be sent in a standard-size envelope (called a "number 10" envelope). Many envelopes have windows in them, so that you can read the addressee's name and address, which have been printed on the letterhead or on the reply form, through the window. This saves time and money in printing the name and address twice.

  Make sure you obtain the right direct mail permits from the post office, including a "business reply" permit for the reply envelope, and a bulk-rate permit for the outer envelope. If you cannot obtain a nonprofit organization permit imprint, at least get a regular bulk-rate permit.

- **The letter.** Your letter to a prospective donor must look, read, and feel like a "real" letter—like correspondence you might be sending to a close friend. It should be printed on your organization's stationery. The tone should be very personal—"Write the way you talk," many direct mail copywriters advise.

  The letter should be single-spaced, double-spaced between paragraphs. And the paragraphs should be very short, consisting of just a few, powerful sentences each.

The length of direct mail letters varies. They're usually at least two pages, sometimes four, and occasionally six to eight when sent to existing supporters. Many direct mail experts believe that longer letters work better, not because they make a donor read every line but because they convey to donors that a message is so important, it takes a lot to communicate!

Good direct mail letters begin with a story or anecdote that puts the issue in perspective, conveying a sense of urgency, success, emotion, or symbolism. The story is usually tied to a need your organization is facing or a problem it is trying to solve.

The appeals you make in your letter should be those you believe will resonate most with your audience. Are people angry over a proposed tax hike? Fearful about the possible construction of a new nuclear power plant? Do they need someone to lobby on their behalf for health care reform? If you know what motivates your readers to care about your issue and join your organization, you'll be able to draft a successful direct mail appeal to them.

You also need to ask for an explicit amount of money. Over and over again, groups that request specific dollar amounts in their letters see the size of the average gift they receive increase.

Many letters contain a final P.S. to make one last point and tie your appeal to a current event or issue.

- **Enclosures.** It has become common practice to enclose an additional note or "buck slip" to give your letter extra emphasis and appeal. Wildlife groups may enclose nature stamps. The Whitman-Walker Clinic, a Washington, DC, AIDS facility, encloses Christmas cards, unsolicited, with its end-of-the-year solicitation. Other groups include a window decal, membership card, bumper sticker, personal notes from celebrities, fact sheets, photographs, surveys or questionnaires that need to be returned to the group (along with your check), and copies of favorable news clippings.

  Of all these potential items, some are more effective in boosting donations than others. According to direct mail expert Amy Leveen, personal notes from celebrities and copies of favorable news clippings work well because they provide a short, pithy way to allow someone else (who may be more

This letter, brochure and response card (see facing page) generated an impressive 4 to 6 percent return when 1,800 pieces were mailed to parents, alumni, and friends of the Baltimore Chesapeake Bay Outward Bound Program, with gifts averaging $50 to $75. The pieces were typeset and laid out using Word Perfect 5.1, Q & A database software, and a laser printer; photos were stock or taken by Bill Hearn, who wrote and produced the package. With a limited budget that precluded purchasing lists, total production costs ranged around $1,200 plus staff hours. Outward Bound offers challenging outdoor courses that help build leadership, team skills and self-esteem among city and other youth and adults.

Reprinted with permission, Parks & People Foundation, Inc.

Parks & People

Baltimore Chesapeake Bay Outward Bound Program
4921 Windsor Mill Road ⊓ Baltimore, Maryland ⊓ 21207 ⊓ (410) 396-0082 ⊓ Fax: (410) 298-3822

October 12, 1992

Dear

The problem simply stated is this:

"There is the <u>decline of fitness</u> due to modern methods of locomotion.

The <u>decline of initiative</u> due to the widespread disease of spectatoritis.

The <u>decline in care and skill</u> due to the weakened traditions of craftsmanship.

The <u>decline in self-discipline</u> due to the ever-present availability of tranquilizers and stimulants.

The <u>decline of compassion</u>, which is spiritual death."

Yet Kurt Hahn, founder of Outward Bound, wrote this more than 30 years ago. Imagine what he would think today, a time of 127 channel televisions, video games, vcr's, city library systems with declining budgets, and the tremendous temptations of a drug industry that is closer to our children then we are?

Who will compete against these challenges? Who will lead the way and show young people that they are strong enough to meet these hurdles head on?

Every young person, parent, teacher, and all the others who care will.

But they need help. No one can do it alone. Not everyone has the resources to or believes that he or she can overcome life's obstacles.

That's where Outward Bound comes in.

As an alumnus you have personally witnessed the difference Outward Bound can make in a life. Whether you sailed or backpacked, traversed the ropes course or a rock face, you gained immeasurably from an experience which engaged both your body and your mind.

This year has been especially rewarding. I am proud of the work we've done with students in both new ways and old.

Baltimore Chesapeake Bay Outward Bound is a program of the Hurricane Island Outward Bound School and sponsored by the Project Committee of the Parks & People Foundation, Inc.

♻ Recycled Paper

Our achievements have been substantial.

Just last month 12 students from Baltimore City finished a six week urban environment course. During the first week they completed a traditional Outward Bound short course. Over the next five weeks they used their Outward Bound developed skills to perform community forestry work. Led by Yale University School of Forestry and Environmental Studies students and Outward Bound instructors, they studied streams and forests, interviewed community people about their parks, reforested patches of community land and cleared streambeds of debris.

Throughout the past year more than 30 courses from public high schools participated in our Peer Leadership Program. Students from those schools cultivated increased self-awareness and refined their leadership skills. Upon their return to the halls of high school they applied enhanced skills to their personal lives and school community.

Courses like these are expensive, but the changes they produce in the lives of our young people are dramatic.

I'm hoping that we can bring the same Outward Bound experience you had to more young people this year, young people who otherwise would not have the opportunity to discover their own inner resources.

But we need your help to do it.

We have increased demand for trips and long waiting lists. Government funding cuts to some of our programs mean we must now raise additional funds to make up the difference.

So please consider sending $35, 50, $100 or $1000. With your help Baltimore Chesapeake Bay Outward Bound will continue to offer more students the confidence to face today's great obstacles and the self-discipline to overcome them.

Sincerely,

Harriette McPherson

Harriette McPherson
Executive Director

P.S. Together we can -- and we must -- take stronger, bolder action to help the youth of our region develop the personal resources they need to face the challenges of our time.

On my trip I learned to never say the word "can't," and if you want something you must try for it, and make it happen, there is always a way. I have never been so cold, hungry, and tired in my life, but what I learned and gained was worth being cold, hungry and tired for, and I would do it all over again if I could.
-Kathleen Fought

My son gained twelve new friends from all over the United States and an unforgettable positive experience. Thank you very much for allowing him the opportunity to experience this kind of leadership.
-Jacqueline Gowans-Maultsby

If it were not for you we would not know how much fun and challenging sailing could be. I hope that you will continue to give others the opportunity such as this one.
-Jamalden Gowans

I have never been so cold, tired, and hungry in my life...

Never in my life have I experienced such a thing. I am very active in my church youth group and we take many wonderful retreats, but this trip was a different kind of wonderful...I learned how to push my limits and abilities...I had the best experience of my life on Baltimore Chesapeake Bay Outward Bound Peer Leadership Group 64 and I will remember it forever, as will, I think, everyone who was with me. Thank You.
-Joanna Tisdale

Baltimore Chesapeake Bay Outward Bound Program
4921 Windsor Mill Road
Baltimore, MD 21207

Yes, I want to help a young person take a big step toward independence and a feeling of self-worth. Here is my contribution to provide the Outward Bound experience to more people.

☐ $1000 ☐ $500 ☐ $100 ☐ $50   Other $_____

Name_____

Address_____

City, State, Zip_____

Area Code, Phone_____

Please make check payable to Baltimore Chesapeake Bay Outward Bound. Contributions are fully tax deductible as allowed by law. Baltimore Chesapeake Bay Outward Bound is a program of the Hurricane Island Outward Bound School and sponsored by the Project Committee of the Parks & People Foundation, Inc.

Parks & People
The Foundation for Baltimore Recreation & Parks

Parks & People Foundation, Inc.

credible) to tell the recipient of the package why he or she should support your group.

Surveys or questionnaires are far less successful in generating revenue, says Leveen, unless they are closely tied to a controversial issue (for example, asking potential donors to fill out a general survey on the environment probably won't raise a lot of money; asking them to return a postcard you can then deliver to their member of Congress requesting support of upcoming legislation on clean air probably will). Urging people to complete a survey or questionnaire for a specific reason is also more effective because it makes donors feel like they're taking part in your work, not just paying you to do the work for them.

Be very clear in all your direct mail that the greeting cards, bumper stickers, window decals, and other items you're offering are free to the recipient. Don't leave people with the impression that they owe you money for items they didn't order.

Finally, there is also a reply card that allows for your name and address and a box in which you check off how much money you are enclosing. It is usually recommended that recipients send checks or money orders, not direct cash.

- **The return envelope.** In addition to the letter and buck slip, your direct mail package should include a "business reply envelope," or BRE, in which to return the donation. The most common BRE is a wallet-flap envelope that incorporates the information usually found on the reply form (including a short statement about the group, a place for the donor's name and address, and a box for the amount of the contribution). Usually, BREs are post-paid to create an added incentive to the recipient to make a contribution. (Some groups add a note on the envelope such as: "Your stamp here will save us money" to encourage donors to use their own stamps.)

CHECK IT OUT: For more information about how to obtain the permits necessary to send direct mail, as well as to help you meet the specifications for printing and mailing direct mail, see *Postal Publication #13* and *The Mailer's Guide*. Both should be available from your local post office.

Mailing Lists

As you prepare your package, think about whom you're going to mail it to. There are two kinds of lists: the "house" list, consisting of past contributors to or supporters of your group; and a "prospect" list, composed of people who have never made a contribution to your group before. You'll need to write completely different letters to the two groups. Previous donors should learn how their past donations have been put to good use, and why their help is needed again. Prospective donors need to be convinced that your group will help resolve a problem the donor cares about.

Renting or exchanging/trading mailing lists is big business; they cost, on average, $65 to $70 per thousand names if you must rent the list. If you're prospecting, choose lists that you have some reason to believe will perform well for you. Mailing to a list of people who have already shown great interest in your issue is one way; exchanging or renting lists from an organization with goals and objectives similar to yours is another. Usually, you will rent another group's list for one-time use only; if you exchange your list with another group, it will be on a name-for-name basis. It is not standard procedure to buy lists you can use forever; few groups are willing to sell their names for indefinite access by another group.

Maintaining your house list is critical. Use a computer to keep it up-to-date, noting how many times each donor gives, how much each gives, and the appeals that generate the most amount of money. Weed out past donors who have not contributed for a given period of time by mailing to them one last time with a "WHYFU?" letter (why have you forsaken us?), asking why they've lapsed, explaining current activities, and suggesting they renew. If they then contribute, keep their names in your active house file. If they send you nothing, put aside their names for a prospecting letter in the future.

Keep donors on your house list informed about your activities through newsletters, annual reports, personal letters to larger contributors, and thank-you letters to all donors. Mail several times a year, and keep track of the size of the donor's average gift, your income per thousand pieces mailed, your percentage of return, and your net income.

Testing

Before you incur the expense of printing an entire prospecting package and mailing it to people you don't know, test the package to determine how effective it's going to be. Choose a random sample of names from the mailing list. Test several different variables, such as different enclosures or two versions of the outside envelope. One of the packages will outperform the other; that's the one you should go with.

Mail the test package to about 10 to 15 percent of the list. Then monitor your returns for three to four weeks. If you break even, continue the process. Mailing to a list of prospects should have a return of approximately .75 percent; mailing to past contributors should bring in 40 to 50 percent.

Evaluating Results

Chances are, you won't make enough money to cover your costs on the initial direct mail package. But this may not be all bad. Many groups feel they can afford to break even or lose money initially if they'll make money renewing members later on. Don't necessarily base your success on how much money you raise in any one mailing. Rather, success may be determined by how you utilize the contributors from each previous mailing to bring in consistent, predictable, and substantial funds. For example, mailing to a list of fifty thousand prospective donors may cost $5,000 and bring in only $5,000 from a total of five hundred contributors. But that same list could generate as much as $10,000 over the next two-year period. Most fund-raisers would consider that a success. And so will your financial advisors—for that is a very good return on your investment.

THE CANVASS

Many groups raise money by going door-to-door. When I was a kid, I used to replenish the coffers of my church youth group by selling candy to everyone in the neighborhood. Bob Forney, Schuylkill County's fund-raising "gopher," raised $500 by knocking on his neighbors' doors and asking for support to send two Committee for a Healthy Environment volunteers to a conference on hazardous wastes in Washington, DC.

Raising money through door-knocking is cheap (since the labor is usually donated by volunteers), effective (it's harder to say no to someone who's looking you in the eye), and empowering (it feels great to take a purse full of money back to the organization and be able to say, "I did this").

On the other hand, canvassing is extremely time-consuming, and it can be equally frustrating if, no matter how persuasive your plea, people just don't want to give.

Before launching a canvass, carefully weigh the pros and cons. (For tips on how to set up a canvass, see Chapter 4, "Building Your Organization.")

HOUSE MEETINGS

A growing number of organizations are using house meetings as fund-raisers. Suppose you have decided to host a meeting in your home. First, invite ten or more people to stop by for light refreshments and to learn more about your organization. Offer background information—a fact sheet you've developed, a recent news story about the issue, an endorsement letter from another respected organization or individual in your community. If you have a videotape (even a tape you made yourself), show it to make the issue more real to your guests. Have along an "expert" who can answer any questions your guests may have.

Your program should last about half an hour. As it concludes, ask people to join or donate. (It might help to "prime the pump" by inviting a volunteer who says he has decided to join and urges others to do so as well.) Reiterate the importance of joining, and thank everyone for participating. To follow up, send a welcome note to all new members, and another recruiting appeal to all those who did not join. Send thank-you notes to all who contribute.

SILENT AUCTIONS

A silent auction is the elegant cousin to the noisy and some-times rancorous exchange you might see in a typical auction house.

This fund-raising activity is an incredible amount of work, in-volving tens (if not hundreds) of hours of labor, tremendous inge-nuity, and the gracious generosity of merchants and businesses. Here's how it works.

At least two people are needed to coordinate the entire auction. They pick a date, find a location, amass a small army of volunteers to help collect the items that will be auctioned, and set the param-eters for what will be donated and sold.

Begin the process at least six months in advance of the event, and perhaps longer, depending on how much money you hope to raise and how many items you need to find. Try to gather an in-teresting collection of appealing items: the most recent auction at Acorn Hill, my son's preschool, included beautiful watercolor paintings, facials and massages, gift baskets of foods, soaps and toiletries, two hours of instrumental music for a party or festive oc-casion, a weekend rental at another parent's cabin, books auto-graphed by the authors, handmade clocks, heirloom wicker doll strollers and other toys, clothing, sheets and linens, various kinds of consulting services, gift certificates to be redeemed at the local food co-ops and bookstores, candles, and many, many more entic-ing objects.

Display the items attractively in a room or gallery where they can be easily examined. Describe each item on a three-by-five white index card or small sheet of paper. People may circulate freely from one item to the next, indicating their interest in pur-chasing an item by writing the amount of money they are willing to pay for it on the card. (If something is unusually rare or special you may want to establish an opening bid on the card; the bidding must then start at that amount.) The next person to come along has the option of upping the ante on the item or moving on.

After a certain time that is prominently posted, no more bids will be accepted. The last name on the card at closing time gets the item at the price he or she wrote down. The successful bidder takes possession of the item when the bid has been paid.

MERCHANDISE

Selling merchandise may be another way to make money—as long as you bring in more than it costs you to buy or manufacture the merchandise in the first place.

There's almost no limit to the variety of products you can sell. Books, calendars, clothing, cookies, fruits, nuts, candy, taffy apples, flashlights, water bottles, vacations, bicycles, and even appliances are among the goods groups have peddled in an effort to make a buck.

Particularly popular are the thousands of "specialty" items available for imprinting with an organization's name and logo. T-shirts, coffee cups, back packs, bumper stickers, and fountain pens are just some of the products organizations have slapped their name on to boost their income (and generate some additional visibility for themselves).

But there's more to setting up a merchandising operation than perhaps meets the eye. First, you must decide what to sell. Many groups choose items that will appeal to their most likely customers, their members. T-shirts and baseball caps bearing an organization's logo and name, bumper stickers brandishing the group's slogan, and buttons that capture the essence of a campaign are popular both because they raise money and because they build solidarity among the members of a group. Schools, churches, musical groups, and civic associations frequently sell candy, cookies, nuts, and fruit, particularly during holiday fund-raising drives, when consumers are either looking for gifts or are in a "holiday mood."

As a general rule of thumb, choose items for which you believe there will be substantial demand and on which you can make a profit. You can strike the best bargains by dealing directly with the manufacturers of the items you want. Check labels of products you want to sell or speak with other groups who are selling similar items to locate the name and address of the manufacturers you want to approach. Send the manufacturer a letter introducing your group and asking for a price list that tells you how much the items cost. Then negotiate with the manufacturer for the quantities of the items you wish to order, how you will pay for them (for example, cash in advance, or payment upon sales), and how you want the items customized.

If you can't locate the manufacturers, seek out an "advertising specialist," a middleman who help groups buy merchandise from the manufacturer and customize it with the group's name, logo, or other imprint. (For example, an ad specialist might help a group interested in organizational T-shirts buy plain T-shirts, design camera-ready artwork, and then get the shirts imprinted.) Advertising specialists are usually familiar with the variety of products that are available and know what prices are competitive. If they have pre-existing relationships with manufacturers or printers, they might be able to help the group get the product at a more reasonable rate.

What can you sell? Here are some examples:

Buttons	Books (especially
Bumper-stickers	cookbooks)
T-shirts	Reports
Calendars	Water bottles
Posters	Mugs or cups
Trees	Games
Candy, dried fruits,	Tools
fresh fruit, baked	Magazine subscriptions
goods	Vacations
Holiday crafts	Flowers
Wrapping paper	Appliances

How are you going to sell your merchandise? Approach your membership at meetings and through your newsletter. To sell to others in your community, you can go door-to-door or set up sales tables at fairs or at meetings of school or church groups. You can also approach local retail outlets, appealing to their community spirit to stock your goods and help you raise funds through sales so you don't have to ask them for a direct donation!

Before you set up any kind of sales or merchandising operation, work with your accountant to create a system to allow you to track your expenses and income and maintain an accurate inventory of the products you've bought and sold. You may need to establish a volunteer committee to handle orders, recruit and train your sales force, and collect money.

☞ **CHECK IT OUT:** For more merchandising tips, see *500 Ways for Small Charities to Raise Money,* by Phillip T. Drotning, Contemporary Books, Inc., Chicago, 1979.

CLASSIC T-SHIRT
Available in white, red, purple, green, black, or navy blue. 100% Cotton. M, L, or XL. $11.95

CREWNECK SWEATSHIRT
Classic crewneck design in white with navy logo or navy with white logo. 50/50 poly/cotton. S, M, L, XL. $29.95

NEON BASEBALL CAP
Adjustable with "BCBOBP" printed on front and Baltimore Chesapeake Bay Outward Bound Program underneath. Hot pink, lime green, yellow or orange. 100% nylon. $10.95

CANVAS RUNNING SHORTS
50/50 cotton/poly with logo. Adult sizes M, L. Red or Navy. $7.50

INSULATED MUG WITH LID
No more cold coffee 10 minutes later. Available in red with white imprint. 12 oz. $6.25

CERAMIC COFFEE MUG
10 oz cup. White with blue logo. $4.95

CANVAS TOTE BAG
This tote bag is perfect for books, toys, or beach goodies. Inner envelope pocket for those hard to find items. 13" wide, 14" tall. Expands to 4". Navy. $10.50

ORDERING INSTRUCTIONS

Send order form with check or money order made payable to Baltimore Chesapeake Bay Outward Bound. SORRY NO CREDIT CARDS OR COD'S.

Mail to:
Baltimore Chesapeake Bay
Outward Bound Store
4921 Windsor Mill Rd.
Baltimore, MD 21207

Questions? Call (410)396-0082.

Shipping and Handling

Please add the appropriate fee to your order:

$1.00 to $25.00   add $4.00
$25.01 to $50.00   add $5.00
$50.00 and up   add $6.00

We ship Parcel Post unless you specify otherwise. For UPS we need your street and house number for delivery.

Baltimore Chesapeake Bay Outward Bound - Order Form

Quantity	Item	Size	1st Color	2nd Color	Unit Price	Total

Name_____

Address_____

City_____ State_____ Zip_____

Phone(Day)[____]_____ (Evening)[____]_____

Item Total	_____
MD Residents add 5% sales tax	_____
Shipping and Handling	_____
Total Amount Enclosed	_____

Here's an example of a flier the Outward Bound group developed to sell T-shirts, mugs, tote bags, and other items to its members.

Here's an example of the astounding merchandising success that has allowed one local group to evolve into a sophisticated $200,000-a-year organization.

Local Hell-Raiser
Our Bodies, Our Selves

It was 1969. Twelve women from Boston's Emmanuel College were attending a workshop on women's health.

Wishing to continue the discussions, the women met weekly in an office on the Massachusetts Institute of Technology campus, where several had husbands in graduate school. Their original goal was to compile a list of "good" doctors for women. But after sharing more horror stories than favorable reports, the women realized that doctors who were informed about women's health needs were hard, if not impossible, to come by. All agreed that they would benefit from a solid research effort on topics such as contraception, pregnancy, and sexually transmitted diseases.

Each woman gathered information on a particular topic, then shared it with the group. One by one, they were struck by the gaping hole in medical research concerning women's issues, and by the lack of published information for women seeking answers to everyday health questions. Forming a "collective" group in which they continued their research and discussion, the women compiled their findings into a document they titled "Women and Their Bodies: A Course."

The fledgling New England Free Press printed this brief report on newsprint and stapled it together, the stark title handwritten on the cover. Almost overnight, the Free Press was swamped with calls for the book (even though not a cent was spent on advertising), which at $.75 a copy was a real bargain. Women were hungry for health and medical information written by women rather than by the seemingly unsympathetic, male-dominated health care industry. The document sold an amazing 250,000 copies between 1969 and 1971.

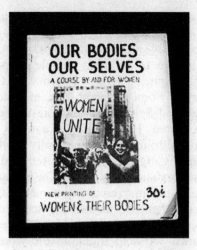

One of the earliest versions of *Our Bodies, Our Selves,* containing invaluable women's health information, diagrams, and photos.

Word of the compilation's success reached larger publishers, and several approached the collective with offers to publish its "best-seller."

But the women, who stress consensual decision making in their hierarchy-free organization, were not pleased by the thought of their grass roots effort being swallowed by a large for-profit corporation. To make themselves "official," they incorporated as a nonprofit and chose the name Boston Women's Health Book Collective. With the help of a woman lawyer, they were able to negotiate and secure an unprecedented contract with Simon and Schuster, the book publishing behemoth, in which the collective retained control over the cover design and advertising copy, set a ceiling on the book's price, and offered a 70 percent discount for clinics and other nonprofit health organizations. The book, renamed *Our Bodies, Our Selves,* hit the *New York Times* best-seller list and stayed there for three years.

Since 1971, this publishing phenomenon has grossed over $1 million. Royalties from the book have been poured back into more programs and additional publishing efforts. *Our Selves and Our Children* came out in 1978, and *The New Our Bodies, Our Selves* appeared in 1985. The collective is constantly working to make its research available to as many people as possible. It maintains the Women's Health Information Center, the largest library of women's health care information in the United States, and fields inquiries from women all over the world. It started "HealthRight," a national newsletter on women's health; founded Community Works, a grass roots fund-raising organization in Boston; and has sponsored conferences and made films. The group's innovation and persistence in overcoming bureaucratic red tape were apparent when it established a health center at a women's prison in Framingham, Massachusetts.

In their own self-help tradition, the women of the collective do their own financial planning and bookkeeping for their annual operating budget of around $200,000. Revenue still comes largely from book sales, but the collective has approached and received funding from several foundations, and the state of Massachusetts helped fund the prison health center. The collective mails a fund-raising letter annually, typically with a solid 10 percent return, receives other donations, and has begun investing in stocks to help batten down its financial future.

ORGANIZING A MARATHON

You can raise substantial levels of funds for your group by organizing a marathon activity that allows donors to pledge a certain amount of money on a per-unit basis. (The "unit" may be a mile walked, a lap swum, an hour danced, and so on.) Volunteers line up sponsors before the event; contributions fulfilling the pledges are collected after the event is over.

Marathon events are appealing because they require very little money up front to be successful. They allow everyone in the organization to participate and make a contribution; they also help build visibility for your group in the community.

Walk-a-thons, bike-a-thons, and dance-a-thons are common, with volunteers getting donors to contribute a certain amount of money—usually less than a dollar—for every mile they cover or hour they spend. But the events don't have to be athletic to work: The National Audubon Society organizes bird-a-thons, in which donors pledge money for each different species of bird its volunteers sight. Aveda hair salons has held cut-a-thons on Earth Day, donating a certain amount of money to an environmental organization for every head of hair it shears.

● ●STEP BY STEP● ●

How to Organize a Marathon

● **Pick a target date.** Try to choose a date that can become an annual event. Consider weather if you're planning an outside function, and set rain dates if necessary. Contact your mayor's office for information about conflicting events. Usually, Saturdays, Sundays, and holidays offer good opportunities to hold events like these because you get so many people who would probably be at work if you attempted to schedule the event on a business day.

● **Select a location or route that is centrally located, easily accessible, well-known, and near adequate parking facilities.** A route for a walk-a-thon should be fifteen to twenty miles long; for a bicycle ride, around forty miles. Choose a safe route; avoid narrow streets with heavy traffic or routes with left turns, especially at dangerous or busy intersections. Walk, drive, or bike the route yourself before making it official to insure that it is easy and safe. The route should also pass by parks, school grounds, or other open spaces with public facilities, or business establishments like gas stations that may give permission for marchers to use rest rooms. End where the route began so those who have driven will be near their automobiles; send a copy of the route and date to the local police department in advance so that you can obtain any permits you need.

● **Line up volunteers who can help recruit people to participate, coordinate the pledges, and collect the money.** You'll have to develop sponsor sheets so that volunteers can write in the name and address of their sponsors on standardized forms. If prizes are to be awarded, describe them on the sheet to provide added incentive to raise money, including background information about your group as well.

How to Organize a Marathon (cont.)

- **Publicize the event to the local media.** Try to line up a local TV or radio station to co-sponsor the event with you. A radio station can broadcast live from the event and increase participation.

- **Solicit donors who care about the issue or are interested in protecting the community.** Ask them to pledge a minimum amount per mile or per unit. Also target local businesses. Rather than get them to pledge a minimum, ask them for a set amount; perhaps, in exchange for a high contribution, the sponsor will receive a substantial amount of publicity. Encourage each business to buy an ad in the program to be used at the awards ceremony. Prices can vary depending on the size of the ad and location in the program.

- **On the day of the event, make sure you have enough volunteers.** They should be recruited from the ranks of your own membership as well as from churches, civic groups, and other community-minded institutions.

- **To ease registration, provide at least five tables per thousand walkers expected.** Space them widely and offer check-in alphabetically so that walkers can register efficiently.

- **Establish checkpoints every mile or two along the route (every five miles for a bike-a-thon).** Open spaces with public facilities—like parks and school grounds—make the best checkpoints, though gas stations work well also. Staff with at least two volunteers. Stamp all cards to verify mileage. Provide water, foot powder, and miscellaneous refreshments.

Be sure checkpoint volunteers have access to a phone and a list of emergency phone numbers. Mark well with large signs and posters. Provide vehicles for those who are unable to complete the entire course. Vehicles should travel the route continuously, picking up the dropouts and returning them to the final/starting checkpoint. Arrange for a doctor and nurses to be present. Have a cleanup crew dismount check-in tables and clean up all litter.

Finally, don't forget to collect the pledges. You can either ask participants to collect the pledges from their sponsor or bill each sponsor directly, in which case you'll need to send a letter and bill with a business reply envelope.

How to Organize a Marathon (cont.)

Monitor the returns. After ten days, send a reminder card to those who have not yet contributed. Follow up by telephone if necessary.

Be sure to send a thank-you note to each contributor. All participants in the event should receive a certificate of appreciation.

WORKPLACE FUND-RAISING

Every year, one small but significant action—taken by thousands and thousands of people—raises hundreds of millions of dollars for important social causes.

The action is payroll deduction, a technique most commonly associated with United Way and adopted by hundreds of other social service, human welfare, and environmental agencies to encourage employees to donate a small portion of their paychecks to worthwhile efforts.

The strategy is brilliant in its simplicity. On an annual basis, a company allows its employees to designate as much money as they wish to be deducted from their paychecks (after taxes) and donated to a charity that is participating in the company's workplace campaign.

In most cases, the charity will have had to meet certain criteria to be allowed into the campaign. In addition to being a 501(c)(3) organization, it may need to keep fund-raising and administrative costs below 25 percent of total support and revenue; may need to be audited annually in accordance with generally accepted accounting principles by an independent certified public accountant; and may need to have been in existence for a certain period of time prior to being accepted into the workplace campaign.

Organizations and institutions that belong to United Way enjoy the lion's share of workplace donations, but getting your organization accepted into the United Way fold can be tough if your work doesn't fall under the strictly "health and human services" rubric around which United Way is organized.

However, more and more organizations are forming independent federations to petition companies to expand their workplaces and allow new charities into their workplace campaign program. It's a win-win proposition, since the more choices employees have, the more donations they make—improving the company's image in the process.

In 1991, the Council of Federations was formed to serve as the cooperative voice of more than five hundred charitable agencies and local organizations wishing to provide employers with an efficient, effective way to revitalize their workplace giving campaigns and respond to growing employee interest in wider choices for charitable contributions.

Charter members of the Council of Federations include the Combined Health Appeal of America, Earth Share, International Service Agencies, the National Alliance for Choice in Giving, and National/United Service Agencies. Their members represent organizations and institutions committed to addressing long-term solutions to a broad range of community, family, health, environmental, international relief, and economic concerns.

If your group is not currently participating in a workplace program or is not eligible for membership in your local United Way, investigate your options. Urge your members to promote payroll deduction programs where they work. If your group cannot participate on its own, apply for membership at an appropriate local or national federation. Then work with the federation you join to encourage employers to expand their giving programs to include your group, promote substantial awareness of your organization among employees, and maximize the contributions you receive.

 CHECK IT OUT: The Council of Federations has developed guiding principles for corporations to use in expanding their charitable campaigns. For a copy of the principles, or for more information about joining a federation, please contact the Council of Federations, 12701 Fair Lakes Circle, Suite 370, Fairfax, VA 22033.

JOINT VENTURES

In addition to asking companies for grants and employee contributions, many nonprofits tap corporate marketing departments, developing joint ventures that create substantial benefits for the organization as well as the company.

One of the best known corporate/nonprofit ventures blossomed in 1984, when American Express collaborated with the nonprofit group managing the restoration of the Statue of Liberty to raise $1.7 million. The alliance was the very model of simplicity and ingenuity: During a period of three months, American Express donated $1 of every item purchased with an American Express card, and $5 from each new card application, to the statue's restoration fund. The fund got to keep the money, and American Express got to reap the marketing benefits when new card applications rose 17 percent and usage of the card itself increased by 28 percent.

As the American Express example proved, corporations profit from relationships with nonprofits in ways that money alone can't buy. Affiliating with a nonprofit gives the corporation instant credibility and access to a group of consumers it might never reach on its own. And joining forces with a citizens organization helps a company generate favorable publicity for itself and build a positive image in the community.

Nonprofits benefit from participating in joint ventures, too. Apart from the obvious opportunity to earn additional income, working with a corporation can increase the credibility of the organization among other companies (which may lead to other partnerships), give the group more clout in the political arena, and generate visibility for the group that will boost its membership recruitment efforts.

In their excellent book *Doing Best by Doing Good: How to Use Public-Purpose Partnerships to Boost Corporate Profits and Benefit Your Community* (Dutton, 1992), Dr. Richard Steckel and Robin Simons describe several forms a joint venture can take:

- **Cause-related marketing** links the benefit to the organization directly to the sale of the company's product or service (for example, a bicycle equipment manufacturer might agree to donate a dollar to a bicycle club for every water bottle it sells).

- **Sponsorships,** particularly of fund-raising and educational events, help raise visibility for the sponsor while giving an organization the financial or promotional backing it needs to get the event off the ground (in Washington, DC, local radio stations take turns sponsoring annual fund-raising marathons for various health charities).

- **Premiums** allow corporations to develop new products or adapt old ones whose sale will either generate income for a group directly or, if given away for "free," will encourage the recipient to become a member of or contributor to an organization. (When Dr. Steckel was helping to pull the Denver Children's Museum out of a big financial hole, he worked with StarKist/9-Lives to develop an educational booklet about pets. Sixty-two thousand copies of the booklet—whose cover included the 9-Lives trademark and whose foreword was written by the company's CEO—were then sold to StarKist for $1.25 each, and the company distributed the premium to families and veterinarians around the country. The museum pocketed a profit, while StarKist established credibility with pet doctors and consumers.)

- **Licensing** gives groups the opportunity to rent their name or logo to a company to apply to a product, in the hopes that such a ploy will boost product sales (hence the intense competition for the Olympics logo).

- **In-kind donations** can include printing, postage, paper, access to equipment, the equipment itself, and a wide variety of services.

How do you go about creating a joint venture with a company? Vanguard Communications faced exactly that question several years ago when American Forests (AF), a nonprofit group based in Washington, DC, asked us to help identify potential corporate partners for a variety of the group's tree-planting activities. Here's what we did:

- **Targeted companies.** We generated an initial list of companies we felt would be interested in trees and tree planting. These included paper manufacturers, automobile manufacturers, oil companies, restaurant chains, and other corporations with an interest in the environment.

- **Identified individuals at the companies.** Using several reference books available in libraries, we located the headquarters of each company and identified the name of the individual in charge of marketing and promotion (sometimes that person is a vice president, other times a director or manager of marketing and/or promotion or public affairs; the title will depend on the company).

- **Wrote letters.** We compiled a persuasive packet of information about the organization and its tree-planting program, known as Global ReLeaf. The kit included a brochure describing the role that trees play in reducing global warming, news clips we had generated about Global ReLeaf, information about American Forests, and a cover letter explaining the many marketing opportunities that existed for any company that affiliated with Global ReLeaf.

- **Made telephone calls.** After mailing the packets to our list of targeted individuals, we followed up each letter with a personal telephone call to make sure that the packet had been received and opened, and to gauge the interest of the recipient.

- **Set up appointments.** We then set up meetings for the Global ReLeaf staff to go in and meet with representatives of several of the companies to begin to talk about the kinds of collaborations that could occur. Global ReLeaf staff remained flexible; some companies were interested in premiums, others in cause-related marketing.

- **Agreed on some potential areas of collaboration.** If, after reviewing its options internally, a company decided to go forward, Global ReLeaf staff secured commitments for collaborations and began to close the deal.

- **Did not let the agreement slip through the cracks.** Global ReLeaf staff continued to monitor the process, providing additional information to the company (particularly in the form of statistics about the benefits of planting trees), working on the publicity plan for the campaign, and monitoring the campaign's success.

In any city or town, there are plenty of businesses with whom you might be able to create a joint venture. When scoping out potential partners, keep your organizational goals and ethics in mind. Don't work with any companies that create a conflict of interest for you just because you need the money—the relationship may end up generating more negative publicity or ill will than the money you make is worth. Do you respect the company you're thinking of doing business with? Is the company respected in the community? Is the company guilty of any legal violations? Will associating with the company in any way impinge upon your credibility with the public, your members, and your donors? Answer these questions honestly before plunging ahead. Once you decide to go forward, however, make as much out of the relationship as you can.

☞ **CHECK IT OUT:** *Filthy Rich & Other Nonprofit Fantasies: Changing the Way Nonprofits Do Business in the 90s* (Ten Speed Press), by Dr. Richard Steckel with Robin Simons and Peter Lengsfelder; and *Doing Best by Doing Good: How to Use Public-Purpose Partnerships to Boost Corporate Profits and Benefit Your Community* (Dutton), by Dr. Richard Steckel and Robin Simons. You can also contact Dr. Steckel at AddVenture Network, a Denver-based firm that works with nonprofits and companies to develop joint ventures and "public-purpose marketing" strategies, at 1350 Lawrence Street, Plaza 2H, Denver, CO 80204.

PLANNED GIVING

If your donor list includes repeat givers or wealthy prospects who are nearing estate-planning age, your fund-raisers should consider asking them to participate in a planned-giving campaign.

Planned gifts (formerly known as deferred gifts) are donations of money, property, or stock that people give using special tax-advantaged techniques like setting up trusts through their wills. Donors receive tax benefits and the security of knowing these designated assets will accrue to their favorite charities when they die. Charities are then able to borrow and plan based on the significant gift they will receive in the future.

But before you establish a planned-giving program, take a hard look at your group's future. Will you be operating, not just ten years hence, but fifty or even one hundred years from now? Your board or fund-raising committee must be prepared to convince your legal and financial advisors as well as your prospective donors that your organization intends to endure and merits such support. A donor who establishes a trust to benefit a charity would be foolish to do so with an advocacy group whose work will abruptly end when its goal is reached—when a disease is cured or a nuclear reactor is shut down, for example.

Once you have decided that your group has staying power, the types of planned gifts fund-raisers can offer to established donors are numerous.

Bequests are the most popular option. Through a will, a person can leave an exact dollar amount, shares of stock, or even an entire estate to a charitable [501(c)(3)] nonprofit group. Donors benefit by making a major gift and by reducing any estate tax burden that may eventually befall their heirs. Organizations then dispose of the assets and receive the money (heeding any restrictions the donor set on how the assets are to be used).

Gift annuities, which are financial contracts written by a nonprofit, generate revenue back to the donor and beneficiaries until the death of the last beneficiary, when the donated funds become the sole property of the nonprofit to use for its purposes.

Charitable trusts receive contributions of cash, stocks, or property essentially in trust for the charity while generating fixed- or variable-interest income for the donor. The contribution, which provides a federal income tax deduction for the donor in the year the gift is made, may be made for life or a limited period of years. When the trust ends, the nonprofit receives the assets remaining in the trust.

Charitable lead trusts are the opposite of charitable trusts: The donor's gift is used by the charity for a period of time, after which the remainder goes back to the donor's designated beneficiary.

Donors can also opt to invest their gift in a diversified portfolio called a *pooled income fund.* The fund will normally have many donors/investors, each of whom receives his or her share of net income. As individual beneficiaries die, the nonprofit managing the pooled-income fund receives the donor/investor's share of the fund.

How to Proceed

Seek advice. Financial planners, lawyers, and accountants can help you set up a planned-giving campaign appropriate for your organization. Ask another charity that solicits planned gifts to recommend a firm. Your group's financial and legal counsel will also come in handy, as will lawyers and bankers retained by donors interested in making a planned gift to your group.

Promote your campaign. Place announcements and articles in your newsletter, mail a brochure explaining the campaign to potential donors, and advertise the availability of your campaign in publications that might reach potential donors. Consider launching your campaign around April 15 when many people are extremely anxious to find alternatives to paying "Uncle Sam" and the Internal Revenue Service, or in conjunction with a major fund-raising drive, such as an effort to renovate a building or establish a scholarship fund.

Acknowledge donors appropriately. Only some will desire public acclaim; be sure to ask. If they do, mention them in your newsletter or annual report and present them with a certificate. Or engrave their name on a plaque that hangs prominently in your offices, adding the names of new major donors to the plaque each year. Make sure to provide an accurate receipt and complete all forms necessary to finalize the gift.

FUND-RAISING
DO's and DON'Ts

- **Do** start raising money as soon as you can. Pass the hat at the first community meeting, and at every meeting thereafter.

- **Do** budget (some) money to raise money.

- **Do** appeal to people's self-interest.

- **Do** give people something to "buy," whether that be a gift for themselves or others or membership in your group.

- **Do** ask again.

- **Do** say thank you. Promptly send thank-you notes to or telephone those who give. Don't forget to express your gratitude to the workers who made your fund-raising event successful.

- **Don't** be afraid to ask.

- **Don't** spend more money than you raise.

- **Don't** accept or pursue a donation that would restrict the choices available to your organization to pursue its own issue agenda. If you take money from a person or corporation you are attacking, make sure you both understand that the donation will in no way affect your adversarial relationship.

- **Don't** overlook "in-kind" contributions— donations of services, equipment, facilities, or supplies—that help stretch your fund-raising dollars.

THE BUILDING BLOCKS

Part I focused on specific steps that will help you establish a sturdy foundation upon which to build your organizing strategy. Those steps include picking a winnable issue, setting short- and long-term goals, identifying the most appropriate targets for your actions, employing basic organizing tactics, building a sound organization, and generating the money and resources you need to keep your campaign moving forward.

With such a foundation in place, you should now be ready to add the "building blocks" that will allow you to pursue your goals with some reasonable chance of success. These building blocks fall into three basic categories: communications, lobbying, and action at the polls.

Chapter 6, "Communications," focuses on the myriad tactics you can use to generate favorable media coverage of your issues and then to use that coverage to pursue your organizing goals. As you'll see from the stories retold in these pages, securing publicity for an issue or a campaign is rarely the raison d'être of the campaign itself. Rather, organizers employ communications tactics to help move along other elements of their organizing strategy.

Such is not necessarily the case with lobbying or action at the polls. Chapter 7, "Lobbying," describes the processes you engage in to create, amend, or defeat legislation locally and nationally, while Chapter 8, "Action at the Polls," presents tactics that will help should you decide (or be forced) to take your issue directly to the public for a vote, to elect candidates to office, or to pursue election yourself. Both lobbying and political action may be used as tactics to pursue a specific goal—or may become organizing goals themselves. The accompanying stories offer examples of both.

*"We are trustees of the future.
We are not here for ourselves alone."*

—JAMES BRYCE, "Trustees of the Future" (1913)

6.
COMMUNICATIONS
GETTING THE WORD OUT

DANIEL SCHORR, THE renowned radio and television correspondent, once observed: "If you don't exist in the media, for all practical purposes you don't exist."

The noblest deeds, the most ambitious plans, the greatest successes may come and go with nary a whisper if they aren't somehow anointed with the scepter of sustained favorable publicity.

Few hell-raising campaigns intentionally avoid the media spotlight. It's more often the case that many hell-raisers fail to get the public credit they deserve because they don't understand how to develop and implement an effective media strategy.

Fortunately that wasn't the case for a band of dedicated wildlife preservation activists in Kansas, whose clever media tactics gave their lobbying campaign the flair and momentum it needed to win. Read on.

ACTIVISTS in ACTION ➤ ➤ ➤ ➤

Saving Cheyenne Bottoms

Cheyenne Bottoms is the largest freshwater marsh in the interior United States. Its 12,290 essential acres emanate from almost the exact center of Kansas. But despite its classification, the wetland rarely receives enough rainfall or runoff to compensate for the amount of water that evaporates from the area. Ironically, Cheyenne Bottoms is dry more often than wet.

When the Bottoms is dry, it takes a patient naturalist to appreciate the critical role an ecosystem like this one plays in nurturing the hundreds of waterfowl species that visit each spring and fall. But, says Jan Garton, co-chair of the Save Cheyenne Bottoms Task Force and a member of the Northern Flint Hills chapter of the National Audubon Society, "when it has water, it's an extraordinary place," whose value is lost on few.

Rain showers can convert the Bottoms into a lush oasis, reviving the heart of the great American plains. Vistas of calm water, bordered by long marsh grasses and cattails that sway to the reedy whistles of red-winged blackbirds, stretch on for miles. Both the water and shore teem with wildlife: turtles, muskrat, deer, even a rattlesnake or two. And everywhere, birds, arriving in stages Nature has spent eons timing: ducks, geese, and cranes in early spring; sandpipers, plovers, and other shorebirds in April and May; blackbirds and orioles in the summertime.

The entire area plays a vital role in the Central Flyway system, the chain of wetlands that provides rest and refueling stops for birds migrating from the northernmost reaches of Canada to the tip of South America. In the spring, 45 percent of all U.S. shorebirds can be found at the Bottoms, and five species of shorebirds send 90 percent of their population through the welcoming Kansas marsh. According to International Shorebird Surveys, Cheyenne Bottoms is the single most important migratory area in the United States, and may be the most important migration stopover in the entire Western Hemisphere.

What makes the Bottoms so important, Garton says, is the birds' migratory habits. "A lot of these shorebirds fly until their energy runs out, a few thousand miles. If Cheyenne Bottoms is not there for them to stop and 'refuel,' they're in trouble." With wetlands disappearing at the alarming rate of nearly 500,000 acres annually, other marshy rest stops have literally dried up. Birds without the energy to keep flying and look for another place to feed could die.

In 1957, the state of Kansas purchased Cheyenne Bottoms and tried to convert it into permanent marshland by diverting water from other rivers. But local irrigation projects were literally draining the region of its water; it looked like the Bottoms was in danger of joining the grim statistics underlying the fact that half the wetlands in Kansas and the United States have been destroyed in the last two hundred years.

The area needed a significant—and expensive—engineering overhaul if it was going to meet Mother Nature's needs as well as those of farmers. But considering how few Kansans even knew the marsh existed, a major campaign would be needed to build support for Cheyenne Bottoms and motivate the legislature to spend money to save it. The Northern Flint Hills chapter of Audubon became involved in 1983, when Garton and other members decided someone needed to save Cheyenne Bottoms before it was too late.

The group knew that task number one was publicity: Their campaign to save Cheyenne Bottoms would succeed or fail depending on the amount of public support it engendered. Deciding to spend the first year working to increase awareness, the activists launched a media campaign in the fall of 1983, even before they sought the help of other conservation groups. A poster and brochure—tried-and-true communications basics—were designed to introduce as many people as possible as quickly as possible to the area and to recruit volunteers. The more unusual and creative elements of this campaign came later.

Art and Design for Cheap

Garton, a patient, organized person who works part-time for United Parcel Service and draws and paints in her free time, became the group's unofficial artist, designer, and writer.

To capture the muted beauty of the Bottoms on a poster, Garton sketched an American avocet, a waterfowl species that frequents the Bottoms, and accompanied it by straightforward text: "Cheyenne Bottoms is drying up. It needs your help. Speak up. Speak out. *We can save Cheyenne Bottoms.*"

The statewide Kansas Audubon Council covered the cost of printing five hundred of the 14-by-22-inch posters. Members distributed the posters, colorfully printed in hues of black, rust, and blue, through the nine local Audubon chapters in Kansas and the state's Wildlife and Parks Department. They also hung them in schools, libraries, the state university biology department, and other places that display informational material— "Wherever there was an environmentally friendly person," Garton said. The following spring, members presented a framed copy of the poster to Governor John Carlin to encourage him to start considering the Bottoms's problem and to let him know that efforts to preserve the area had crystallized.

Jan Garton joins representatives from five Audubon chapters in presenting the Save Cheyenne Bottoms poster to Governor John Carlin in March 1984, six months into the campaign.

Garton also created a brochure to provide more details about the Bottoms and motivate people to join the campaign. The simple but organized four-panel design consisted of crisp line art and typeset text with boxed quotes from authorities such as a Kansas Fish and Game biologist. The simple brochure, printed in brown ink on inexpensive yellow recycled paper, encouraged readers to help preserve the wetland by contacting legislators, joining a conservation group, or buying federal duck stamps at the post office. Interested readers could fill out a tear-off coupon to return to the Save the Bottoms office (also known as Garton's home) to receive more information. The outside back panel consisted of a self-mailer with a return address and the Save the Bottoms logo.

If the volume of brochures printed and distributed is any indication of success, then the Bottoms piece was a real winner. When a substantial first print run of ten thousand copies (again funded by Audubon) was quickly exhausted, high interest in the area eventually warranted two more printings.

Since they had not yet compiled a concrete mailing list, the Audubon activists relied on displaying the brochures in public places for passersby to pick up. They dropped them off at libraries and schools—outlets that, more than supermarket counters, would attract people likely to be interested in the cause. They also put copies in the Wildlife and Parks Department offices and in offices of environmental organizations. In a particularly insightful move, the activists contacted the state's tourist agency and sent brochures to visitors' centers on the Kansas Turnpike so people visiting the state would read about Cheyenne Bottoms and feel motivated to help. Garton said that twenty or thirty people on the Bottoms mailing list became involved because they noticed the brochure at a visitors' center, and a number of them sent money or volunteered to work.

Later in the campaign, Bottoms bumper stickers and mugs were produced and given away or sold. The group also produced an attractive and inexpensive newsletter, which they used to keep coalition members and other supporters informed about the Bottoms and their efforts to protect it.

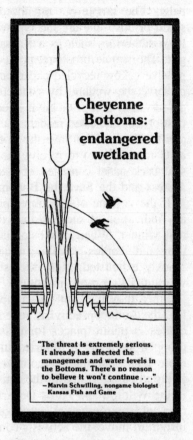

This poster, brochure, and bumper sticker were used to educate the public about Cheyenne Bottoms.

Hunters and Hikers

The activists spent the winter of 1983 building the Save Cheyenne Bottoms Task Force. The most important function this coalition served was to "unite a number of environmental and conservation groups that had never worked together in the past, and had often viewed each other as adversaries," Garton said. Groups with hunting interests, for example, usually clash with Audubon and other wildlife preservation organizations, but these differences were transcended in the interest of conserving the Bottoms. "There were some rivalries, jealousies, suspicions. We got to know each other, got to unite around a cause," she said. Eight groups joined forces, including the Kansas chapters of the Audubon Council, Ducks Unlimited, the Ornithological Society, the National Wildlife Federation, the Natural Resources Council, the Sierra Club, the American Fisheries Society, and the Wildlife Society. They set two goals: to preserve a natural wildlife habitat, and to maintain the economic benefits surrounding communities derived from recreational and sporting use of the Bottoms.

Coalition members began writing to their state legislators that winter. Though they did not yet have a specific proposal or bill drafted, they wanted to lose no time in educating their target audience. "We weren't telling them to do this or that about it, we were just saying please help save the area," remembers Garton. Believing that a concentrated, broad-based educational effort would effectively inform legislators and mobilize the public, activists began planning a Saving Cheyenne Bottoms conference to be held in September 1984.

The Opening Pitch

Garton took charge of publicizing the conference. She secured lists of all the daily and weekly newspapers in the state via the reference section in her local library and the West Central Regional Audubon office in her hometown of Manhattan, Kansas. Hoping to boost attendance from the public and the media by getting reporters to feature the Bottoms before the conference occurred, she wrote a press release that, in one of the campaign's largest mailing efforts, she and her fellow task force members sent to every paper in the state. The results were stunning. During the week prior to the conference, newspapers and even a Wichita television station filed stories about Cheyenne Bottoms and advertised the event. Seven or eight reporters attended the conference and wrote stories, and another TV station and the Kansas affiliate of National Public Radio toured the Bottoms and ran stories. Environmental and other interest groups were contacted about the conference and encouraged to attend as well.

However they heard about it, two hundred people showed up at the two-day gathering at a community college near Cheyenne Bottoms to witness firsthand the dry, cracked mud floor of the marsh and to listen to the eerie silence—there were no birds. No more compelling evidence was needed to convince conference attendees that the marsh needed help. State wildlife officials who were involved in managing the Bottoms discussed the increasing conflict over how the region's water resources should be used, while wildlife groups described the area's unique ecology. Each speech went into a booklet that was given out to registrants as well as to reporters to make their reporting job easier and, it was hoped, yield more comprehensive, accurate stories.

Focusing the Campaign

Ironically, the sophisticated water management information presented at the conference convinced the task force that it needed even more facts about the Bottoms. A professional study of the area and its inadequacies, along with suggestions for renovations, was essential if any progress was to be made. Securing state funding for this study—in a state where public money had never been used to preserve wildlife—became the crux of the Save Cheyenne Bottoms Task Force's battle for the next two years, and the basis for a stepped-up publicity campaign.

Now was the time to nurture press contacts. The editorial writer at the *Eagle-Beacon* in Wichita, one of the state's major papers, was sympathetic to the cause, so the task force sent press releases and fact sheets to him regularly and could usually count on an editorial. In addition to the newspaper lists, Garton maintained a list of contacts at the state capitol and a list of "unique" outlets such as National Public Radio. A writer for one news service whose work ran in several state papers saw the press release, attended the conference, continued to call Garton for information, and wrote several stories. Garton described screening his stories to decide if he was in fact a supporter, and then sharing information regularly with him once she decided he was. "We learned to be cautious. You don't tell people your strategy, you play just like a politician. Don't tell the media what you really think; save your feelings for your inner group," she said. The key, she summarized simply, is never to say anything you don't want to see in the paper.

A Pillow Fight

In February 1985, the activists had a brainstorm: Why not present Save Cheyenne Bottoms seat cushions to the state lawmakers they were lobbying for the study money? The media event would give reporters a highly photogenic opportunity to cover the campaign, allow activists to interact on a personal level with their legislators, and give the legislators something to smile about. As it happened, the Denver Broncos football team had canceled an order with a local vendor for orange stadium cushions before the cushions were printed, so the vendor happily unloaded them on the task force at a significant discount. Though it was not the color of choice, orange was so bright and noticeable that it proved to be a terrific background for the Save the Bottoms logo and graphic—a bird rising from a marsh. The cost of purchase and printing was covered in part by the cushion sponsor program: Bottoms supporters were encouraged via the group's newsletter and word of mouth to spend a minimal $3 and sponsor a cushion that they could present to their legislator in person. Out of 165 legislators, 60 to 70 were handed a cushion by one of their constituents; the task force sponsored the rest.

While Garton coordinated media outreach, a volunteer phoned each legislator and set up a time to deliver the cushion. Garton mailed a press release to reporters who covered the state legislature announcing the presentation and accompanying press conference. On February 12, the activists gathered at the capitol and gave their respective policymakers their pillows—telling them: "We'll save your bottoms if you save ours!" Volunteers then met with each legislator to discuss the importance of funding the proposed study. The group was rewarded with highly favorable newspaper coverage, stories on two local TV stations, and several radio interviews.

In another strategic media move, Garton requested and was granted a half hour interview on a Kansas radio station that received good statewide coverage. The show's agricultural focus attracted many farmers who, in need of the Bottoms's water for irrigation, opposed renovating the wetland. "It targeted an audience we wouldn't have had a chance to contact otherwise," Garton explained. "They needed to know our view of the issue."

Getting the Word into the Streets

Though few Kansans seemed to oppose preserving the wetland, it still took the Save Cheyenne Bottoms Task Force a year and a half to convince the state legislature that the area was worth any expenditure—let alone the $200,000 it was projected to cost. When a bill to fund what amounted to only a portion of the study passed in the spring 1985 legislative session, the group had to mobilize again in 1986 to squeeze even more money out of the government to finish the study. The now-seasoned band of advocates hit on another clever idea. To put more pressure on legislators and give the entire campaign a "shot in the arm," they needed a single event that would reach millions of people in towns across the state. Why not organize a two-hundred-mile relay run from Cheyenne Bottoms to the capitol using a vial of Bottoms water as a baton?

On March 22, 1986, the first runner and supporters gathered at the Bottoms for the starting gun. Each of the forty-five participants ran between one and ten miles before handing the vial off to the next runner. Two hundred miles, three days, and many tired feet later, the final runner sprinted up the capitol steps and presented the vial of water to Governor Carlin. Like the cushions, the run was timed to precede the spring legislative session, so members of the coalition presented each legislator with his or her own small brown jug of Bottoms water and again met with them to urge their approval of funds to complete the study.

The task force held another press conference at the capitol shortly after the race was finished. They had sent press releases to the capitol reporters and to media in the towns along the run's route, and response was tremendous, with local papers printing photos of their townspeople jogging alongside the designated runner. "It gave people a chance to show that they were supporters, building a network of activists working together," said Garton. The event generated a chain of coverage that followed the race across the state to Topeka, where the larger city media took over.

At the end of the spring 1986 legislative session, the Kansas state legislature approved Governor John Carlin's recommendation that $60,000 be allocated to complete the Cheyenne Bottoms feasibility study. Task Force members breathed another sigh of relief the following fall, when the study was finally completed.

The untiring activists continued to keep a watchful eye on the Kansas Wildlife and Parks Department to insure that the study's recommendations included a comprehensive renovation plan for the marsh. They did; following the study's recommendations, the Wildlife and Parks Department began overseeing the renovations to the marsh and the creation of additional marshland. Renovations (to compensate for lack of rainfall, as well as ongoing irrigation projects in the area) include digging canals and pools to allow water to move into the marsh more freely, constructing dikes, and installing water distribution structures such as pump stations. The improvements, scheduled to be completed around 2001, will enable Cheyenne Bottoms to handle the water it receives from rainfall and runoff more efficiently so that, even during drought years, wetland habitat will be available for wildlife.

The Bottom Line

The elements of this publicity campaign—poster, brochure, media events, bumper sticker, newsletter—were consistently deployed in a tactical way to keep the task force's overall strategy moving forward. Regular press contact, well-conceived press releases, and clever publicity hooks kept Cheyenne Bottoms in the Kansas spotlight for three years; without it, the state would not be improving Cheyenne Bottoms today.

The Cheyenne Bottoms campaign left its imprint on Kansas in other important ways, too. For one, Kansans today are a good deal more savvy about wetlands and their importance to the region and the continent. But perhaps more important, there is now a strong, experienced environmental voice in the state that did not exist before.

WHY IT WORKED

Organizers of the effort to save Cheyenne Bottoms ran an impressive media campaign that succeeded on three levels. It informed and mobilized the public, educating people about the issue and getting them to take action at the right time. It shaped the policy agenda in the state legislature, forcing lawmakers to deal with a problem they initially knew nothing about. And it strengthened the group itself, transforming a disparate band of environmentalists into an effective coalition ready to take on other fights.

The campaign was able to accomplish so many challenging goals because it revolved around a well-conceived and well-implemented strategy that never lost its focus. Here's a quick review of the steps you should take to develop a media strategy for your organization.

Developing a Media Strategy

In devising an effective communications strategy, ask yourself the following questions:

- **Does the media campaign reinforce your group's overall agenda?** Your media strategy should fit hand-in-glove with your organizing goals and objectives. If you have created an organizing blueprint (see Chapter 2), refer to it frequently to keep yourself focused on the point of your organizing efforts. If given a choice between two events or activities, choose the one that will move you most readily toward your goals.

- **What organizational resources can you commit to your media program?** Define resources broadly—not just in terms of money, but also in terms of volunteers, staff, a newsletter or other publications you produce, computer capabilities, and of course, time. There is no simple prescription for the percentage of resources you should devote to your organization's media efforts; it will probably increase when you are staging specific media events and other labor-intensive activities, and decrease when you are primarily maintaining your program. At a minimum, communications should always be included in overall organizational strategic planning.

- **What time constraints do you face?** Be realistic about what you actually have time to pull off. Because the Kansas legislature meets for only ninety days each year, the Bottoms activists had a substantial amount of time in between legislative sessions to plan their media events and build relationships with reporters that they could tap when votes on funding for their study were pending. On the other hand, because so much time did lapse between legislative sessions, Jan Garton and her colleagues had to make sure their media events kept the Cheyenne Bottoms story alive for lawmakers. Use the media to get your message out at the times most critical for mobilizing volunteers and pressuring elected officials.

- **Are you reaching your target audiences?** Obviously, the media are one target audience; you'll know if you're successfully educating reporters by the frequency with which and the way in which they cover your stories. But you also use the media to reach other target audiences, such as legislators, government officials, and the public. You'll get a sense of how much support you have among legislators by how they vote; among government officials by how responsive they are to your concerns about the enforcement of laws and regulations affecting issues you care about; among the public at large by the number of people who attend events you organize, contact their elected officials, give you money, or join your group.

The Cheyenne Bottoms Media Strategy

Here's how the Cheyenne Bottoms media strategy unfolded:

GOALS

- To protect Cheyenne Bottoms by getting the state legislature to allocate government funds for a study on the disappearing wetlands.

- To conduct the study and renovate the wetlands.

ORGANIZATIONAL CONCERNS

- Inadequate staff and resources; solve by building coalition to mount campaign and augment resources available to win campaign.

TARGETS

- State legislators.
- Reporters.
- Public at large.

TACTICS

- Develop posters, brochure, bumper stickers to educate public.

- Stage media events to educate reporters, alert the public, and pressure legislators (those events eventually included a conference and presenting legislators with Cheyenne Bottoms seat cushions and, later, vials of Cheyenne Bottoms water).

- Build lasting relationships with reporters: disseminate news releases, hold news conferences, feed reporters story ideas.

- Develop a newsletter to keep coalition members informed.

TIME FRAME

- Three years (until study was completed).

The activists modified their tactics as the campaign evolved. For example, when they began, they had no way of knowing that it would take them two and a half years to get the study completely funded, or that they would need to plan multiple media events to influence legislators and the public. Still, what never changed was their understanding of the role the media could play in helping them achieve their goals and objectives. The question was never about whether they would generate publicity—but how.

☞ CHECK IT OUT: *Strategic Communications for Nonprofits* is a set of nine guides explaining media and communications techniques and technologies that contain over four hundred pages of case studies, how-to's, and strategic approaches reflecting the successful communications experiences of many nonprofit organizations. To order the complete set, contact the Benton Foundation, 1634 I Street, N.W., 12th floor, Washington, DC 20006, $50.00.

Your local library will also carry manuals and books on public relations that will provide additional information on dealing with the news media. Many of the larger nonprofit organizations in your area are likely to have materials as well. Check hospitals, community colleges, United Way offices, and your local Chamber of Commerce. Also, many community colleges and adult education classes offer courses in public speaking and media relations. Consider signing up for one, if only to hone your skills.

HOW THE MEDIA WORK

"Just by reporting it, how they report it and how they play it—emphasizing one angle over another, using it at the top of page 1 or as the lead story on the evening newscast—the media can help shape or crystallize or accelerate public response to a story," wrote reporter David Shaw in "Media Impact: Why Some Stories Have It, and Others Don't" in the *Los Angeles Times* (October 26, 1992).

They can also bury your story so that the public never has a chance to give it a second thought. You can improve the odds that reporters will give your story the kind of coverage it deserves by understanding how reporters and editors decide what to cover and what to ignore.

What Is Newsworthy?

Your chances of getting in the papers or on TV are greatest when your story is considered "newsworthy." To most editors and reporters, this means your story has the following characteristics:

- **It is new.** (Remember the Mothers of East L.A.? No one could remember the last time they witnessed a pack of mothers in white scarves marching across a bridge to oppose a prison.)

- **It is timely.** In other words, it connects to another local, national, or international story to make yours seem more relevant. (You will make your story seem more timely if you link or "peg" it to the release of controversial reports, an upcoming vote, or some other event that you expect to garner substantial media attention—for example, the spectacular sight of thousands of birds arriving at the marsh gave Cheyenne Bottoms supporters a wonderful publicity opportunity to focus attention on the region's ecological significance).

- **It is colorful.** It can be accompanied by compelling visual images, either photographs or videotape.

- **It involves substantial numbers of people.** Stories that impact hundreds, thousands, or millions of people usually get more play than those that affect just a few.

- **It revolves around a crisis, conflict, or catastrophe.** The media thrive on other people's misfortunes—and from a reporter's point of view the more awful, outrageous or unbelievable it is, the better.

- **It involves controversy.** For the purposes of generating media coverage about an issue, it is always better to have two sides than one. Some activists have been known to suggest names of opponents reporters might interview in order to play up the controversial elements of a story and enhance media opportunities.

- **Celebrities are participating.** A "celebrity" can be the latest Hollywood hunk or starlet, a well-known politician, or the local football or basketball hero—in other words, someone who enjoys relatively good name recognition and has been in the news before. Celebrities add some glamour to events; they're particularly useful for getting access to interview programs. But they can also be quite independent and uncontrollable, and use your venue to deliver a message about their own agenda. Make sure the celebrities you recruit stick to *your* message *exactly*. Agree to the ground rules of their participation in advance. At the least, give them a list of talking points on paper that you want them to address. It is not unreasonable to give them a prepared statement to read, and to restrict freewheeling interviews until you're absolutely confident the celebrity will accurately reflect your point of view.

Whether trying for broadcast or print coverage, on radio or TV, in newspapers or a magazine, keep in mind these essential elements a story must contain to be considered newsworthy. If, upon reviewing your story, you don't believe it meets at least a third of the criteria listed above, don't waste your time and resources trying to get reporters to cover it. Go back to the drawing board, reconsider the elements you have to work with, and rework your story idea until you feel you can "pitch" it to reporters more successfully.

Media Outlets

A media outlet is the vehicle that delivers news and information to your target audience. There are many categories of media outlets, and you need to be familiar with all of them. They include television stations, radio stations, daily and weekly newspapers, magazines, and local cable television stations.

TELEVISION

Of all the media available to work with today, perhaps none has a greater impact than TV. According to the Roper organization, 65 percent of Americans identify "the boob tube" as their main source of news and 49 percent consider television the most believable news source. Claims the Washington, DC–based Benton Foundation, which promotes the use of media by public interest organizations, more than 92 million U.S. households, or 98 percent of all homes, own television sets—more than those with indoor plumbing or telephones!

Given the pervasive role television plays in providing information and shaping public opinion, its importance in positioning issues and winning a campaign can't be ignored. Even if you don't watch TV yourself, assign one or more "media monitors" to keep track of the quantity and quality of the media coverage you're generating.

Many television reporters are "general assignment" staff—they cover any breaking news story rather than just one topic or "beat." At larger TV stations, TV reporters will specialize in beats like consumer affairs, health or city hall; all stations, regardless of size, usually have one beat reporter who covers sports and another who does the weather.

While television is more immediate—and frequent—in the delivery of news than newspapers or magazines, it is also much briefer. Essentially, broadcasters compress various news events into short segments they then string together to give a brief overview of the day. The average story lasts no longer than one minute and forty-five seconds and is accompanied by a video image; even very short newspaper stories contain more words. The average half hour broadcast of evening news contains only about twenty-two minutes of actual reporting, including the weather and sports reports (the rest is commercials and station identification messages), providing far less information than what's offered in the entire daily paper. In TV, the visual is everything.

Still, television offers activists many opportunities for coverage. Most stations deliver news four times a day: in the early morning (sandwiched around the network morning shows like "Good Morning America" and "The Today Show"); at noon; in the early evening, around suppertime; and then again at 10:00 or 11:00 P.M. If an important story is "breaking," newscasters may interrupt regularly scheduled programming to give viewers a news update.

In addition to its news programming, a station may also offer morning talk shows, which usually air at 9:00 or 10:00 A.M. weekdays, and occasionally at 7:00 or 8:00 on Saturday or Sunday mornings. These talk shows, which on average last half an hour, provide activists with an excellent opportunity to deliver their message—unfiltered by reporters or editors—directly to the audience (see page 254 for information on how to arrange appearances on these programs).

Nearly two out of three local stations also run editorial comments, either during regularly scheduled newscasts or during feature programs. Activists can lobby station managers to influence the content of these editorials and viewers can appear on-camera to respond to an editorial or comment on an issue of public importance, delivering a "guest editorial" the way they would write a letter to the editor of a newspaper.

Some news stories are simply recounted by one of the news "anchors," people who sit at a news desk and read news reports live. But because TV is a visual medium, most stories are told by a reporter who has taken a camera crew to the scene of the story and obtained compelling video images and interviews with the subjects involved.

Though many people work at a television station, you'll primarily be concerned with those who are involved in the production and delivery of the news:

The **news director** is responsible for the overall news program and determines what beats reporters will cover as well as which particular stories they'll cover.

The **assignment editor** helps determine which stories the station will cover and assigns reporters to do the stories.

The **news producer** is responsible for getting the story onto film or tape; the producer works with the reporter to select images the camera crew shoots and helps write the reporter's script. The producer also enforces deadlines to make sure a news segment gets produced in time to make it on the air.

The **reporter** actually goes out with the camera crew, conducts interviews, and works with the producer to put the story together.

NEWSPAPERS

Newspapers report stories in much greater detail than either television or radio. On the whole, they contain more information about local issues than national ones; stories are either written by locally based staffs, "stringers" or free-lance reporters, or picked up from wire services and news syndicates.

There are approximately 1,600 daily newspapers published in the United States every day, with a combined circulation of 62.6

million readers. Most cities will have at least one daily newspaper and several weekly papers, as well as an array of special-interest newspapers or newsletters.

According to the National Newspaper Association, approximately 63 percent of all American adults read a newspaper every weekday. The number pumps up to 68 percent on Sunday. In addition, nine out of ten people read at least one paper every week.

In addition to the newspapers published in English, ethnic communities may publish a weekly or (more infrequently) daily in their own language: Chinese, Korean, Spanish, Arabic, and so on. For example, African Americans in Washington, DC, can read the *Washington Afro-American;* Hispanics can peruse *El Diario de la Nacion;* and Koreans can thumb through the Washington edition of the *Hankook Ilbo,* published in Korea.

Newspaper people are fond of saying that their news and editorial sections are as separate as church and state. That's because the news sections are supposed to carry unbiased reports that reflect all sides of a story, while editorial writers are allowed to express their opinions about a story. Reporters are expected to remain neutral, reporting on the details, like who was involved in an event, what the event was, why it was held, and what the opposition had to say about it. Editorial editors and writers usually take sides, making judgments based on their analysis of the impact of the story on people, their community, or country. Unless they change jobs, reporters don't write editorials, and editorial writers don't cover stories.

On the editorial side of the paper, you'll want to know three people:

The **editorial page editor** is responsible for the editorials that appear in the newspaper. Larger papers may employ several individuals to write editorials; they serve on the editorial "board." At smaller papers, the editorial page editor may be the person who writes all the editorials as well. Where an editorial board does exist, the editorial page editor convenes meetings of all the editorial writers to discuss potential editorial topics and assign individual editorial writers to write a piece.

There is also an **editor of the op-ed page.** An op-ed is a column, sometimes called an opinion editiorial, that is written by a guest writer and expresses an opinion on a particular topic; it usually runs opposite the editorial page.

And most papers have an **editor of the letters to the editor** section of the paper who determines which letters to run (see pages 226 to 231 for tips on how to arrange editorial board meetings and write and place letters to the editor and opinion editorials).

Columnists are individuals who are allowed to express their point of view in a feature that the newspaper carries on a regular basis. Columnists either find their pieces on the op-ed page, facing the editorial page, or in the features section of the paper.

If editorial writers and editors are the "church," news editors are the "state." Most newspapers have an **editor** for each section of the paper: sports, health, food, life-style, metropolitan news, national news, and so on. Editors of these sections assign reporters to cover stories based on their knowledge of a particular "beat" (or category of news) or availability of reporters in the newsroom.

Beat reporters cover specific topics, such as those described above; **general assignment reporters** cover an array of topics from day to day, depending on how many different stories need to be reported upon and the availability of other reporters in the newsroom.

Newspapers offer an encouraging array of opportunities for activists in need of publicity. Through letters to the editor and op-eds you can express your point of view or respond to others. Hard news reporters will cover breaking news stories, while feature reporters and columnists will prepare more in-depth pieces with a stronger human interest angle. You may be able to "sell" your story to the health editor, metro desk, real estate section, city desk, and even the food writer, depending on how you "pitch" it.

(Remember the news about alar, a pesticide used on apples? After the story became a "60 Minutes" sensation, different versions of it appeared in many sections of local newspapers: food sections did stories on organic produce, business writers reported on the economic impact banning the chemical would have on apple growers, health editors examined the links between pesticides and cancer, and news reporters went to supermarkets to document the effect the story had on consumers.)

WEEKLY NEWSPAPERS

Don't overlook weekly newspapers, which are sometimes perceived as "lesser" newspapers in the community. Many citizens read their community newspapers from cover to cover. Even when you can't get your story into the main paper, the weeklies may cover it religiously. Keep track of their deadlines, circulation, advertising rates, and the conditions under which they'll accept your photos so that you can employ weeklies from the beginning to build your case. Use them particularly to announce meetings, recruit members, and promote events.

RADIO

According to the Radio Advertising Bureau (RAB), Americans average about three hours of radio listening per day. Two out of three Americans listen to the radio during prime time; radio is their first source of morning news. The RAB estimates that 53 percent of the work force listens to the radio at work, and 77 percent of adults listen in their cars.

Almost every city in America hosts several radio stations. All-news stations in each market specialize in twenty-four-hour coverage of local, regional, and national news events. Other stations devote themselves exclusively to a certain type of music: classical, rock, funk, and country are among the most common. Still other stations offer a mix of news, music, and talk shows.

The kind of news coverage a station produces is frequently a function of the station's size. Whereas large stations may attend press conferences, for example, smaller ones without reporters to dispatch often prefer to do on-air or recorded interviews via telephone. Small and medium-sized stations might be willing to take pretaped comments (called "actualities") from your group over the telephone; larger stations probably will want to conduct their own interviews.

Many radio stations offer news updates once an hour, giving you more opportunities for coverage on radio than with any other medium. But remember: In addition to having a solid news story, you can increase your chances of radio play by having something that *sounds* terrific. Good sound effects—like the wail of a train whistle, the chant of demonstrators, the roar of engines, or a colorful "quotable quote" (a ten-to-fifteen-second phrase or sentence that sums up your position in a memorable way)—will improve your chances for radio coverage significantly.

A local radio station has the fewest number of reporters available to actually go out and cover stories. But it offers you several in-studio options for coverage:

- **Talk shows.** Whether live or pretaped, talk shows let you deliver your message directly to the listening audience; appearances usually are scheduled through the host of the program or its producer (see page 254 for tips on how to set up an interview).

- **Public service announcements.** These outlets give you a free forum for disseminating your message—if you meet certain criteria (see page 232).

- **The news.** Radio news is usually read on the air by an announcer. But occasionally announcers interview newsmakers live at the radio station or over the telephone.

Radio can use the voice of your spokespeople in four main ways: as an actuality, in an interview segment, on a talk show, or in a public service announcement. Actualities and interviews are usually produced by the news department, while talk shows may be scheduled through the program's host or staff, and public service announcements are arranged through the public service director.

An **actuality** is essentially a prerecorded quote from the newsmaker that is used to enhance a radio news story. It is short—between ten and twenty seconds, and no more than seventy-five words. It may be excerpted from a taped interview, a speech, or a statement at a news conference. Some groups have gotten more sophisticated, taping comments from their leaders on important issues and then calling radio stations to ask if they want to take the comment right over the telephone. (Check with audio equipment stores for the cables you need to connect your tape recorder to the telephone.)

If you are recording an actuality for your group, make sure it conveys a complete thought. Pause between statements so that radio personnel will be able to edit your comments into the news piece easily.

An **interview segment** is basically what it says it is: an interview between a reporter and a spokesperson. It can be live or taped, and if it is taped, it can be aired in its entirety or edited into a shorter piece. Interview segments are usually aired as part of a news program.

A **talk show** interview can be given either in the studio or by phone. Interviews may last as little as five minutes or as long as an hour. They may be punctuated by listener call-ins, and questions may range from complicated to simple, from hostile to supportive.

In addition to pursuing radio for news coverage and interview opportunities, many groups recruit local radio stations to broadcast live from the scene of high-visibility fund-raisers, demonstrations, or celebrations. Throughout the week leading up to the event, the station offers its listeners T-shirts, mugs, or other paraphernalia. And one of the station's disc jockeys emcees the happening itself.

MAGAZINES

Many cities and regions of the country now enjoy the publication of a magazine specifically aimed at their geographic locale (most newspapers also publish a Sunday magazine). These magazines consist almost entirely of feature stories with high human interest value and are usually accompanied by photographs.

Several other types of magazines offer activists publicity opportunities as well. The national newsweekly magazines—like *Time* and *Newsweek*—and the general interest monthlies—like *Good Housekeeping* or *Life*—appeal to a mass audience and cover a wide range of topics. "Industry" or "trade" magazines cover a narrow topic that is usually of interest to only a limited segment of the public (for example, the *Journal of the American Medical Association* is a "trade" publication read primarily by doctors, medical researchers, and others involved in the health industry).

The magazine's **senior editor** functions much like a TV news director; this person is responsible for overall coverage, assigning free-lance writers to tackle some stories and staff writers who work directly for the magazine to others. Sometimes it is possible to generate magazine coverage by first pitching your story to a writer, who will convince the editor it's worth doing.

If you already have a relationship with an editor or writer, you can probably have a conversation in person or by telephone to discuss your story idea. But more likely, you'll have to write a "query" letter, asking the editor if the publication is interested in the idea, then send a story outline and either photographs if you already have them or a list of potential photos. Some magazines ask you to include a self-addressed stamped envelope with your query so that they can respond to you more quickly. Other editors may call you by phone to discuss the idea.

Most activists use a combination of these outlets to disseminate information about their issues and build support for their campaigns. The group you'll read about in the following story relied on T-shirts, bumper stickers, brochures, and pamphlets to build community support for their campaign to protect their community. But they sparked national interest in their work—particularly among important federal officials—through the local and national newspaper and television stories they generated, as well as the consistent coverage they received in a trade magazine.

Local Hell-Raiser
Not in Our Backyard

Members of the Creek and Cherokee nations who live on the banks of the Arkansas River near Gore, Oklahoma, 135 miles east of Oklahoma City, had always been uncomfortable with the nearby Sequoyah Fuels uranium-processing facility. The place had had numerous accidents throughout its twenty-two years, and many residents worried that their air and water were being contaminated with radioactive fallout.

When the plant announced plans to open an underground storage facility for hazardous waste in 1985, twelve people in the community sat down together to consider their options. All felt strongly that the project should be stopped—and that they would have to be the ones to stop it. Forming Native Americans for a Clean Environment (NACE), they launched a vocal and visual campaign against Sequoyah Fuels.

"We made a conscious and focused decision to close it, despite the economic hardship local residents would perceive such an action as causing," said Lance Hughes, who is part Creek Indian and executive director of NACE. The plant employed three hundred people from the surrounding area, of whom around fifty were Native Americans, and few people felt they could afford to lose their jobs. Hostile workers would later call the activists Communists, and even harass Lance and other members with death threats. But the group pressed on.

Early in the campaign, the activists began issuing

news releases to alert reporters in the region to the corporation's plans to build the storage facility.

In addition to generating substantial favorable coverage in the local media, Hughes worked with a reporter from *Nuclear Fuels,* an industry trade publication widely read by policymakers, nuclear regulators, and industry professionals. The reporter wrote an article every two weeks for two years about the deplorable shape Sequoyah Fuels was in, using information provided by NACE.

NACE published a ten-page newsletter, passed out information on the nuclear industry at fairs, and printed posters and postcards. One of many T-shirts the group sold displayed a photo of a nine-legged frog, a freak genetic foul-up attributed to the contaminated soil and water.

This grisly nine-legged frog was found by a boy in Gore, Oklahoma—downstream from Sequoyah Fuels.

Dan Agent

In 1991, it looked like the Nuclear Regulatory Commission (NRC) might renew Sequoyah's operating license. NACE responded by issuing a damning environmental assessment report of Sequoyah Fuels that made local news. (Fund-raising among private foundations, church groups, and potential members generated $13,000 to cover the cost of printing the report.) Besides issuing a press release announcing the report, NACE offered copies to the public via its newsletter and at public meetings. They also made it available to the

NRC. The report, of course, was covered in *Nuclear Fuels.*

In its own investigation of Sequoyah, the NRC had found that water on the plant grounds contained 35,000 times the level of uranium considered safe by federal law. Soil samples revealed almost 40 percent pure uranium. The facility was illegally discharging tons of uranium into the Arkansas River, the ground, and even into the company's administrative offices through the air-conditioning. Through its newsletter, fact sheets, and opinion editorials, NACE wasted no time in reminding the public that minute amounts of uranium are toxic, and that the metal can't be recovered from a contaminated substance.

An explosion at Sequoyah in 1986 that killed one worker and injured thirty-six others helped capture the attention of the national media. A front-page story in the *Washington Post* kicked off a string of stories in the *New York Times, St. Louis Post Dispatch,* and the *San Francisco Examiner* that helped focus an uncomfortable national spotlight on Sequoyah Fuels. All the while, to avoid losing momentum, NACE kept in frequent contact with reporters who were following the story.

"It involves a lot of phone work, faxing, mailing to reporters . . . we wanted to make sure that our information was credible and timely. We needed to break down the secrecy of what is going on here," Hughes said of the group's ongoing publicity campaign.

A *New York Times* article about raffinate, a hazardous waste–based fertilizer the nuclear industry was attempting to market, inspired NACE to videotape areas where the substance had been applied. The compelling footage showed dead cows and the burn line where the "fertilizer" had been sprayed on trees. NACE sent the video to the largest eastern Oklahoma television station, where it headlined the nightly news. The same station had already produced an award-winning series on Sequoyah, based on the criminal investigation of the facility that was conducted by the Nuclear Regulatory Commission from August 1990 through October 1991 and that had resulted in the plant's first shutdown, though only for six months.

On November 17, 1992, a toxic gas leak that injured thirty-four workers proved to be the proverbial straw that broke the camel's back. Seven days later, and seven years after NACE began its fight, General Atomics, the plant's owners, shut it down.

The group's activism earned the surprising praise of Ivan Selin, the chairman of the federal Nuclear Regulatory Commission, who acknowledged his agency had been pushed in the right direction on nuclear safety issues because of the pleas and protests of "watchdog" groups like NACE.

REACHING THE MEDIA: THE MECHANICS

Engaging in various media tactics is the most exciting and fun part of the communications job. But in order to implement any communications strategy well, you need to have certain mechanical functions in place. For most groups that means identifying someone or a group of people to do the work, targeting reporters, and developing materials that present your group and your issue in the most favorable light.

Form a Media Committee

Developing and implementing a communications strategy for your organization is hard work, but as the old saying goes, someone's got to do it. Since many groups can't afford to hire an employee just to coordinate media outreach, they frequently end up forming a media committee.

The committee should be led by a coordinator who is experienced at working with reporters, writing press releases, and has a basic understanding of how the different media work. (Jan Garton, who has a master's degree in journalism, found her education invaluable in her efforts to publicize Cheyenne Bottoms. She knew how reporters operated, understood print and broadcast deadlines, and could anticipate what the media would be looking for in her stories.) If no one with such experience belongs to your group, find someone who is willing to learn and has demonstrated leadership skills.

Ask others to join the committee who will help contact reporters, schedule interviews, and track the success of the media campaign. Everyone should help identify story angles that can be fed to reporters at the appropriate time.

Though the committee will do most of the "hands-on" work of generating stories and creating a positive relationship with the press, the entire organization should develop an appreciation for the media effort. Those who are responsible for refining the organization's policy positions should help craft and implement media strategies as well, acting as spokespeople and contributing insights about the effects media coverage will have on the organization or a specific campaign. Once a year, the organization as a whole should examine the brochures, pamphlets, and news releases the group produces. Do they contain too much copy? Is the copy too strident—or too weak? Are the graphics compelling? Compare the materials you produce to others you like even more, and recommend changes that will improve your organization's image and efforts.

Here's an example of how the media committee of an ad hoc coalition effectively squelched an effort to market cigarettes to blacks in Philadelphia.

Local Hell-Raiser
Nipping It in the Bud

Uptown cigarettes looked cool. The slick gold-and-black foil packaging that encased the twenty cigarettes was meant to reflect hip city nightlife, while the sparse ad copy promoting the product nonchalantly suggested: "Uptown. The Place. The Taste."

Uptown tasted cool, too. It had a distinctive menthol flavor, a tang some studies showed black smokers in particular would find appealing.

In fact, everything about the product was designed to attract blacks, a minority group suffering from the highest cancer rate in the country and a disproportionately large number of smoking-related deaths. And that deliberate marketing tactic outraged a lot of people, especially in Philadelphia, where the cigarettes were scheduled for a test launch before being moved into national distribution.

R. J. Reynolds Tobacco Company (RJR), the cigarettes' manufacturer, defended its effort to entice black
smokers into choosing Uptown by claiming that blacks
had the right to select one cigarette brand over another.
But opponents from Philadelphia's African American
community, the health industry, and other concerned citizens believed the clever campaign would actually encourage more African Americans to take up the deadly
habit. In their view, RJR had concocted a spurious rationale to justify an insidious campaign. The launch had to
be nipped in the bud.

Early in January 1991, just a few weeks after RJR
announced its intention to test-market the new cigarette
in the City of Brotherly Love, representatives of the
National Black Leadership Initiative on Cancer, the
American Cancer Society, the American Lung Association, the Committee to Prevent Cancer in Blacks, and
others formed the Coalition Against Uptown Cigarettes.
Their goal: to insure that the cigarette was not introduced nationally. Their strategy: to deflate the Philadelphia market test scheduled to begin the first week of
February. Charyn Sutton, a public relations specialist
based in Philadelphia, volunteered along with represen-

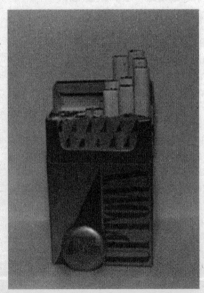

When R. J. Reynolds tar-
geted this cigarette explicitly
to African Americans, it
generated a groundswell of
opposition in Philadelphia
and created national news.
Rev. Jesse W. Brown, Jr.

tatives from each member group to form a media committee and spearhead the group's unorthodox effort to fight back.

Members of the coalition concluded they had to channel their outrage at the unfair targeting of blacks, protesting the cigarette launch but avoiding campaigning against smoking in general so they would not alienate smokers. They hoped to appeal to that side of a smoker's psyche that regrets his or her own addiction and doesn't want more young people to get hooked. Their program had to oppose marketing this deadly product to a high-risk group, but without casting blacks as poor or uneducated.

With these considerations in mind, the coalition opted to combat the tobacco industry's sophisticated, polished media campaign with one that was just the opposite: Though carefully orchestrated behind the scenes, it would appear spontaneous on the surface, and would target primarily local media, even though national outlets were clamoring for the story.

Their first priority was to mobilize community support against the cigarette. The coalition recruited doctors, members of the clergy, city council members, and marketers, all African Americans, to condemn RJR's campaign, circulating comprehensive fact sheets internally to keep its spokespeople fully informed about the issue. The group made a point not to condemn target marketing per se because many black agencies engage in the practice themselves. Instead, they challenged RJR's decision to market a deadly product to such a susceptible audience.

To attract business leaders in the community as well as those concerned about health, they positioned the issue as a business story rather than as one focused only on health. When the *New York Times* requested a list of the coalition members, the coalition waited until it could send a roster that included not just the health-oriented members but all interest and community groups and individuals that opposed Uptown cigarettes—reinforcing the grass roots image the activists were carefully cultivating. "The media got excited about the 'real' people—that this was really a grass roots organization, not a bunch of people pushing press releases," Sutton said.

The coalition did the unheard of when it turned down an offer to appear on ABC-TV's "Good Morning America": they would have been positioned against black marketers, and did not want to divide the black community in any way. A better way to mobilize Philadelphians to join the protest, they knew, was to encourage coverage by local media, including the *Daily News,* a Philadelphia tabloid with a large African American readership.

In the span of a few short weeks, the Stop Uptown campaign attracted a wave of local editorial backing as well as the attention of national media outlets. Dr. Robert G. Robinson, a local researcher who worked in tobacco control, asked U.S. Health and Human Services Secretary Louis Sullivan to address the issue during a trip to Philadelphia, which the secretary, himself a black physician, gladly did. (Reporters later wrote lead stories in which they noted the secretary's "scathing indictment" of Uptown cigarettes.)

The day after Secretary Sullivan's statements, officials at R. J. Reynolds decided to pull Uptown's test market. The date was January 19, less than two weeks after the coalition's first meeting. The company expected the story to be buried in little-read Saturday newspapers. But the papers held the victory story until Sunday, the most-read issue, and gave it significant front-page coverage. Television networks featured the story nationally on Monday.

"It was really a landmark in terms of marketing, community mobilization, and media," Sutton remarked when the campaign was over. She said that the group was prepared to distribute damning fliers if RJR persisted in distributing the cigarette to stores in February, but the tactic was not needed.

Meanwhile, RJR went back to the drawing board to try to figure out where it had gone wrong. The mistake was not a little one. One industry source estimated that the failed launch cost the company over $5 million—not to mention the negative PR it suffered.

Create a Press List

A press list is perhaps an organization's most essential media tool. In one handy, organized place, it helps you keep track of reporters and editors so that you can phone them and mail information efficiently and without delay.

To develop your list, identify reporters, columnists, and editors at your local newspapers, magazines, radio stations, and television stations who cover a beat that pertains to your issue. Some of the most common beats include environment, science, health, transportation, crime, the arts, and city hall. If the media outlet is small and doesn't have a reporter specifically assigned to your issue area, ask for a general assignment reporter who can be interested in your story. You may need to talk to the assignment editor to get the right name.

Make two lists: a mailing list, consisting of first and last names of reporters or editors, the name of the outlet, and its address; and a telephone and fax list, consisting of the name of the reporter, the outlet, and the phone and fax number. While you'll use the first list for mailing press materials, you'll need the second list so that you can easily follow up all your mailings with telephone calls, a task you *absolutely must* undertake to insure that reporters received your materials and are planning to cover your story.

(Likewise, give your office and home phone numbers to reporters. Many media deadlines extend far beyond standard office hours; if a story breaks late, reporters may need to reach someone in the evenings and on weekends. If your office has an answering service or machine, always include a number where a contact can be reached during off hours.)

You can develop media lists by painstakingly going through your telephone directory and looking up the address and phone number of each outlet. But a quicker way is to use any of the media directories kept in the reference section of your local library. Some books to look for include: *Bacon's Publicity Checker, Gebbie Press All-In-One Directory, Broadcasting Yearbook,* and *Editor and Publisher.* You might also be able to borrow media lists that have already been created by other groups, such as a local labor union, the United Way, a local hospital, or one of the organizations participating in your coalition.

OUTLETS YOU SHOULD PUT ON YOUR PRESS LIST

- Daily newspapers
- Local radio stations
- Local cable systems
- Associated Press and United Press International wire services
- Local magazines
- Weekly newspapers
- Local bureaus of national newspapers and magazines
- Local TV stations
- Local affiliates of national TV networks
- Organized newsletters
- Policymakers, legislators, legislative staff
- Other "VIPs" (large contributors, foundations)

REPORTERS YOU SHOULD PUT ON YOUR LIST

- Beat reporters
- Editorial page editors
- Editorial writers on your issue (if they exist)
- Assignment editors
- Columnists
- Feature editors and writers
- News directors at radio and TV stations

Develop a Press Kit

A press kit is a compilation of materials that help "sell" your issue to the media, as well as to possible sponsors, contributors, and even members. It gives credibility to your organization by presenting information that makes your group seem solid and reliable. And it represents you as you would like to be represented in the press.

Press kits can be extremely versatile. Though used primarily to introduce you to the media, they can serve several other functions as well. Armed with the proper cover letter, they can be sent to foundation executives who are considering your grant request, to legislators you are trying to lobby, to public officials at agencies you're trying to impact, and to other organizations with whom you would like to work.

Usually, press kits are sent out to introduce a group or issue; it is not necessary to give a reporter a complete organizational press kit each time you meet. Besides, depending on how many elements your press kit includes and whether it's copied or printed, press kits can be expensive. So send your kit to a reporter just once; don't resend it unless the information you're forwarding has changed from what the reporter has already received. And don't use complete press kits to respond to general queries from the public at large; send a less-expensive brochure or fact sheet instead.

Your press kit should contain the following components:

- A kit folder, one with pockets so that you can insert news releases, fact sheets, your business card, and other materials.
- A press release about the issue or activity for which you're seeking coverage.
- Your group's most recent newsletter or annual report.
- Fact sheets on your issues.
- Selected press clippings.
- A standard one-page description of your organization.
- Brief biographical profiles of your spokespeople.
- Charts, visuals, or photographs as appropriate.

If you send a photograph, it should be a professionally photographed 8-by-10, black-and-white glossy, not a silly-looking snap shot. Type a brief photo caption on a sheet of white paper and rubber cement the caption to the back of the photograph. The caption should expand on whatever information is already apparent in the photo, listing the correct names and titles of every person in the picture. Make sure you have permission to reproduce the photographs from everyone who is pictured.

Outlets may use photographs of book or report covers or of people who are the principals in an organization, not your "action shots" from a press conference or media event. (In those circumstances, you're better off coming up with good "photo opportunities" that will entice the media to send their own photographers to your event.)

In all materials you produce, strive to make them visually interesting. Print in two colors (black can be one), write in short paragraphs, and use subheads, graphic illustrations, boxes, and bullets to break up text in a way that can be easily followed.

TACTICS

Many tactical tools are available to help you get positive media coverage and sustain media interest in your issue. The following chart offers a brief summary of the many media tactics available to activists and suggests how to use them effectively. The rest of the chapter explains each of these tactics in greater detail.

Media-Grabbing Tactics	How to Use
• News releases	• Explain position, describe an event.
• News advisories	• Invite reporters to attend an event.
• News conferences	• Explain breaking news to reporters.
• Letters to the editor	• Respond to editorial coverage or demonstrate community support for an issue.
• Editorial meetings	• Influence the editorial position of a newspaper or radio or TV station, especially before an important vote or decision.
• Editorial mailings	• Explain status of issue or campaign and ask for editorial support.
• Opinion editorials	• Express your opinion on your issue and establish your credentials as a spokesperson.
• Special events	• Demonstrate strength, rev up media, put pressure on targets.

Media-Grabbing Tactics	How to Use
• Public service announcements, signs, fliers, posters	• Educate the public; tell people where they can call or write for more information, or how to attend meetings or events.
• Advertising	• Lobby; when coverage from other sources is inadequate and you have the budget to go directly to the public.
• Calendar listings	• Turn out people to meetings, rallies, other events.
• Interview shows	• Explain position directly to public.
• Community cable	• Educate the public, explain your position.
• Video	• Provide your own footage to the media, legislators, or the public.
• Newsletters	• Communicate with members.

Press Releases

A press release serves many purposes. It may explain a new report, announce an upcoming event, reveal your position on an issue, provide background information, or supplement an ongoing news story. It expresses your point of view as you would like to see it reiterated by the media in a headline and in the way the story is reported.

Some smaller newspapers (particularly weeklies) may print a news release verbatim, actually cutting off the bottom of the release before they send it to be typeset for the paper to accommodate their space limitations. That practice is frowned upon at larger media outlets, which use news releases to help decide whether to cover a story, but then go out and get the facts themselves. Because editors may spend only a few minutes scanning the first few paragraphs of the release, you can increase the impact the release can have by following standard journalistic procedure and putting the most important information first (putting the most critical information at the top of the release will also make sure your key points make it into the paper should an editor at a smaller outlet take a pair of scissors to it!).

The diagram below is called an "inverted pyramid." It represents the way you should present your information in your news release. The cardinal rule is: Put the most important information about your story at the top. Subsequent paragraphs should contain useful information that substantiates your opening paragraphs, as well as quotes from your spokespeople. But if you want to pique a reporter's interest, put the essential news first.

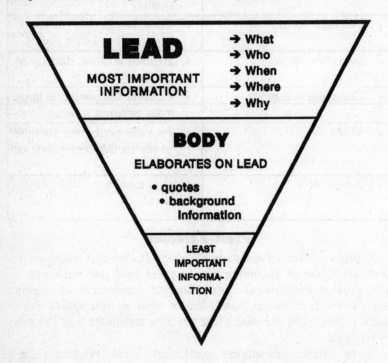

What elements of your story should you consider "essential"? These usually revolve around the "Five W's"—what (is happening), who (is involved), where (is it happening), when (is it happening), and why (is it happening). The most newsworthy of the W's goes first, the next most newsworthy W goes next, and so on until you've laid out all of the W's in the most newsworthy way.

To determine which of the W's is the most newsworthy, review the criteria at the beginning of this chapter. Then make some judgments. For example, if the "who" is famous enough (like the President of the United States or the Pope), almost anything he or she does will make news. But for most activist groups, their biggest news will revolve around the "what." "Environmentalists said today that . . ." is not very newsworthy. "The Dallas water supply is contaminated, according to a report released today by the public health department" is.

While your press release might seem extremely newsworthy to you, it is likely to be just another piece of paper to a news editor, who could receive two hundred releases on any given day. Pressed for time and forced to meet strict deadlines, the editor may not read more than two or three paragraphs before deciding to either toss your news in the trash or give it to a reporter for follow-up. So use releases sparingly but effectively. When reflecting on the Cheyenne Bottoms campaign, organizer Jan Garton said that they could have issued more releases and maintained an even larger presence in the media, but she felt that the press paid better attention to their occasional, well-planned releases about concrete events than they would have to more frequent releases, not all of which might have been important.

You can increase the chances that your release will lead to a story by making it look and sound like news right from the top.

Contact: Jan Garton, Chairperson
Public Involvement Task Force
SAVING CHEYENNE BOTTOMS
219 Westwood
Manhattan, Ks. 66502
(913) 539-3004

FOR IMMEDIATE RELEASE

CHEYENNE BOTTOMS STUDY APPROVED

In one of its final actions, the 1985 Kansas legislature approved a funding
measure for a major study of the Cheyenne Bottoms wildlife area near Great Bend.
The study will take eighteen months, and will look into the geology, biology,
ecology and hydrology of the area before recommending actions to the Governor
and the legislature on the best ways to restore the area as a manageable wetland.
Financing for the study will come from the Kansas Fish & Game Commission, which
will be the contracting agency, as well as from the general fund and the Chickadee
Checkoff donations.

Legislation supporting a feasibility study was backed by several of Kansas'
major conservation, environmental and sportsmen's organizations, including the
Kansas Natural Resource Council, Kansas Audubon Council, Kansas Wildlife Federation,
the Kansas Chapter of the Sierra Club, Kansas Ducks Unlimited Council, Kansas
Wildlife Society, and Kansas Chapter of the American Fisheries Society.

"We're very pleased that so many people were involved in initiating and
following through on this proposal," said Darrell Montei, chief of the Game
Division in Kansas Fish & Game. "This is the kind of support that will be needed
if the natural resources of the state are to remain for future generations to
enjoy."

Cheyenne Bottoms is a 19,000-acre marshland of critical importance to
migratory bird and wildlife populations. In the late 1940's and early 1950's,
water appropriation rights from the Arkansas River and Wet Walnut Creek were
granted Cheyenne Bottoms to provide a permanent, reliable source of water for
managing the marshland.

Today, the Arkansas River no longer flows for over 150 miles, and is
nearly out of water at Great Bend. Lack of reliable inflows has caused the
Bottoms to dry up by fall. Decreasing wetland habitat in the United States,
Canada and Central America has resulted in population declines in many species,
especially in duck populations.

Cheyenne Bottoms has been identified by the International Shorebird Survey
as one of the top three wetlands east of the Rocky Mountains in attracting numbers
of shorebirds and importance of species.

- 30 -

This news release from the Save Cheyenne Bottoms Task Force announced that the
Bottoms' feasibility study had finally been approved by the Kansas legislature.

• •STEP BY STEP• •

How to Write a Press Release

• **Write a catchy title and lead sentence that "hook" the reporter or editor into your story.** Try to be somewhat objective in the way you present your information. Let the news speak for itself.

• **Put the most important information first.** Include who, what, when, where, why, and how in the first paragraph. Be accurate with all names, numbers, statements, and quotes.

• **Keep it to two pages.** The release should be double-spaced, with wide margins. Write in the active voice, which is easier to read and requires fewer words. Keep sentences and paragraphs short. Copy the second page on the back of the first to save paper.

• **Issue the release on your organization's stationery.** If you don't have stationery, put your organization's name, address, and logo at the top of the paper. Write "For Release" in the upper left-hand corner, followed by the date you wish the information to be available for coverage. In the right-hand corner, provide the name of one or two contact people, along with home and work phone numbers where they can be reached.

• **Include a quote from someone representing your organization** by at least the third paragraph of the release to identify your organization and add interest to the text.

• **Type the word "more" or "over" at the bottom of the first page; type "- 30 -" or "###" at the end of the release.**

• **Include one or two sentences that describe your organization at the end of the release.**

• **Mail, fax, or hand-deliver releases** at least one week before you want the story to appear. If your story is complicated or you're releasing a detailed document, distribute it earlier to give reporters more time to review it. You can ask reporters not to release your story before a certain date or time by writing the words "Embargoed for Release" above the release date and time you've chosen.

How to Write a Press Release (cont.)

• **Follow up by telephone** to see if a reporter intends to cover your story. Talking to reporters by phone also helps you build relationships with them that may help you get more and better coverage over time.

☞ **CHECK IT OUT:** The *Associated Press Style Book and Libel Manual* can help answer your questions about capitalization, punctuation, abbreviations, and other stylistic questions, and offers instructions on how to avoid libel. Written for journalists, it is valuable for anyone who is working with newswriting-style text. To order, send $6.75 ($4.75 for AP member publications) to Stylebook, AP Newsfeatures, 50 Rockefeller Plaza, New York, NY 10020.

The *Chicago Manual of Style* is a more in-depth grammar, style, and word-use reference that includes information on foreign languages and mathematics in type and on copyrighting. Check any bookstore or your library; or send a check for $40, plus $3 to cover shipping and handling, to the University of Chicago Press, 11030 South Langley Avenue, Chicago, IL 60628, 800-621-2736.

● ●STEP BY STEP● ●

How to Pitch a Story

To get a story covered, you have to "pitch" it to either the assignment editor or a reporter; in other words, you need to convince the editor or reporter that the story is newsworthy. What should you do?

- **Send out a news release or information packet.** The material should meet the criteria for newsworthiness described on page 190.

- **Follow up with a phone call.** Try to reach every person to whom you mailed the information. Attempt to secure a commitment from each one to at least consider the story, and to attend an event if you are organizing one.

- **Arrange interviews.** If the reporter is interested in your story but cannot attend your event or has other time constraints, try to schedule a one-on-one interview with the reporter and your spokesperson at the reporter's convenience.

- **Think visually.** Television reporters in particular are not interested in what they call "talking heads": one or several people who do nothing more entertaining in front of the camera than move their mouths. You'll have more success if you offer reporters an interesting picture to accompany your words. Remember the success of the Save Cheyenne Bottoms Task Force? Both media events they organized—the presentation of the seat cushions and the run across the state—generated substantial print and television coverage because their images were unusual, interesting, and entertaining. Help reporters and cameramen set up shots. If necessary, provide them with photographs, slides, or video footage they can use in their papers or on the air.

Ironically, even though television is more pervasive than newspapers, this broadcast medium often takes its cues for stories from newspapers. In other words, a TV reporter or editor may be reluctant to prepare a story on a topic until he or she reads about it in the daily paper! If you're having a hard time convincing television to cover your issue, fax the reporter or editor any newspaper clips you've generated about it, accompanied by a pitch noting how current the issue is.

A final note about pitching a story: Occasionally you may be in the position of having to interest reporters in more than one story at the same time. How do you juggle your desire and need for coverage with a reporter's natural reluctance to do more than one story on your organization in any given week (or even month)?

First, try to avoid the situation by developing a well-planned media strategy that reasonably times your events and news releases so that they don't fall on top of each other. Convene your media committee regularly to discuss upcoming activities and to anticipate unexpected media opportunities and decide ahead of time how to handle them in the event conflicts occur.

If you're still faced with several competing events, reevaluate the news stories remaining on your plate. Prioritize them again for newsworthiness to make sure they all merit serious news attention (the last thing you want to do is develop a negative reputation among reporters as someone about whom they can complain, "There's not enough ink in the world to satisfy you"). Hold off on promoting the less important stories to avoid undercutting yourself.

Now take a look at your press list. Can you parcel out your story ideas so that different reporters get different stories? For example, suppose you are launching a new program to educate pregnant women about the causes of infant mortality at the same time you are preparing a detailed report about the incidence of infant mortality in your community. Review your press list. Then get the hard news reporter to look into the conclusions of the report, and urge a features reporter to do a special about your new educational program.

During those periods when you simply can't avoid going back to the same reporters several times in the course of a few days or weeks, be up front with the media about what the future holds. Call reporters and let them know that you'll be in touch with them frequently, convey to them the newsworthiness of all of your stories, and ask them for suggestions on how you can avoid overloading them and still get them the information they need to stay apprised of your issue. Most reporters will appreciate the advance warning, and you'll come off looking like a savvy media professional.

Accept the challenge
Help prevent child abuse

NEWS ADVISORY

April 24, 1991   Contact: Kim Walker/Karen Lee/
                                     Genene Valleau
                                     373-7717

<u>FOR IMMEDIATE RELEASE</u>

On Monday, April 29, at 10:30 a.m., approximately 600 teenagers in juvenile corrections facilities and on parole around the state will complete physical and community challenges to raise money for child abuse prevention. Many of these kids have been abused and neglected and have gone on to commit crimes. They want to give something back to the community. They want to prevent other children from suffering.

Join former Dallas Cowboy Claxton Webster, Trailblazer spokesman Nick Jones, former Gov. Goldschmidt, Dept. of Human Resources director Kevin Concannon, and Children's Services Division administrator Bill Carey as they participate in Mission: Possible events at juvenile corrections facilities statewide.

Community businesses supporting Mission: Possible include PacifiCorp Group, First Interstate Banks, Elks clubs statewide, Kelly Services, ProImage, Roth's Fo⁻ iliners, Fred Meyer, United Grocers, Steinfeld's, Pepsi, Franz, and Farmington Thriftway.

The Portland Trailblazers have donated an autographed ball and two autographed T-shirts to raffle off. All proceeds from Mission: Possible will go to the Children's Trust Fund of Oregon. Mission: Possible is coordinated by Children's Services Division.

A list of the Mission: Possible community challenges is attached, along with contact names and numbers. A brochure of Children's Trust Fund programs around the state is also attached. Students will be released for photographs and interviews at every location.

- 30 -

◆   198 Commercial SE, Salem, Oregon 97310   Phone: (503) 373-7717   ◆

This news advisory invited reporters in Oregon to Mission: Possible fund-raising and media events.

219

News Advisories

A news advisory differs from a news release in that it acts like an invitation to an event. Rather than contain all of the details about the event itself, it only focuses on the most important information—the lead of the story. It should be one page long, and should explain in one or two sentences what the event is about, where it is, when it is, what time it is to take place, and who will be participating. It should also contain the name and phone number (home and work) of someone in your organization who can be contacted for more information.

News Conferences

A news conference is a gathering of reporters convened to provide a story to them all at the same time. Beat and general assignment reporters for print, radio, and TV are most likely to attend news conferences, though columnists might as well.

Reporters despise news conferences unless they offer real news. A press conference that consists of nothing more than "talking heads" needs a stronger, more sensational peg to draw the media out than a colorful, photogenic event like a marathon. (However, in smaller cities, where there is less competition for hard news, it is easier to be considered "newsworthy".) So it is essential that you truly have something newsworthy to discuss before you convene a news conference (review the criteria on page 190). Releasing a controversial report, filing a lawsuit, or announcing a boycott, picket, or some other dramatic action all make for good news conferences, especially if they are spiced up with good sound and visual images.

If your topic doesn't warrant a news conference, consider several alternatives: Issue a news release by mail or fax, then try to arrange one-on-one interviews with reporters who are particularly interested in your topic. Or organize a "background briefing" over an informal breakfast or lunch to discuss the story.

Sometimes, even though you've organized a good, solid press activity, you'll get overshadowed by a more stunning event—such as a local fire or plane crash. In that case, hand-carry your press release and other materials to reporters and attempt to schedule one-one-one interviews and in-studio interviews on the spot or the following day.

• •STEP BY STEP• •

How to Organize a News Conference

• **Determine that your event merits a news conference.** Review the criteria described on page 190 to make sure the event is newsworthy.

• **Choose the right location.** If you are planning to picket a factory, hold the news conference outside the factory gates (get a permit if necessary). You can also hold news conferences at your own offices—if you have a room large enough—or at your city's Press Club (though you may need to be a member or have a member sponsor you). The location should be safe, easy to reach, relatively close to the center of town, and offer access to parking. If you need advice about desirable sites, ask a friendly reporter for suggestions. Take precautions if you're holding the conference outside. Though an outdoor activity may work for you symbolically, bad weather could put a real damper on the event; in your news advisory, identify a rain location to which reporters should go if it's raining thirty minutes before the event is scheduled to begin.

• **Choose the right day.** Tuesday, Wednesday, or Thursday are better days than Monday or Friday. Reporters may forget about the event over the weekend, and on Friday they may be leaving work early, or may be preparing stories for the weekend. Weekends aren't often a good option, either, because so few reporters are on the job (although you'll have less competition for news). If you do have to hold an event on a Saturday or Sunday, make sure you alert the "weekend assignment editors" well in advance. And if you opt for a Monday, remind reporters on Friday and again first thing Monday morning.

• **Hold your event between 9:30 A.M. and 1:00 P.M.** Any earlier than that, and reporters will have a hard time getting there—unless you're inviting them to breakfast. Any later, and reporters will miss their deadlines for filing stories.

• **Notify reporters in advance.** Send out a written news advisory so that it reaches reporters at least a week before your event. A word of caution: Don't "scoop" yourself if a reporter

How to Organize a News Conference (cont.)

reads your advisory and wants to do your story before your press conference or event. The point of an advisory is to get the media to cover your event, not to get them to do an early story—and spoil your chances of coverage with everyone else. Politely tell eager reporters they must wait for the event to get all the details. Putting the words "Embargoed Until" with the date and time of your event on your advisory will alert reporters that the story is not available to be released until that date and time.

- **List the event on the "daybooks."** A daybook is a computerized calendar of media events that is maintained by the local newswire service and fed electronically into newsrooms. All assignment editors regularly consult the newswire to find out what is happening during that day or in that week. In most cities, the daybook is published by the local Associated Press and/or United Press International bureaus. You can also pay a fee to a placement service such as PR Newswire (check your telephone directory for a local phone number), an agency that will send your announcement to newsrooms the same way the daybooks do. Fax the daybooks a copy of your news advisory, then call to make sure they received it and will list it.

- **Follow up by telephone.** Phone the reporters to make sure they received the news advisory and plan to attend your news conference. If they don't plan to attend, try to set up a one-on-one interview with your spokesperson so that you don't lose the opportunity for coverage. (Many reporters now allow you to fax news releases to them. If you can get access to a facsimile machine, you can get your materials into a newsroom yourself.)

- **Prepare a press kit.** The kit should include a press release summarizing your story in news format, a verbatim statement of what speakers will say, brief biographical profiles of each speaker that include their full name and correct title if they have one, fact sheets or backgrounders that further substantiate your position, and other materials that help make your case compelling. The press kits should be distributed to reporters as they sign into the news conference; reporters who do not attend

How to Organize a News Conference (cont.)

the news conference can get a kit afterward by mail or special delivery.

• **Prepare yourself.** Prepare for the news event thoroughly. Anticipate tough questions. Make a list of the kinds of questions you're likely to get asked and practice short, quotable answers. Hold a mock press conference with others in your organization.

• **Think visually.** Having good pictures to show will greatly enhance the likelihood that television will cover your event. Unusual charts or graphs, videotape, slides, large mounted blowups of photographs, maps, ballots, letters, or petitions will work well. All slides for television should be horizontal, in color, and graphically simple. Color labs and graphics shops can inexpensively produce charts for you on their computers.

To alert assignment editors that you have good photos, the advisory should include the words "Photo Opportunity Available" followed by a brief description of what the "photo op" is. (You can also include a description of the visual in the body of the advisory. A group enticed reporters to one press conference by beginning: "Surrounded by a jumble of dead and broken trees in a massive clearcut, an Arkansas Congressman will introduce a bill to protect . . .")

• **Take pictures.** If you can, hire a photographer. If you can't afford it, have someone in your organization take pictures. Though large daily newspapers probably won't want to use your photos, smaller weeklies might. The pictures will also come in handy for your own internal public relations purposes, such as inclusion in annual reports, newsletters, promotional brochures, and fliers.

• **Prepare the room.** Don't pick one that's too big, or you'll look like you've had a flop even if a lot of people show up. And avoid one that's too small, or you'll have disgruntled reporters standing outside in the hall complaining. If you can afford it, offer reporters coffee, tea, and juice; otherwise ice water will do. Equip the room with flip charts or whatever other technical equipment you need, a podium with a microphone, a "mult box" that allows several reporters to easily record the event by plugging into the podium microphone, risers in the back of the room

How to Organize a News Conference (cont.)

for TV cameras, and a table in the front for radio reporters to set up on. Attach a sign bearing your group's name and logo to the podium so that it appears in all camera shots. A quick-print shop can probably make you a 15-by-19-inch sign for less than $50; the sign can be reused many times. Ask the hotel or Press Club to list your event on the daily events board in the lobby. If you're holding your event at a more exotic location, you may need to post your own signs.

• **Set up a sign-in table right outside the room.** Ask one of your volunteers to register reporters and hand out press materials. Record the name, outlet, and phone number for each reporter who attends so that you know who did and didn't show up. Signing in reporters also gives you a chance to introduce yourself to those you don't know and get reacquainted with those you do.

• **Get started.** Begin the conference approximately five minutes after the designated time. Introduce yourself, welcome all those in the room and thank them for attending. Introduce your other speakers, then read your own remarks if you have any. If not, the other statements should begin.

• **Let a limited number of people speak on your behalf.** The last thing reporters want to hear is one person after another deliver a speech. Organize the key points you need to present, and have them presented by no more than four people. If the conference is being convened by a coalition, let other people sit on the stage behind the podium; they can be called upon to answer questions if necessary. They can also prepare written remarks to include in the press kit. All speakers should have written statements prepared, which they should read regardless of how many (or how few) reporters attend the event. No delivered statement should be longer than five minutes. When reading the statement, look up from the paper occasionally to make eye contact with reporters, and to be more effective in delivering the best "sound bites."

• **Answer questions.** After all the statements have been made, open the conference up to questions from reporters. Ask all speakers to answer questions at the podium, where the mi-

How to Organize a News Conference (cont.)

crophones are located, not from their seats, where their comments may not be picked up. Expect additional questions "one-on-one" after the news conference, especially from the electronic media (radio and TV stations), who may ask you questions to get you to reiterate key points or your best quotes.

• **Close the news conference.** Try to limit the conference to no more than an hour. Thank all those who came, and offer to answer any additional questions or provide more information if needed.

• **Follow up again.** When the conference is over, fax or deliver by hand press kits to any key reporters who planned to attend but didn't. Return to your office or home in the event that reporters need to contact you for additional information. Follow up with any reporters at the news conference who asked for information you couldn't provide.

• **Get clips of stories.** Monitor the media for stories based on your news conference. Keep copies of stories that appear in newspapers or magazines. If you can afford it, ask a local TV and radio monitoring service (check your telephone directory or the Press Club for names) to tape radio reports and TV news stories for your archives. These might also help you compile a promotional videotape of your group at a later date.

Usually, you can't get reporters to travel far out of town to cover a story, unless they're on the campaign trail with a presidential candidate or their beat is foreign affairs and they have no choice but to travel halfway around the world.

So when the Washington, DC, organizers of the 1980 Year of the Coast decided to kick off their event, they were taking a risk. On a morning when everyone they knew (including reporters) would rather sleep in (January 1), and on one of the coldest days of the year, they organized a symbolic beach walk along a coastline an hour away from the city.

Remarkably, the idea worked like a charm. Despite whatever hangovers reporters, crews, and activists might have been feeling after ringing in the New Year, a crowd of media and participants showed up at Sandy Point State Park on the shores of the Chesapeake Bay to announce their plans to wage a year-long campaign to protect America's coastlines.

What compelled editors to send reporters to a beach on the first day of the year? Frankly, there wasn't a lot else going on that day and editors still needed to fill their news "holes," the spaces allotted for stories, not advertising.

The canny strategy resulted in a big payoff for the organizers. All three local TV stations attended the event, as did reporters and photographers from the *Washington Post,* the *New York Times,* and National Public Radio, giving the campaign a national launch as well.

Letters to the Editor

The "Letters to the Editor" section of newspapers and magazines provides a forum that lets you deliver your message directly to a large audience. Through letters you can explain how your issue relates to items being covered in the news, correct or interpret facts after an inaccurate or biased article has appeared, respond to an editorial or op-ed piece, or rebut a news or feature story.

Letters to the editor can also be used to spur editorial writers or news editors to cover an issue they had thought their readers weren't interested in.

According to several surveys, more people read letters to the editor than any other section of the paper.

A letter to the editor should be brief, clear, and to the point. You can write it either as an individual or as a representative of your group. Be sure to include your address, as well as phone numbers at work and at home. Type your letter, or handwrite clearly. Try to limit the text to one page, as short letters are more likely to be printed than long ones, and be polite. Make sure any facts and figures you use are accurate. If your letter isn't printed, don't be discouraged; keep trying.

To generate a number of letters, organize a letters-to-the-editor committee and give the members outlines of sample letters. Stagger the mailing of their letters a few days apart when responding to a timely topic, or a week or more apart when trying to maintain general interest in the topic. (For more tips on letter writing, see Chapter 3.)

COMMENT/OPINION

OUR READERS WRITE THE BOYERTOWN AREA TIMES — SEPT. 1, 1988

What we are missing without library

Editor:

I was not a resident of Boyertown for the "first" library but as a resident for the last 24 years, I have missed having a community library. I am the leader of the Girl Scout troops involved in this venture and would like to address the facts presented in a letter appearing last week.

When the girls proposed the idea of starting a library we did not just think it was "a neat idea." We did some research and visited all the libraries in Berks County. Tiny villages like Bethel have libraries! Morgantown's library is in the other half of a municipal building housing the cinder truck. That facility circulates over 1400 books a month. We are not talking a half a million dollar building here. We are talking of people who want to read and are willing to share whatever space they can find. Most of Morgantown's equipment is second-hand, with homemade furniture.

Tax dollars are really the bottom line in the formation of a library and the library committee is well aware of that. However, state tax dollars we pay are now going to other communities because we cannot share in the "library pie" since we have no library. Based on size and staff and municipal support, we could begin sharing those revenues by 1989, if we had a location.

The "first" library probably failed because of lack of material that would interest the public.

Through the use of Access Pennsylvania, any book can be in Boyertown within two days for a reader's use: so the size of the collection need not be large . . . definitely not worth $350,000.

Yes, you can get material from the high school library if you can find a parking space or edge your way through the halls during school hours. The library staff there is in favor of a public library. Yes, the bookmobile comes to town if you can remember when. Yes, you can go to Pottstown or Reading to their libraries if you can find a route without construction.

I do not know the Economy League or anything about them but they will probably tell you that the highest salary for a full time librarian in Berks County is $18,000. Most libraries do not begin with full time paid staff, but volunteers . . . ideal positions for senior citizens.

Speaking of senior citizens, most libraries contain areas of large type books for which our future library has been given a monetary donation already. In fact the library committee has received several donations.

If you or your organization want more information about our plans, we would be happy to show some slides of the Birdsboro Kutztown, Wernersville, and other neighboring libraries with facts on their use and support to show what we are missing living in the Boyertown area without a library.

Somehow my Girl Scouts, their families, and I do not feel a $1 or $2 per capita increase yearly in our tax dollar could be called a "burden" when we read where other tax dollars go.

Today, everyone is paying for and watching videos. . . shouldn't we consider the value of reading too?

Please consider signing a support pledge or attend a committee meeting for this most worthwhile cause. . . for a library benefits everyone.

Marian Borneman, leader
Girl Scout Troops 192 and 698
146 Crest Drive
Boyertown

Troop leader and Library Committee member Marian Borneman took it upon herself to tell *Boyertown Area Times* readers about all a library had to offer their community.

Reprinted courtesy the Boyertown Times

Editorial Meetings

Editorials have clout. Whether they correctly reflect public opinion or not, editorials tend to be perceived by policymakers as representing the majority view of the local area. Thus, legislators and policymakers read them to get a sense of which way the political winds are blowing and to determine whether or not a position they're taking is going to make them popular or get them hanged (figuratively speaking, of course). And the public turns to them to help make up their minds. Just as you strive to generate news (and, to a lesser extent, feature) coverage about your campaign, you should make it a priority in your media strategy to generate favorable editorials.

Try to meet with editorial writers in person. Request the meeting by letter, followed by a telephone call. Ask to meet with the editorial writer who most frequently addresses your topic, though at larger papers the editor of the editorial page and other members of the editorial staff (also known as the "editorial board"), along with reporters who cover the issue, may sit in, too. (At smaller newspapers, editorial positions are determined by the paper's publisher or managing editor, with whom you can also attempt to meet.)

At the meeting, make a solid presentation of your story, conducting yourself in a friendly but businesslike way. Expect tough, probing questions—editorial writers consider it their duty to thoroughly explore both sides of an issue. Emphasize not only why your issue is important but also when an editorial on the topic could have the greatest impact. Unlike reporters, editorial writers are expected to take sides; the editorial meeting gives you an unusual opportunity to lobby editorialists to adopt your point of view and express it in the paper.

You can generate an editorial on your TV or radio station in much the same way. Contact the station to determine if the station manager, the public affairs director, or the public service director of the station is responsible for editorials. Then send a letter explaining your position, along with background information on the issue. Follow up with a personal meeting or telephone call to stress the importance of an editorial favoring your position.

For both print and electronic media, send a short thank-you note after the meeting. Include any new details that reinforce your position. Then monitor their coverage. Have your volunteers ready to follow up any editorial with appropriate approving or disapproving letters to the editor. A favorable editorial merits a phone call from you to the writer thanking him or her personally.

If the editor will not meet with you, attempt to have a lengthy telephone conversation. Keep all editorial writers regularly informed with news releases, background papers, fact sheets, and other materials you develop for the campaign.

Editorial Cartoonists

Editorial cartoonists, whose drawings caricature issues, politics, and politicians on the editorial page, should not be overlooked as another avenue of publicity. Though few cartoonists agree to attend meetings, they can still be lobbied via the mail and, occasionally, telephone. Their cartoons, reflective of their sometimes substantial knowledge of and interest in an issue, often carry as much zing as an editorial or op-ed piece.

Don't assume that a cartoonist's position reflects that of the newspaper that employs him or her. Cartoonists frequently pen their drawings independent of the paper's editorial stance; in fact, cartoons can effectively counter negative editorials just as much as they augment positive ones. Add editorial cartoonists to your press list and alert them by fax or phone when a cartoon favorable to your issue would be particularly timely.

Editorial Mailings

In addition to keeping editorial writers apprised of your issue through the normal flow of information you send out, you may want to develop specific mailings aimed at editorial writers when a vote or decision is imminent and an editorial favorable to your position is essential.

Send a one- or two-page "editorial memorandum," addressed to editorial writers and editors, with the date and your name clearly indicated. Explain the tactical situation you're in and explicitly ask for editorial support at this time. Attach a breaking-news release, a relevant fact sheet or backgrounder, and copies of favorable editorials from other papers.

Follow up by telephone.

Op-Eds

Op-eds can help raise public awareness on issues, inform elected officials and policymakers, suggest solutions to important problems, and educate the media. Op-eds are longer than letters to the editor (750 words as opposed to one or two paragraphs). They are usually printed opposite the editorial page, and generally are widely read.

Unlike straight news stories, op-ed pieces allow the writer to express a point of view. Still, they must be very well written and to the point, offering a fresh perspective on a timely issue. To write an op-ed, develop an outline of the most important arguments about your issue you want to make. Use anecdotes or vignettes that put the story in a context readers can relate to and that make the op-ed more appealing. Offer solutions to the specific problems that concern you rather than whine about the problems alone.

Even though you are writing to state an opinion, the more your point of view is bolstered by solid facts and figures, the more likely it is that your piece will run. Quote experts and authorities whose opinions substantiate your own.

Editors are inclined to run an op-ed piece if it is written by someone who is well respected in the community or is in a position of authority vis-à-vis the issue being discussed. So to enhance the chances that your op-ed will be printed, co-sign it with an individual editors will recognize and respect.

Unfortunately, most editors will not "commission" op-ed pieces.

One meeting could save your life

By Nancy Ricci

The author is a Yalesville resident, a graduate of Lyman Hall High School and the State Academy of Hairdressing. She and her husband James have three sons.

I'm tired of hearing vital statistics on drunk drivers. Why don't we just stick with human feelings; the love and compassion we all share.

Drinking has its feelings; it's great to sit back, relax and have a drink, but before you know it that feeling brings on being drunk. You have that right, but NO ONE has the right to mangle or kill someone. That's forever. When you get in a car your feelings are numb, your reaction time is longer, red lights become blurred, on-coming head lights are blinding, your weaving pattern gets easier for you can't follow or see the lines. Why don't we just think before we drive? No one ever thinks of themselves getting into an accident or even wants to.

When you're tucking in that special someone tonight, think about it. Look at your baby's or child's eyes, their smiling innocent faces full of love and life. The tenderness you feel is so very special; they're a part of you. No matter how old they are, they're special. You know they're safe, you would never ever want them harmed. You give them hugs and kisses and tuck them in for a quiet peaceful night of beautiful dreams. How would you like to tuck them into a casket and finally into their graves? Never ever able to see, hear or feel them? There are no more tomorrows and no more beautiful dreams for you. Hold someone special, right now! Close your eyes, you still have the sensation of touch — death takes that! You see blackness and emptiness — death brings that! It never goes away. Sure, you learn to live with it, but the pain and emptiness last forever, they are feelings!

Constant

When you see that empty chair, bike, car, bed, whatever it is is a constant reminder of what was. But even if you give away their belongings, you can't erase your feelings. You can't bury them!

We can't sit back and rob our loved ones of the right to live. We have to help one another. Think before you drive drunk. Hand over your car keys — it's not a weakness, it shows your strength and compassion for others. Never let those feelings die! By making others aware, you make them able to help you. Don't let yourself wake up some morning to discover that you mangled or killed someone the night before. Your feelings were dead because you were drunk and you might not be able to even remember the night before. You are lucky to be alive — it's your right! What right do the victims or their families have? None. Let our hearts reach out. Have the feeling of working for a safe community. Remember it can happen to you or yours!

Personal

I once thought, "Us? Never! — it's something you read about." But at age 19, my brother was killed thanks to a drunk driver. God, it hurt!

We never expected that to happen again. It

Nancy Ricci

could never happen again. Yet four years ago my sister and her friend were killed by a drunk. "My God!" we thought, "What's happening! When will it ever stop!"

It won't stop unless we do something about it. I'm scared constantly. Parties and socials are always going on. We all need to learn to think before we drive drunk.

RID — Remove Intoxicated Drivers — is trying to educate and make people aware of our drunk driving problem. We are trying to make it a whole community affair. Our Mayor Dickinson and Assistant Chief of Police and other local groups have given us their support. We are working with the SADD (Students Against Drunk Drivers) chapters of Lyman Hall and Sheehan High Schools. We will be the adults behind them if they need our help in their programs.

We are having a town-wide meeting Sept. 25, at the community room of our town Library at 7:30 Tuesday night.

Nancy Winialiski — area RID coordinator and Director of Gaylord Hospital's Alcohol Rehabilitation Unit, will be opening speaker. Elenor Sutton of Lyme, who is on the board of directors of RID and was formerly of Citizens Concerned about Drunk Drivers, will speak and show the movie "Kevin's Story."

SADD — Students Against Driving Drunk — Advisors Nancy Morand and Frank Stupakevich will represent Lyman Hall and will show a commercial made by the students. Pam McCarthy Kromble and students from Sheehan will also speak.

If you can attend only one meeting a year, this will be one of the most important. The life you save may be yours or that of someone you love.

Please come.

Nancy Ricci's passionate appeal to take action against drunk drivers sounded from the op-ed page of the *Meriden Record-Journal*.

Reprinted courtesy the Meriden Record-Journal

In fact, they will not even tell you if they will consider running what you want to write until you write it and they have a chance to review it.

Don't just mail off your op-ed piece willy-nilly. Write a cover letter to the op-ed page editor explaining why your commentary on the topic is so important. Send the letter, the op-ed, and your press kit to the editor. Call at least twice to urge the editor to print your piece and to remind him or her about its timeliness.

If your op-ed is rejected, rewrite it into a shorter version and re-submit it as a letter to the editor.

Special Events

Special events—like fairs, celebrations, demonstrations, and fund-raisers—can generate substantial positive publicity for your group. For an in-depth description of how to organize successful events, see Chapter 3.

Public Service Announcements

Public Service Announcements (PSAs) are advertisementlike messages made on behalf of nonprofit groups (for a description of what constitutes a nonprofit group, see Chapter 4). Most radio and television stations and some newspapers and magazines run PSAs free of charge as a community service, though they are under no legal obligation to do so.

PSAs can be used to announce a meeting, invite the public to a rally or celebration, build membership, raise money, or provide information. They will not be used if they attempt to lobby for a piece of legislation, elect someone to office, or are otherwise political in nature. In other words, you can use PSAs to invite people to a candidate's forum—not to tell them which candidate to vote for.

Although they can be useful in reaching a mass audience, PSAs suffer from several limitations. You cannot determine when the spots will appear or what kind of story they will accompany. (Frequently, they run on broadcast outlets in the wee hours of the morning, when the fewest people are watching or listening). They're very labor intensive to produce and place. And they have very limited (if any) impact on policymakers. To get substantial placements you must convince an outlet that your message is especially important to its local audience. In fact, some organizers feel that you should only turn to PSAs when you've done everything else to organize and publicize your issue.

Still, if you decide to produce PSAs, these general tips should help you make the most of your placement efforts for both broadcast and print outlets.

BROADCAST (RADIO AND TV)

Although exact guidelines for the acceptance of PSAs vary from station to station, some general rules hold:

- The sponsoring group must be a nonprofit organization.
- All announcements must identify the sponsoring groups.
- Finished or preproduced PSAs must be provided to the outlet at least three weeks before you want them to run, earlier if possible.
- PSAs that promote a specific event or activity are more likely to be used than those that address long-term issues or revolve solely around encouraging people to join a group.

According to The Advertising Council, a consortium of advertising executives from around the United States that helps nonprofit clients develop and place PSA campaigns, shorter broadcast PSAs and smaller print PSAs get used more frequently than longer, larger ones. The Ad Council's surveys of outlets that carry PSAs show that ten- and twenty-second spots garner the most air time, while print PSAs that take up an eighth, a quarter, or other fraction of a page receive more placements than do full-page PSAs.

When you approach the radio and television stations in your area for air time, keep in mind these following suggestions:

• •STEP BY STEP• •

How to Get Your PSA on the Air

• **Telephone the station.** Ask to speak to the public service director. Give your name, the organization you represent, and briefly explain the campaign. Ask if you can make an appointment to come in and discuss your request. Always attempt to make your case in person.

• **If you are asked to send your PSA by mail, draft a letter and send it along with your PSA.** The letter should explain why PSAs on your issue are important to your community. If the public service director agrees to meet with you at the station, prepare the letter anyway and submit it as part of your "formal" request. If the station agrees to air the spots, you will not need the letter. If you are turned down, ask that a letter of reply be sent to you, giving the reasons for the station's rejection of the spots.

• **Provide the public service director with a "PSA Campaign Kit" that includes the following items:**
    * a brochure, fact sheet, or press kit describing your group;
    * a copy of your press release announcing the purpose of the campaign;
    * a letter from the IRS or some written material regarding your group's tax status if the station requires it;
    * scripts of the radio spots and videocassettes of the television spots if they are preproduced; announcer copy for spots you want read on-air;
    * a "bounce back" postcard that outlets can complete listing when the spots aired or will air to help you keep track of outlets that use the spots for future PSA campaigns.

• **Follow up the meeting with a letter of thanks.** Even if the station refuses to air your spots now, send a letter expressing your desire for reconsideration in the future.

• **Monitor stations to insure that your spots are being carried.** If they don't appear with the frequency you were promised, contact the public service director and find out if there's a problem.

PSA DIRECTOR CONTACT SHEET

Use the worksheet on the facing page to help you secure PSA placements promoting your issue. Try to reach all the TV stations, the most popular radio stations, and the major newspapers in your area. Aim for ten outlets. To discover which radio stations in your area are most popular, consult the *Radio Market Report,* a series of books that offer statistical profiles of radio listeners in over 250 cities or metropolitan areas. Included is the gender, age, and demographics of listeners and the frequency with which they tune in—from this may be derived what stations are most popular among a given group in that area. The books are published several times a year by the Arbitron Company and are sold only to stations and other Arbitron clients, but you will likely find that a set has been donated to your local or university library. You can also check *Broadcasting* magazine, which has published analyses of the Arbitron data. To find phone numbers, consult your local telephone directory or the library. Contact each outlet and ask to speak with the public service director personally about the campaign. If no public service director is employed, ask to speak with the person who screens PSAs. If possible, arrange a meeting to discuss the campaign, and secure a commitment to run the spots.

PSA Director Contact Sheet

Outlet	Public Service Director	Address & Phone Number	Outcome
(TV)			
(TV)			
(TV)			
(TV)			
(radio)			
(radio)			
(radio)			
(radio)			
(newspaper)			
(newspaper)			

PRODUCING YOUR OWN PUBLIC SERVICE ANNOUNCEMENTS

Radio PSAs may be submitted to stations in two forms: a written script (known as "announcer" or "live" copy) that will either be read live on the air by the announcer or taped for reuse by the station; or a preproduced tape of your message.

There are several advantages to submitting radio PSAs as live copy. It is less expensive to submit a written script than to tape a spot. And announcer-read copy can be much more effective since the announcer's voice is familiar to the station listeners.

Approximate word count for PSAs:
    10-second spot: 20 words
    20-second spot: 50 words
    30-second spot: 75 words
    45-second spot: 110 words
    60-second spot: 150 words

When writing a script, follow these guidelines:

- Get the attention of the audience. Be punchy and upbeat.

- Develop interest by showing how listeners will benefit. Ask for action (tell where to go, what to do, when to do it). Suggest the listener call you for more information. Include the name, address, and/or phone number of your organization.

The following PSA copy, read live by radio announcers, helped solicit books for the Boyertown Community Library. Local Radio Station WBYO ran this PSA twenty-three times.

ATTENTION BOYERTOWN AREA RESIDENTS

Senior Girl Scout Troop 688 requests donations of books for the newly formed Boyertown Community Library. We need books for preschool children and adults in particular. We are applying for the Colgate Youth of America grant and need your support by your donations of these books.

Presently we have wastebaskets of appropriate books located in local beauty shops, barber shops, and day care centers.

If anyone is interested in donating books or would like books delivered to a new location, contact Boyertown Community Library, Box 641, Boyertown, PA 19512, or call 367-9167.

Thank you for helping our troop's efforts in helping form a new public library.

Check the requirements of your local stations before you begin to write your PSAs; many radio stations prefer to have their deejays read PSAs to keep the message flowing with the rest of the program. Call the station to determine its preference before producing a spot that will not be used. Ask the following questions:

- Will the station accept both tapes and scripts, or only scripts to be read by announcers? Does the radio station want a cassette tape or reel-to-reel spools?
- How many copies of the scripts/tapes does the station require?
- How much lead time does the station need to fit PSAs into its schedule?

Television PSAs are very expensive to produce. Film and videotape, camera crews, on-camera "talent," editing, and reproducing the spot could cost you several thousand dollars, even if you produce a simple spot. Before you begin to rack up costs, ask your community cable station if it can help you produce a PSA. Local cable stations frequently offer training in video production skills to members of the community, providing access to cameras, editing equipment and studio space, as well as staff assistance on the production itself. When preparing to distribute your PSAs, ask television stations what format they prefer to receive them in: one-inch, three-quarter-inch, or beta. (You can submit live copy and a slide to a TV station if you don't have a videotape to distribute, but it probably won't be used much.)

PRINT

Most newspapers and magazines are reluctant to forego advertising revenue to run free public service announcements. The nature of the medium also works against public service advertisers: "The problem with print media," says Jim Laseter of Wray Ward Laseter Advertising in Charlotte, North Carolina, "is that a newspaper is an expandable medium; a publisher can decrease the number of pages at will if there is not enough material to fill them. Radio, on the other hand, has a set amount of air time that it has to fill."

Large-circulation magazines that are distributed nationally (like *Time, Life,* or *Ebony*) receive a flood of print PSAs every week and tend to choose those that are slickly produced and pertain to an issue that's perceived to be of national importance. Smaller local or regional magazines may be interested in your issue, but their available ad space is tighter; most run PSAs on a space-available basis and may have to choose only those that represent a publisher's pet cause or charity.

The same rules generally hold true for newspapers. Most of the large daily papers do not run PSAs at all (unless they are produced by The Advertising Council), while others print only PSAs about issues that the publisher is personally interested in. Smaller newspapers vary in their PSA policies. The *Burlington Free Press* in Burlington, Vermont (circulation 54,000), does not run free PSAs but offers about a 30 percent discount for religious groups and charities that purchase ads. The *Ithaca Journal* in central New York State (circulation 24,000), on the other hand, runs a few public service advertisements per day, and tries to include all that the paper receives.

The Phoenix Newspapers, Inc., publisher of three papers in that city, takes public service a step further by choosing a community issue it feels needs to be addressed and building a public service advertisement campaign around it. "PSA space is limited today," said Laura Atwell, manager of community and corporate affairs for the newspaper group, which has a combined circulation of around 500,000. "There are few opportunities for grass roots groups to get into the paper." She told of one campaign, launched in 1990, that was spurred by the unusually high number of child drownings that had occurred in Phoenix's backyard pools. The newspaper group's community and corporate affairs department sought out a coalition of companies, officials, media, and parents that had organized to fight the problem, and worked with them to create awareness ads that ran in all three papers for three years.

Newspapers and magazines have slightly different requirements for the artwork they'll accept as a PSA. Newspapers generally accept black-and-white only, and artwork should be printed at approximately a 65-line screen. The line screen designation signifies the density of the tiny dots that form the image. Newspapers and magazines use different printing technologies, and artwork must be prepared accordingly. Magazines prefer color ad "slicks"; in fact, the more colorful or visually striking the ad is the more likely they are to place it. Magazine PSAs should be printed at approximately a 110-line screen. And remember, all print PSAs should be developed in several different sizes—full-page, half-page, quarter-page, eighth-page—to increase the chances they will be used.

Drugs Don't Work.

For Drug Free Workplace Info, call: 488-2236

Making your workplace drug-free can also help you make a profit. All you have to do is start a "Drug-Free Workplace Program" and you'll receive a 5% discount on your worker's compensation premium. Any Florida employer who complies with Florida Department of Labor guidelines is eligible. So clean out your office today. Because drugs don't work. For anybody.

BUSINESS AGAINST NARCOTICS AND DRUGS PROGRAM (B.A.N.D.)
VENICE AREA CHAMBER OF COMMERCE

PARTNERSHIP FOR A DRUG FREE FLORIDA

Businesses can have their own drug problems. The Partnership for a Drug-Free Florida got together with chambers of commerce to run this print PSA campaign in 575 local media outlets. Other state partnerships are designing PSA campaigns using Florida's as a model.

What can you do to increase the likelihood that newspapers and magazines will carry your PSAs?

- **Build relationships with publishers.** Help publishers understand why your issue is so critical to the health and welfare of the community.

- **Make your PSAs relevant.** The more your PSAs focus on solving a local problem or addressing a community issue, the more likely they are to be placed.

- **Get to know the outlet in which you're trying to secure placements.** Do you meet their PSA requirements?

- **Submit PSAs in the proper format.** You may be asked to provide PSAs in a variety of sizes (full-page, half-page, quarter-page, vertical, horizontal, etc.)

- **Treat advertising campaigns like fund-raising drives.** Appoint a committee to do the work, plan the drive carefully, and build solid relationships with those who have the power to agree to run your campaign.

- **Use alternatives.** If your PSAs just aren't getting placed, try community calendars and other publicity techniques discussed in this chapter.

Local Hell-Raiser

A PSA Campaign Brings People Back to the Park

Meridian Hill Park opened in the nation's capital in 1936 with 11½ acres of lush gardens, imposing statuary, and spectacular fountains designed to rival Paris's breathtaking Tuilleries. The site of popular starlight concerts and lighting displays for the next twenty years, the national park hosted Pearl Bailey and the Von Trapp Family Singers, among many other celebrated entertainers. But lack of funding and upkeep by the National Park Service in the 1960s gradually surrendered the park to the ravages of crime. Several attempts to revitalize the area, which had come to be called Malcolm X Park, failed; before long, gang activity, drug dealing, and murder had supplanted the wholesome entertainment that had made Meridian Hill so special.

When a local boy was killed there in a drug-related execution in January of 1990, concerned citizens took matters into their own hands. Residents of the four ethnically diverse neighborhoods that bordered the park formed a crime patrol and began policing the area to intimidate the derelicts. They badgered the Park Service to make the sanctuary safe again. And they formed the Task Force to Save Meridian Hill—later, Friends of Meridian Hill, Inc. (FOMH)—so that they could take direct action and achieve more immediate results.

The determined activists began to overhaul the park, cleaning up both its lawns and its image. FOMH doggedly nagged the Park Service to repair lights, install more trash bins, and step up park patrols. On Earth Day 1990, volunteers picked up garbage and planted trees. The hell-raisers also launched a lobbying campaign to pressure the Park Service to spend more money to restore the Meridian Hill grounds and chase out the riffraff.

To draw people back to Meridian Hill, the heartened group scheduled an appealing lineup of cultural events. A booming July 4 celebration in 1990 featured the first starlight concert in fifteen years and paved the way for the annual Save Our Park Day, regular concert seasons, community picnics, even a birthday party for the regal statue of Joan of Arc that graces the center of the hill on which the park is built.

The activists have been applauded for the substantial drop in crime their activities engendered. But for many, Meridian Hill's scary reputation lingers. Though it again has its "regulars"—one local man reverently speaks of the park as "the most beautiful place in the world"— many Washingtonians still choose to take the long way around rather than walk their dogs through Meridian Hill.

To erase these last vestiges of fear, FOMH decided to advertise the rebirth of Meridian Hill in a public service advertising campaign featuring the park's most stunning feature: its cascading fountains, reputedly the largest in the country.

Since the park was accessible by the newly completed "green line" of the city's Metro subway system, the group approached Metro for free advertising space, promising an ad that would promote both the park and the subway. Metro generously agreed, donating large illuminated panels in twelve different locations throughout the subway system that would normally cost a group $600 a month per panel.

Now the group needed the actual PSA. Just as it began to search for an inexpensive but striking photo of the fountains, a Baltimore photographer serendipitously phoned offering the activists free photos he had taken of the park. A sympathetic advertising agency agreed to produce the PSA at the discounted cost of $3,500, which was still a substantial sum to the nonprofit advocates of Meridian Hill. Luckily, a FOMH member (who was a former employee of the local C&P Telephone company) managed to raise every precious dollar by sending a letter to the president of C&P and following up diligently. Steve Coleman, FOMH president, explained the group's good fortune in fund-raising and public relations in a word: ambition. "Groups can 'arrange' to be lucky," he said. "People come out of the woodwork to help."

FOMH's good luck continues. As the group prepares to unveil the beautiful PSA, whose copy—"Experience the Renaissance: Take the Green Line to Europe"— celebrates Meridian Hill's cultural diversity and lives up to its promise to Metro officials, a national lighting company has tentatively offered to illuminate the cascading fountain permanently, which is now shrouded in darkness at night. Originally, Coleman said, the huge fountain was designed to be lit so brightly that it would make the park visible from the air.

The PSA will have a wider audience than those who ride the subway in Washington, DC. FOMH plans to mail reproductions of the PSA, coupled with news clippings about the park's makeover, to travel guidebooks such as Fodor's and Michelin's. In the past, these books have warned tourists to avoid Meridian Hill if they valued their wallets. FOMH hopes that the editors will update their information with the positive news about

Meridian Hill, giving the city's many tourists a greener oasis to visit the next time they come to the nation's capital.

EXPERIENCE THE RENAISSANCE

MERIDIAN HILL

This lit subway PSA encouraged riders of the Washington, DC, subway system to "Experience the Renaissance" and visit the city's Meridian Hill Park.

Advertising

There are times in a campaign when you need to bypass the "media gatekeepers" (publishers, editors, producers, and reporters) and take your message—undiluted—directly to the public. There are also times when your issue is just too controversial to be promoted adequately in a PSA campaign. When instances like this occur, provided you can raise the money, you may opt to buy advertisements.

Paid advertising is appealing for several reasons. It gives you much greater control over when your message runs, where it runs, and what it says. Paid ads can thus be far more effective in promoting a controversial issue or urging people to take a position on a contentious topic. While PSAs can be used to announce an event (for example: "Attend a public hearing to discuss question #3") and relay information, paid ads (in many, but not all, cases) can tell people what to do ("Be on the safe side: Vote *for* question #3").

Activists also turn to paid ads when the timing of an issue is critical and they can't wait for a PSA placement, when they need to counter an opponent's advertising or respond to attacks, when they want to reframe the way an issue is being discussed, or simply to thank their supporters.

BROADCAST ADS

Depending on the city or "media market" in which you live, radio and television time can be extremely expensive. The cost of an ad will depend on how long it is, what time it runs, and how often it runs. A sixty-second radio spot in a major market such as Philadelphia, aired between 6:00 A.M. and 6:00 P.M., would cost between $800 and $1,000. For a smaller suburban station, the same spot could cost as little as $20. Rates for TV ads vary much more. A thirty-second overnight or early-morning spot on a Philadelphia station would cost around $200, an early news spot would go for around $2,000; prime time placements during specials or sporting events would cost $10,000 or more. In general, the smaller the ADI (area of dominant influence) or geographical area that the TV signal covers, the cheaper the ad. Cable TV is usually cheaper than broadcast TV in the same ADI; you may also be able to buy print ads in the viewer's guide.

PRINT ADS

The price of placing print ads also varies, depending on the size circulation of the publication you're buying space in and the size ad you plan to run. Some newspapers and magazines offer discount rates for charities, though you'll need to verify that your organization meets the outlet's definition of what constitutes a charity. Most likely, you'll have to show proof that you're a 501(c)(3) organization in order to receive the outlet's charity rates; even then, you may be excluded if your group is deemed too controversial.

Advertising rates at newspapers vary widely. Compare the following September/October 1993 rates at four dailies of different sizes. Rates and circulation are for weekday papers; Sunday rates and circulations are typically slightly higher (a column inch is one column wide and one inch deep).

- The *Houston Chronicle,* Houston, Texas (circulation 423,000). Standard rate is $156.96 per column inch; charity rate is $90.12 per column inch.

- The *Salt Lake City Tribune,* Salt Lake City, Utah (circulation 111,000). Standard rate is $56.14 per column inch; no charity discounts.

- The *Salina Journal,* Salina, Kansas (circulation 29,000). Standard rate is $11.04 per column inch; no charity discount.

- The *Idaho State Journal,* Pocatello, Idaho (circulation 20,000). Standard rate is $13.02 per column inch; charity rate is $7.88 per colum inch.

Rates at magazines are going to fluctuate even more wildly, depending on the magazine's circulation and the size ad you buy. Check with the magazine's advertising sales department to get current rates for the magazines you're interested in.

Though buying ad time or space can be expensive, producing the ad will also cost you money. To help keep your advertising budget under control, try to find a local advertising agency that is willing to donate its creative services on a *pro bono* basis, either because it cares about your issue or wants to beef up its image as a socially conscious firm. But remember, even an agency that gives you free advice in writing ad copy and designing the spot will probably still charge you to produce the ad (to shoot and edit the videotape for a TV spot, to tape record an audio spot, or to design and lay out a print spot). Before production begins, get a detailed budget on what it will cost, and make sure it is something you can afford.

Outdoor Advertising

Everyone knows about Smokey the Bear and McGruff, the trenchcoated dog determined to "Take a bite out of crime." You may have seen these two mascots and their messages in public service announcements on television or in newspapers or magazines, but you are just as likely to have met them through an ad panel in the bus station or on a billboard.

Outdoor or out-of-home advertising outlets—which also include subway stations, bus stop shelters, and telephone kiosks—are an effective but frequently overlooked option for both public service and paid advertisements—especially those designed to deliver quick, visual messages.

The owners of outdoor advertising outlets, just like the management of print and broadcast media, will donate space to nonprofits and community groups when it is available—that is, when they can't attract paid advertising for it. Sometimes the space is less than prime, or perhaps the advertising or media company is having a slow month or needs positive PR. (Billboard companies frequently donate space to public interest groups, but there are some obvious environmental conflicts—see the discussion later in this section.) Usually, if you are given free space for an outdoor PSA, you must still pay to have the ad produced and installed or pasted into place.

Outdoor ads that run at the same time as broadcast PSAs and carry the same image and/or text copy, can have more than twice the impact of either option alone: Hearing and seeing your organization's name in two completely different contexts helps people remember what you have to say.

TRANSIT AND RETAIL OUTLETS

Many opportunities exist to post your message:

- The exteriors and interiors of buses, on bus stop shelters and in bus stations. Of the country's 60,000 buses, 35,000 carry advertising.

- Both inside and outside subway cars; in elevators; along escalator corridors and walkways; and on the platforms of subway or train stations. The large, "backlit" panels you see in the stations are called dioramas.

- Airport displays.

- Mall displays, and ads in supermarkets, drug stores, and convenience stores (ads are frequently seen on shopping carts or illuminated alongside a large public clock).

- On telephone enclosures and outdoor kiosks belonging to ticket agencies, quick film developers, banks, and other businesses.

- In stadiums and other sports arenas, at ski resorts, and in movie theaters.

Public gyms and health clubs, doctors' offices, and blood donor stations often provide framed wall space for health-related messages. Other miscellaneous media forms available in some markets include painted walls, truck advertising, displays on college campuses and on military bases, air banner towing, and rooftop taxi ads.

PAID OUTDOOR ADS

If you can't convince a mass transit company or business to post a PSA, don't rule out purchasing the space. As you will see in the next Local Hell-raiser story, outdoor advertising offers one way to spread your message that can give you a big bang for your buck. (The Inland Empire activists actually chose to purchase their busboards because their time-sensitive campaign could not rely on waiting for advertising leftovers.)

Transit and other outdoor ads typically costs much less— considering the number of people reached, the longer display time and the potential for repeat viewings—than newspaper advertising or broadcast. Outdoor ads appear longer than newspaper ads and are more inexpensive to produce than broadcast spots. The subway ads prepared for Friends of Meridian Hill only set them back about $3,500 in production costs. For that, they got twelve large backlit ad panels. (The price included design and materials—labor was donated.)

Some companies have contacts with a wide assortment of out-door advertising outlets and will help plan your entire promotional campaign, as well as negotiate and arrange the purchase of space. Outdoor Services, Inc., is the largest of these firms, with seven offices across the country. But such companies are expensive, especially for small campaigns. You can do the work yourself by looking in the telephone directory under "Advertising—Outdoor."

Outdoor campaigns—whether using purchased or donated space—require the same planning as those designed for print or broadcast media. Once you have come up with a general message for the campaign and made the decision to place your ads out-doors, adhere to the following steps to develop an eyecatching campaign:

● ●STEP BY STEP● ●

● **Identify the type of audience you are trying to reach—the "demographics" of those at whom you are aiming your message.** Are they citydwellers or suburbanites? What are their socioeconomic status and purchasing patterns? Also study the space available in your media market. The different boroughs of New York City, for example, offer a wide variety of outdoor media opportunities: in Manhattan, subway, bus, commuter trains, and urban panels are most available and effective; Staten Island, on the other hand, has abundant billboard space.

● **Is your objective to reach a broad range of people once, or a more targeted group repeatedly?** Choose your medium accordingly: An airport ad panel reaches many travelers, but it will only be seen once or twice. A subway ad reaches many of the same riders every day.

● **Decide how long you want your campaign to run,** depending on the way different outlets are scheduled. Supermarket ads are sold or donated in four-week cycles, truck panel ads for six months or a full year. The outdoor vendor can tell you how many postings you need and how long to run them to ensure your message permeates an area.

● **Time of year is important, so plan ahead.** Time slots in late October, November, and early December are sold many months in advance for holiday promotions.

● **Avoid jarring political imagery in your ad.** Your city's mass transit department may not want its buses plastered with graphic depictions of animal testing or abortion. Your cause may be controversial, but your image doesn't have to be.

● **Although outdoor ads are less expensive than broadcast spots, factors such as weatherproofing, lighting, size, and full-color processing make their production complicated and costly.** The vendor will tell you exactly how to prepare the ads and who can do it locally, but don't hesitate to shop around for someone cheaper. The Friends of Meridian Hill, while based in Washington, DC, worked with a Florida art production company that had a slightly longer turnaround time but a significantly lower price. Nonprofits and community groups frequently can get discounts.

BILLBOARDS

Many environmental organizations believe that billboards, regardless of the message, deface the countryside. Organizations such as Scenic America, a national group based in Washington, DC, is among the organizations striving to restrict or ban billboards along roads and highways in many states. Scenic America's Frank Vespe notes that billboard industry lobbyists are always on the lookout for social interest groups—especially environmentalists—who use billboards so the industry can showcase this usage and promote its cause by saying: "See, public interest group X is in favor of billboards!" Beware of billboard companies that gladly give up free space and ask for a letter of thanks or support in return; that letter may be used by the company or by lobbyists to ballyhoo billboards. Using billboards for advertising or PSAs will also make it difficult for you to join any local efforts opposing more offensive activities of the billboard industry, such as cutting down trees to improve billboard visibility or carrying alcohol or tobacco ads aimed at young people.

While it is easy to sympathize with Scenic America's argument, environmental and other organizations may not want to reject out of hand billboards as a potential communications medium—though you can still oppose the construction of new billboards. They just need to weigh the tradeoffs. Dave Crandall and the Inland Empire Public Lands Council produced arresting but tasteful billboards that complimented other outdoor ads in a striking campaign to protest clearcutting in Washington State's Colville National Forest. Crandall said that even environmental groups in the state overwhelmingly supported the campaign's creativity, although some noted that they wouldn't launch a similar campaign because they disapproved of billboards. "Whether we like it or not, billboards are there and they are a form of advertisement," he said. He felt that the cost-effectiveness and high impact of their campaign outweighed the small amount of money they contributed to the industry.

Billboards are donated on a space-available basis or may be purchased by the month, for as many as four months at a stretch.* Billboards line the roadways in 8,000 U.S. cities and towns in ev-

---

*To request donated billboard and sign space, contact a billboard leasing company such as Gannett Outdoor [which can be reached in New York City at (212) 297-6400], which owns billboards around the country.

ery state except Hawaii, Alaska, Maine, and Vermont, where they are banned. Posters are normally rotated to a new location every thirty days.

You can choose from different sizes:

- Bulletins, the largest billboards, measure 14 feet high by 48 feet wide, are illuminated, and have their messages rotated every sixty days.

- "30-sheet posters," or smaller billboards about 9 feet high by 21 feet wide, are found along commuting routes.

- "8-sheet posters"—5 feet by 11 feet—can be seen in neighborhoods and can target harder-to-reach audiences who are less exposed to other media forms.

Local Hell-Raiser
Tell It Like It Is

In a state like Washington, on an issue like clear-cut logging, it's not hard to be outspent or outmaneuvered, particularly when you're an underfunded environmental organization and your nemesis is the well-heeled timber industry.

So the Inland Empire Public Lands Council was putting a lot at risk when it launched a five-phase campaign to educate Tom Foley, a U.S. Representative from Washington State and the powerful Speaker of the House, about the need to protect Colville National Forest from logging practices that could reduce the majestic stands of ancient timbers to a clear-cut tract of barren land.

Dave Crandall, executive director of the council, a nonprofit group working to protect forests in northern Idaho and eastern Washington, described Colville as one of the worst managed forests in the entire country, in response to a slick industry advertising campaign intended to reassure the public that the timber industry was environmentally conscious. "We wanted to get the true message out," said Crandall. "The environmental community often just talks among themselves; we needed to

Spokane commuters saw this billboard on their way to work for two months during the summer of 1992. Considering the number of out-of-town drivers who passed the ad as well, the Inland Empire reached far more people—and spent a lot less money—with the billboard PSA than it would have on a full-page ad in the Spokane newspaper.   *EPLC Photo Archives*

reach the broader public" (and thereby educate Representative Foley), he said.

Outdoor advertising, besides being an obvious choice for a forest preservation campaign, allowed the council to use its strong grass roots volunteer base. People were needed to hang doorhangers and pound in yard signs. Outdoor ads would also enable the council to reach many Spokane residents while staying within its limited budget—all the while assuming that in retaliating the industry would outspend them ten to one.

Julia Reitan, a Sierra Club organizer who happened to have a background in advertising, designed the artwork for the ad that the group conceptualized: a startling black-and-white photograph of a clear-cut forest with the words "FOREST SERVICE APPROVED. YOUR COLVILLE NATIONAL FOREST: A CLEAR-CUT SHAME!" stamped across it in red.

Their first outlet was billboards. The council placed

the alarming image on eight different billboards in Spokane throughout May and June of 1992. According to the billboard vendor, roughly 139,000 people saw at least one of the billboards during that time. The cost ($4,300), Crandall remembered, was less than a one-page, one-hit ad in the *Spokane Spokesman-Review,* the city's daily newspaper—which, for $5,400, would have reached only 101,000 people.

Also in May, 250 volunteers hung 40,000 door hangers—bearing the same image as the billboard—in one day. "We just blanketed this place," Crandall said. Television crews gathered for a speech at the council's office, then followed the door hangers on their routes. The council invited the media to attend or cover each component of the campaign, trying to get as much free media as possible.

On August 11, the ad was posted on busboards and traveled more than half the bus routes in Spokane for the next month, at a cost of about $3,000. Grass roots help was again called upon when the council launched a postcard campaign, repeating the image and message and asking concerned citizens to mail the cards to Congress, where legislation was in the works to save ancient forests.

Mid-August also saw council volunteers knocking on door after door, asking people to display the clear-cut ad—which the council had printed on yard signs—on their lawns. By this time, Crandall said, everyone had seen the ad somewhere and was aware of the problem: People passing through neighborhoods stopped the activists in the streets, wanting signs for their yards. The team installed a thousand signs in a single weekend, and was able to place fifty-three of them on one of the city's busiest streets.

The timber industry responded by buying up the same billboard space in the following months and running its own ad (that showed a young forest and read "The National Forests: A Legacy of Caring"). And industry employees systematically drove pickup trucks through neighborhoods and removed yard signs.

But the campaign's impact on the public and the legislature had already been made. The offices of Speaker Foley and Washington Senator Brock Adams were flooded with calls from citizens concerned about the clear-cutting. Foley, after flying over Colville to view the barren stumps and discarded timber the clear-cutters had left behind, approached the House Appropriations Committee and secured money for an inventory of old-growth trees in Colville, due to be completed in 1994; a health study of eastern Oregon and Washington forests, which found the trees to be ailing; and an analysis designed to recommend a new management system for those forests, also to be completed in 1994. Although the council won't take full credit, Crandall felt that blanketing Spokane with stark photos of the devastation undeniably helped persuade Speaker Foley to take some action.

Interview Shows

Practically every radio and television station produces a talk show or community service program that airs at least once a week. The shows will either be live, taped in advance and edited, live with phone-ins, or interviews via phone. Some stations also take portions of a taped interview and use them in news or feature segments throughout the week.

In many regards, you treat interview shows the same way you treat reporters who cover news beats. You keep lists of the shows and who their producers and hosts are. Popular shows in big media markets will have a producer (sometimes called a "booker") who books guests for the hosts. Programs in smaller media markets may be arranged entirely by the person who hosts the show.

Send information in advance—a cover letter explaining what the issue is and who you're recommending as a guest, and a press kit describing the issue in greater detail. The press kit should also include a one-page biographical profile establishing your guest's credentials.

Just as you need to convince an assignment editor that the topic is newsworthy, you'll have to convince the booker or host that the topic is current and that your guest will be relevant and entertaining to the show's audience.

• •STEP BY STEP• •

How to Book a Talk Show Interview

• **Contact the person at the show responsible for booking interviews.** Telephone the station and ask for the name and title of the "booker." Tell the booker who you want to book and why the issue is important.

• **Send the booker compelling information.** The materials should include a cover letter explaining what the issue is, the expert you're recommending, and a press kit describing the issue in greater detail. The kit should also include a one-page biographical profile that establishes your spokesperson's credentials and a written set of questions for the interviewer to consider asking.

• **Call back to schedule the appearance.** If the booker is not interested in your spokesperson, find out why. If your spokesperson is scheduled for an appearance, find out if anyone else will be joining him or her.

• **Before the appearance, arrange to provide the audience with the address and phone number of your group.** Television stations can mechanically superimpose this information on the screen.

If possible, your organization's media coordinator should accompany interviewees to the station on the day of the interview.

WHAT SHOULD YOU SAY?

It's important to recognize that the best spokesperson is not a slick public relations professional who knows it all. The most effective communicators are those who believe in what they are promoting, whose sincerity and commitment to the subject are readily apparent.

The tips on the facing page apply to interviews with both the print and broadcast media. But a special section is devoted entirely to television and radio interviews, because in some ways, these are more challenging than dealing with print reporters. Not only do you have to be concerned about what you say and how you say it, but you also have to take into consideration how you look and sound—how you will come across "live."

HOW SHOULD YOU SAY IT? MASTERING THE SOUND BITE

A "sound bite" is a colorful quote or succinct one-liner that summarizes an opinion or position in a clear and simple way. "We'll save your bottoms if you'll save ours!" captured the essence of the Cheyenne Bottoms campaign; "If you're a mother and you live in East L.A., you're a Mother," helped explain who the opponents of the East L.A. prison were.

Examples of Effective Versus Ineffective Quotes	
Unmemorable Statements	**Quotable Quotes**
• "Oppose the Incinerator."	• "Ban the Burners."
• "Preserve the Cheyenne Bottoms Marsh."	• "We'll save your bottoms if you save ours."
• "We will be in trouble if we stay frightened."	• "The only thing we have to fear is fear itself."
• "Unless we resolve our differences, our association will fall apart."	• "A house divided against itself cannot stand."
• "Do your civic duty."	• "Ask not what your country can do for you; ask what you can do for your country."
• "I love New York."	• "I ♥ N.Y."
• "I am a candidate for office."	• "My hat is in the ring."
• "A small number of people is responsible for our victory."	• "Never in the field of human conflict was so much owed by so many to so few."
• "Re-elect the President."	• "Four More Years."

Delivering effective sound bites is essential if you want to make it into television news, where competition for air time is fierce. But even in print media, where space is not as much at a premium, the spokesperson who can convey his or her message in a brief sentence or two—and sound lively while doing it—will have a much better chance of having a quote used than someone who rambles on and takes several minutes to make a point.

For both print and broadcast reporters, captivating quotes and zinging one-liners are more likely to be used than lengthy explanations and lackluster discussion. Punchy lines and quotable phrases are a favorite of all reporters.

How can you insure your quote is quotable?

- **Avoid jargon.** Give specific examples that your audience can relate to.

- **Use analogies.** Simplify complex issues by relating them to common everyday examples. For example, environmentalists have helped many people understand the causes of a condition known as global warming by comparing the earth's atmosphere to a giant greenhouse that is overheating, with no way to cool off.

- **Be personal.** Talk about why the issue is important to you personally, not just the amorphous "public at large."

- **Talk to your audience.** Let someone know how he or she will be affected by your issue.

- **Be brief.** Divide long ideas into short sentences—for the public, they're easier to remember; for editors, they're easier to edit.

- **Be direct.** Nancy Ricci, a citizen lobbyist you'll read about in Chapter 8 who helped pass laws to reduce drunk driving, was fond of saying, "A collision involving a drunk or drugged driver is a *crash,* not an *accident!*"

WHO WILL YOU BE SAYING IT TO? (KNOW YOUR AUDIENCE)

Knowing who is going to see you, hear you, or read about you will help you determine the most effective way to communicate your message. You may emphasize certain issues if you are on a college public radio station and address others if you are appearing on a local TV talk show.

Before granting an interview request, familiarize yourself with the type of outlet your interview will be in. Ask the reporter, producer, or host for information clarifying the interests and concerns of its audience. Pinpoint the kinds of concerns, biases, and interests their audience has. Then work with your group's media committee to determine how to characterize your issue in a way that will generate the most empathy from that audience and motivate it to act.

Understanding the reporter's personality is also important. To the extent possible, get to know a little about your interviewer before the interview takes place. Watch or listen to the program you're going to be on in advance; if you're going to be talking with print reporters, read a few of their columns or stories first.

PREPARE FOR THE INTERVIEW

Before the interview takes place, decide what you want your overall message to be; know what you want to say and how you want to say it. Anticipate questions (especially difficult ones) and have answers ready. And rehearse, but not too much—you want to be relaxed and confident, not robotic and spouting clearly rehearsed lines.

Line up supporters in advance of the interview who can both attest to the viability of your arguments and pose questions typical of those the other side would raise. Be ready with sharp answers that bolster your position and deflate the opposition.

———————————— **The Interview Checklist** ————————————

___ **Always be honest and forthright—don't conceal information.** If you don't know the answer to a question, say so. Then get back to the reporter with the information as soon as possible.

___ **Never answer a question you don't understand.** Ask the reporter or host to restate it.

___ **Be concise.** Long answers are often ineffective, unless you are giving background for a print interview.

___ **Be informative, accurate, and helpful.** Be consistent. A seasoned reporter can easily tell when you are fudging facts and figures. Try not to be argumentative or defensive.

___ **Use news to create news.** Try to relate your message to news of the day.

___ **Avoid distractions.** Rapid hand movements, repeatedly clearing your throat, foot tapping, eye rolling, and repeated use of "you know" and "uh" can be very disconcerting.

___ **Don't forget that being nervous is normal.** Try to redirect your fear into excitement and enthusiasm.

___ **Use humor during the interview to lighten the tone.** Making people smile helps them connect with you in a more personal way, making you seem less distant.

___ **Don't forget that** *you* **are the message.** Focus your words, voice tone, facial expressions, and body language.

Get to interviews early so that you can get a sense of the questions the reporter is planning to ask. Take along a "cheat sheet" to help you keep track of points you want to make and facts or figures you don't want to confuse.

If, after the interview, you discover you gave the reporter some incorrect information, call him or her and correct your mistake. Thank reporters with a quick phone call or note if they do run a positive story or program.

Always remember that, like you, reporters are just people trying to do their job. When possible, help them do the best that they can by supplying useful information and resources. If a story contains inaccuracies, politely correct them in a letter or phone call and continue to try to build a productive relationship with the reporter that will result in a better story next time around.

TAKE CHARGE OF THE INTERVIEW

Just because you have agreed to an interview doesn't mean you have waived your right to courtesy and respect. Here are some ground rules to help you take charge of the situation.

• **Ask about the show's format.** Is it taped or live? Will there be a studio audience? Will there be call-in questions from the public, and will they be screened? Can answers be lengthy, or must you hold to a short, "hard news" format? What kind of questions will be asked? Will other guests be on at the same time? In what order of appearance? How long will your interview last? If it is a live show, you probably will be given an approximate time; if it is taped, you have the right to know how much of your interview will be used. Knowing the length of the interview ahead of time helps you prepare appropriate answers to questions you anticipate.

• **Know what you want to say.** If your goal is to promote a program or particular point of view, steer the conversation to your issues before time runs out. Be prepared. Take notes with you on the set or to an interview (but don't read from them on the air; it will make you look like you don't really know what you're talking about).

Speak up. Don't wait for the interviewer to ask the questions that will lead to the message you wish to deliver. Take advantage of a pause in the interview to make your points. If you are interrupted before you have had a fair chance to answer the question, you have a right to complete the answer. Do so politely, but firmly.

Remember: Your duty is *not* to follow in whatever direction a reporter wants to head, but to convey the message *you* have chosen through the reporter.

- **Use visual materials.** Slides, film clips, photos, charts, and videotapes are often welcomed by producers, who like to enhance their shows with visual variety.

- **Monitor the reporter's "reserve" or "cutaway" questions.** Since most broadcast interviews are shot with one camera, reporters tape their questions for the camera's benefit after the interview is completed. Later, the questions are edited into the interview to give the impression two cameras were used. Sometimes, a reporter changes his questions from those you were actually asked. If you have any reason to believe that the reporter may misrepresent you when he or she tapes the cutaway questions, ask the reporter in a low-key way if you can watch while the questions are being taped. Object should those questions differ from the ones you were asked during the actual interview.

- **Obtain a videotape copy of what was aired.** Take a VHS tape with you to the studio if you would like them to make a duplicate on the spot. Radio stations may be willing to provide you with a cassette of the program. Also, many cities have private media monitoring services that will tape any show on the air if you contact them ahead of time and tell them when it will be on. A cheaper alternative is to ask someone in your office or family to tape the show.

"OFF THE RECORD"—SHOULD YOU OR SHOULDN'T YOU?

The best rule of thumb is very simple: When it comes to media interviews, there is no such thing as off the record. There should be no item under discussion that can't be on the record (or, as Jan

Garton put it: "Never say anything you don't want to see in the paper"). In fact, when dealing with a reporter you are much better off by assuming that anything you say can end up in the newspaper or on a program—so speak and act accordingly. There are no "social occasions" for reporters; you're always fair game.

The best way to handle off-the-record interviews is simply never to give them. But if you must for some reason go off the record, be very explicit with the reporter about the exact ground rules under which you are providing information. The chances for getting "burned" are great: You must establish exactly what kind of information you are providing. Here is a brief rundown on several of the off-the-record terms you should clarify with the reporter:

- **Background briefing.** A briefing provided only for the reporter's education, to offer general background that will help the reporter cover the story or subject. The material in this briefing is probably not intended for immediate publication or airing. Some or all of the material may be confidential and off the record.

- **Deep background.** A briefing or interview intended to be used as part of a story. However, the source of the information, either by name or organization, may not be attributed to the story. Often, no quotes are allowed.

- **Not for attribution.** A briefing or interview in which the source's name may not be used but everything else is on the record.

- **On the record.** The best interview of all. When in doubt, always speak on the record.

 ACTIVIST'S NOTEBOOK

So You're Going to Be on the Air

SPECIAL TIPS FOR BROADCAST INTERVIEWS

While most people view interviews with print reporters as fun or challenging, the notion of being on camera or live on the air can still strike fear in the heart of even seasoned spokespeople. There's something about being under the lights that can reduce an eloquent and articulate person into a babbling mess; it's called stage fright.

Fortunately, the more such interviews you do, the easier it becomes, and veteran spokespeople actually use stage fright to their advantage: a little bit of nervousness can heighten your presentation and make you more enthusiastic and exciting.

There are some excellent remedies for stage fright. One is knowing your subject—and knowing it well. The most articulate speaker will be a poor interviewee unless he or she knows the subject inside and out.

Another is to focus on the reporter or host who is conducting the interview. Don't think about the viewers or audience; visualize yourself in an interview situation with a single individual who is interested in what you have to say.

Here are some specific tips to keep in mind when you are going to be on the air:

- **KISS (keep it short and simple).** Even if your appearance is live and unedited, long rambling answers will put your audience to sleep. When your interview is to be edited for a news show or documentary, try to keep answers to seven to fifteen seconds.

- **Listen and wait.** Make sure you understand the question before you answer. If you need to think about your answer for a moment, do so. If the interview is being taped and you're unhappy with your answer to a question, start over.

- **Don't get angry.** If you don't like a question, pause before you jump into a hot retort. If it's asked on a live show, state simply that the question is unfair or that the tone makes it difficult to have a productive discussion. Then make a point you want to make.

- **Be alert.** Assume that a camera or tape recorder will be on at all times, anywhere in the studio. Don't say anything, even jokingly, that you don't want to come back to haunt you.

Once you get to the studio, let the producers do all the fussing. They'll show you to your seat, adjust your microphone, and ask you to say a few words to adjust sound levels.

- Sit upright in your seat, leaning slightly forward.

- Look at your interviewer—not the camera—throughout the interview.

- Use your hands within reason, and don't be afraid of mannerisms, but avoid fussy, nervous, or exaggerated movements.

- Glance at your notes if you need them for reference, but don't read from them.

TELEVISION TIPS

- **Moderate gestures help make you interesting.** Smiling, pointing, and nodding are good devices when used sparingly to drive home a major point.

- **Everything you do will be magnified.** Sit still; don't swivel in your chair, scratch your head, or wiggle your legs back and forth.

- **Act as if you are on camera at all times.** You may think someone else is being pictured, but even if you see only one camera, there may be others. Don't slouch; do keep your feet together.

- **Be sincere and enthusiastic throughout.** Try to stay animated; otherwise you'll put viewers to sleep.

- **Dress carefully and conservatively.**

Women should:
- Wear a simple tailored dress or a suit and blouse, free of fancy frills, in solid colors.

- Avoid white or very light colors because they can produce a halo effect; black, navy blue, royal blue, and red are considered best.

- Avoid busy stripes and patterns that will draw attention to your clothes rather than to what you are saying. Large pieces of jewelry or sparkling jewels reflect studio lights and tend to flare and distort the picture.

- Wear everyday makeup in natural tones.

Men should:
- Wear simple clothes. Gray, brown, or blue suits with off-white or pastel-colored shirts are fine. Stripes, checks, or sharply contrasting patterns should be avoided.

- Choose ties and handkerchiefs in muted colors. Ties with large figures or designs appear too busy.

- Select rings, tie clasps, and cuff links that do not sparkle or glint off studio lights.

<u>RADIO TIPS</u>

- **Focus on your words.** Be very clipped and precise. Every "uh" and "er" is magnified on radio. Also, use a normal conversational tone. Make listeners feel like you're talking to them.

- **Remember: You can't win them over with a smile.** Gestures don't help either. Nodding yes or no is a waste of time, but a lot of people answer that way nonetheless. Sounding friendly on radio is very important, although it's harder to sound friendly than to be friendly. Smiling while you talk may help give you the right quality to your voice, but remember it's the voice that counts.

- **Be personal.** As far as the listener is concerned, you are sitting right in his or her living room or car. Don't shout; don't preach; don't give a speech. Use a gentle, conversational tone.

Community Cable Access

Cable television has been installed in more than 25,000 communities around the United States, reaching 60 percent of all U.S. households each year and giving activists access to yet another important communications tool.

Each local cable system reserves at least one channel for noncommercial use by individuals and nonprofit organizations. Thousands of these cable access channels are available on more than 1,400 cable systems in the United States. Major cities like Boston, Chicago, Dallas, Tampa, Tucson, Atlanta, and Portland have several access channels, and there are more than 440 access channels in California alone.

In fact, in some communities, cable television is delivered to more homes than the daily newspaper.

Local cable systems usually have much to offer a citizens group. They provide a way for citizens to keep track of what is going on in their community itself, as the channel usually covers local school board meetings and other community events that are obviously not carried on commercial television stations.

Also, local cable systems are usually willing to loan their equipment and personnel to the public so that groups or individuals can make and air their own programs. In fact, many cable systems offer training programs that certify citizen producers for free or at very low cost.

The public access channel offered by the local system may be managed either by the local cable operator, a school or community institution, or an independent nonprofit corporation. The public access center might be based at the cable office, at the public library, in a school, or even in a local shopping mall. The public access center usually includes a television studio, training and meeting rooms, and equipment storage and checkout areas. Equipment, training and access time on the channel are usually provided on a first-come, first-served basis.

Even if you don't want to produce your own programs, you can use your public access channel to organize. Every system has at least one text channel, which conveys information in words and numbers on the TV screen. The text channel acts as a sort of community bulletin board. It lists community events, job openings, and other announcements.

Notify your local cable system that you would like to submit a message for the community bulletin board. You may need to complete a form requesting a message of up to thirty or forty words; fill it in, mail or fax it back, and call to confirm that your message will run at least four times an hour for several days or a week.

In addition to submitting your own text messages and producing your own programs, you can submit preproduced programs or films to your cable system that further your issue. As long as they are of broadcast quality, and contain no obscene material, the station should run them.

You can also appear on an access program or series if the system produces its own talk show.

As with any promotional event, you'll want to target the programming you produce to the right audience. Build viewership by promoting your cable programs heavily in the community newspaper and in fliers posted in local shops and at the library. Whenever possible, feature local people in programs about important issues in the community.

Most cable programs can be aired several times; make sure yours is, too.

Here's just one example of the ways in which you can put your local cable channel to work for you:

LOCAL HELL-RAISER
"At Issue" on Cable

Every Wednesday night at 7:00 P.M., the League of
Women Voters (LWV) of Bucks County, Pennsylvania,
reaches an estimated 150,000 households with a half
hour of local issue programming they write themselves—and enlist their local cable company to produce.

The effort started in 1983, when a group of LWV
volunteers decided to use their local cable system's production resources to produce a documentary called "At
Issue." Since then, the group has produced over thirty
shows on topics ranging from the environment, women,
and health care, to housing and the homeless, hazardous
waste, teen pregnancy, and senior citizens.

Some of the videos, like "Cheap and Delicious,"
which demonstrates how to shop and cook on a limited
budget for people living on public assistance and using
food stamps, can be bought for $15. Others, like "Jus-
tice for Kids" and "Laws for Kids," about the juvenile
justice system, are being shown in high schools as well
as by the Bucks County Probation Department as training films.

"At Issue" costs the League around $10,000 a year,
money that is primarily spent advertising the program
and buying videotape (production facilities and equipment are free through the local cable systems). The sum
seems large, but LWV has been able to secure grants
from local businesses and some national corporations to
help cover the costs of the program.

Shirley Hart, chair of the Bucks County LWV chapter, believes that the cable show's benefit in the community is apparent. "Our program called 'Poor and Sick in
Bucks County' has been exceedingly significant in
terms of expanding low-cost health facilities in the
county. One hospital has given us credit for its entire
change in policy," she said. And people are watching
"At Issue" regularly: When one program wasn't aired
for technical reasons, the station received calls from
viewers seeking assurance that it hadn't been dropped.

The chapter's latest project involves a new program and video called "Raising Healthy Children" that provides information on health resources available to the area's poor parents. Recognizing that school nurses are often the most accessible health care providers for many kids, a local branch of the Pennsylvania Nurses Association donated $500 for LWV to distribute a copy of the video to every school nurse in the county.

Using Video

If you do produce a program, don't put it away after it's been televised a few times. Send it to schools, use it for fund-raising, and urge elected officials to watch it.

Indeed, many groups are foregoing films and slide shows altogether in favor of videos, showing their programs before groups of activists, at public hearings and board meetings, and to individuals seated in their own homes. Some activists have been known to back a station wagon into a supermarket parking lot and run a VCR and TV off the car's cigarette lighter. Others rent booths at the local county fair and show videos while they are handing out fliers explaining their campaign.

Enterprising organizations are taking advantage of the relatively low duplication costs (sometimes only $3 per tape) to send out videos the way they previously distributed direct mail. Some groups transfer entire slide shows to videotape; with over 90 percent of American households equipped with a VCR, they want to take advantage of a ready medium to help them educate, communicate, and organize.

Few community groups have the budget to go on location and produce a state-of-the-art videotape. But you don't need to look like "60 Minutes" to produce a video that will meet your fundraising and organizing needs. In fact, home video has become increasingly popular as a way to organize and inform. Some of the most powerful videos are those produced by amateurs. Remember the footage, shot on a home video camera, that showed Rodney King being brutally beaten by four Los Angeles police officers? The tape, though of extremely poor quality from a professional point of view, sparked a national introspection on race relations in America and was admitted as evidence when the police officers were put on trial. (In Chapter 7 you'll read about Terri Moore, an activist whose home video exposed unhealthy garbage-dumping practices in her town—and helped her lobby for new legislation to regulate the dumpers.)

☞ **CHECK IT OUT:** EarthKind, the environmental arm of The Humane Society of the United States, has launched a new project to loan video cameras and equipment to environmental organizations interested in documenting ecological problems in their communities. Called Eyes of the Earth, the project is working with activists on every continent, including North America. For more information and an application to participate in the program, contact EarthKind at 2100 L Street, N.W., Washington, DC 20037.

Video News Releases

A video news release (VNR) is the television version of a press release. Corporations, professional associations, government agencies, and lobbying groups often hire production or public relations firms to produce expensive pieces that look and sound like part of the evening news. Ranging in length from 90 seconds to 1 minute and 40 seconds, they're distributed to TV stations for free in hopes that they will be aired in local newscasts.

The biggest drawback is that a piece like this must be professionally produced at a cost of about $10,000. And even after you spend that kind of money, there's no guarantee you'll get on the air. You're better off either going the home video route or convincing a television station to produce its own piece.

If you do opt for a slick VNR, make sure your tape includes extra sound bites (interview segments) and "b-roll" (extra footage) so stations can edit the story to fit their particular format. Some stations will have their own reporters write a new script and narrate the story over the footage you provide. Identify everyone speaking on camera, the sponsor of the VNR, and the reporter separate from the actual VNR so stations can see and use their own titles. Use news writers, producers, professional camera crews, and news announcers to make the story look like "real" news.

You can distribute your video news release on a cassette tape or via satellite if you have the budget and you are aiming for national distribution. Either way, notify each station that the tape is coming with a written telex to a specific individual, and follow up with a phone call.

Newsletters

Newsletters are primarily a communications tool intended to keep members of an organization apprised of what the organization is up to. They don't have to be slick and snazzy to be effective, though given the proficiency many activists now enjoy with computer graphics, many newsletters are at least nicely laid out and free of typos. They do need to contain essential information that makes readers and members feel that the organization is doing valuable work, motivating readers to become more involved.

But beware! Consider newsletters public domain; don't include any information in them that you wouldn't want a reporter, officeholder or an opponent to read . . . and use against you. Use newsletters to boast about your work and convey the strength of your position, not to debate strategy or reveal any weaknesses your group may have.

Information should include highlights from any victories you've enjoyed, analysis of any defeats you've suffered, updates on activities the organization has launched, announcements about upcoming events or actions, and useful insights into the issues the group is working on. Many newsletters also provide a letters-to-the-editor section so their members can write in.

As much as possible, the newsletter should highlight your group's accomplishments, making members feel proud to belong and keeping volunteers motivated to do more. In addition to being an effective organizing tool, your newsletter can help you raise money by alerting members when a special fund-raising opportunity is about to come their way.

Most grass roots groups publish their newsletter monthly or bi-monthly, although those on tighter budgets may opt for a quarterly or even semiannual publication.

If you publish a newsletter, distribute it widely. Mail it to members, and make copies available in your local library and among various civic associations, schools, and churches in your community. Exchange your newsletter with other interest groups who also publish one.

And if possible, integrate some light humor and appealing cartoons into the newsletter to keep it from getting too serious or boring.

Meal Times

MEET PROJECT OPEN HAND'S VOLUNTEERS

by Reshima McKelvin
Project Open Hand Volunteer

In the Project Open Hand kitchen, you can see them peeling potatoes, assisting the cooks, or assembling the nightly meals. At the Food Bank, they shop for clients and assemble their grocery bags. They drive all over San Francisco, Berkeley and Oakland to home-deliver 800 bags of groceries weekly and 1,550 meals daily to men, women and children with AIDS who depend on Project Open Hand. These people are, of course, Project Open Hand volunteers, and those are just a few of the many tasks they perform for the organization.

As a non-profit organization, Project Open Hand relies heavily on its volunteers, says Executive Director Steve Burns: "We're one of the most volunteer intensive organizations I have ever seen. Without them, we'd literally have to shut our doors tomorrow, and a lot of people with AIDS who depend on us for their food would be hungry."

Currently, there are about 1,500 volunteers in San Francisco and 200 in the East

who dedicate part of their free time to Project Open Hand. Some come in five days a week, others for a few hours. They come from a variety of backgrounds, different kinds of people working together for a common goal — making sure people with AIDS are fed.

There are as many reasons to volunteer as there are volunteers. Erwin Affolter, who has clocked in over 1000 hours in the kitchen, says, "After I retired, I knew I wanted to do something, and I had heard of Project Open Hand because they were feeding someone I knew from work. This gives me something to do, and I've always done office work so it was also a change."

Other volunteers say they chose Project Open Hand because of the issues surrounding nutrition and people with AIDS. Jean-Claude Kouri, who has worked at the Food Bank for two years says, "Food is one of the the best ways to make a contribution to

fighting the disease."

Chuck Arnold says he chose to come to Project Open Hand because "It was time to fulfill a promise I made to myself in remembrance of a special friend who lost the battle to AIDS, as have many other friends, that I would get more involved to help win the fight against this epidemic. Project Open Hand would be my 'Avenue of Attack.'" A retired case records administrator for the State of California, Arnold alternates his time working in client services and development.

In addition to people who volunteer at Open Hand because of friends who have AIDS, there is also a large number of volunteers who have AIDS themselves. Burt Huguley, who

(See "Volunteering" p. 11)

See back page for exciting details!

7th on sale

During the organization's fledgling days, in-house designers and writers used PageMaker computer software to create Project Open Hand's newsletter. Recently, a public relations firm has begun helping the organization with its newsletter pro bono.

Other Publicity Tools

Here is a list of other tools you can add to your communications arsenal:

PROCLAMATIONS

A proclamation is a document usually issued by a mayor, the city council, the state legislature, or a governor to officially kick off a campaign or to commemorate various weeks, months, days, or years. Hold a news conference at which these "newsmaker" personalities issue the proclamation.

INVITATIONS

Invite local pastors, politicians, and community leaders to join your board, participate in your coalition, or attend an event when appropriate.

CHURCH BULLETINS

Mail promotional information to pastors two weeks in advance to include in their bulletin the Sunday before an event takes place. Send educational material, diagrams, quizzes, and op-eds that can be included on a space-available basis.

SIGNS OR BANNERS

Make and hang over the site of your event several weeks in advance.

FLIERS

Print fliers on bright paper and distribute them in your neighborhood (but do not put them in residential mailboxes, which is against the law). Ask other community groups to hand out the fliers at their events the weekend before your event, include as stuffers in other programs, and hand out at busy intersections. You can easily make your fliers top-notch by using a computer desktop publishing graphics package.

POSTERS

Hang them in shops, train stations, sporting goods stores, bike shops, doctor's offices, and at the YMCA and YWCA. Post on bulletin boards in schools, churches, bars, theaters, and youth centers.

COUPONS

Add coupons offering a discount on appropriate merchandise or access to special facilities or services to your own newsletter, fliers, letters, or invitations.

DISPLAYS

Set up a simple display at the local library or YMCA to develop interest in your issue.

SIGNS AND MARQUEES ON BUILDINGS

Write the head of the firm that owns the building and ask for free use of the sign, particularly before a big campaign that is going to garner a substantial amount of participation from city residents.

OTHER ORGANIZATIONS' NEWSLETTERS

Ask for a paragraph plug on an event you're organizing or an issue you're trying to publicize. Promise to reciprocate.

CARTOONS AND OTHER GIMMICKS

Be creative! Take some cues from the Minneapolis League of Women Voters, which invented four unique cartoon characters to bolster participation in the city's school breakfast program.

Local Hell-Raiser
The Breakfast Queen

When the League of Women Voters of Minneapolis, Minnesota, learned that one out of eight kids in Minnesota was coming to school hungry, and that only one quarter of Minneapolis's public schools served breakfast, the group jumped into action with a major publicity campaign promoting the benefits of a good breakfast for children. Clearly, parents were unaware that their kids could get a fast, economical breakfast at school.

The group invented four funky cartoon birds to promote the breakfast program, featuring them in a colorful brochure. With the slogan "Break for Breakfast to Be Your Best," the piece was translated into five languages, sent home with kids, and distributed throughout the community.

The League then organized School Breakfast Week, during which it held a puppet show for kids starring "the Breakfast Queen" that was performed by students from the Minneapolis School for the Arts and later turned into a video that ran on the local cable station. Bright posters and stickers of the breakfast characters were given out to kids. The mayor, superintendent of schools, city council members, and state officials kicked off a major media event by having breakfast with children at an elementary school. One network station covered the breakfast and plugged the program on that evening's news, and articles and editorials ran in both state and local papers.

A year after the publicity effort began, participation in the state's breakfast program had jumped from 25 to 80 percent. "We noticed a tremendous increase in the number of kids who were eating breakfast," Karlynn Fronek, the project manager, reported. "Minnesota has been especially successful—one of the top ten—in increasing the number of low-income kids participating," she added, citing a rise of 33.2 percent from 1990 to 1991.

This fun-loving, brightly colored brochure gave parents reasons to feed their children breakfast and to take advantage of the city's economical and dependable school breakfast program.

© LWV Minneapolis, 1989

Computers

Computers have become an indispensable communications tool that help you get more work done faster and more accurately than ever before. You can use them to prepare annual reports, newsletters, fact sheets, and press releases; maintain correspondence and handle other word processing tasks; keep track of your budget; send out bills; network with other groups; do research electronically; and track media coverage of your issues. If your group does not use computers currently, find a way to get access to them. Perhaps a local business is acquiring new, more sophisticated computer equipment and would be willing to donate its older models to you.

If you can afford to purchase equipment, consult with a sales representative from one or several computer suppliers. They can help you identify what the computer equipment is likely to cost. The basic package includes a computer hard drive unit, keyboard, monitor, printer, and the software you need to set up your various computer functions.

Usually, the company that sells you the equipment will install it, perhaps for a fee; it can also give you a service contract to maintain your equipment for annual checkups and fix parts should they break.

Many high-tech companies have established programs to help nonprofit and community groups by donating equipment and software, including Apple Computers, IBM, Microsoft, and the Lotus Development Corporation. (For example, Microsoft gives free software to any charity, offering word processing, spreadsheet, and Windows programs, which makes an IBM-compatible computer function like a Macintosh.)

Many of these companies distribute their products nationwide via Gifts in Kind America (GIKA) an Alexandria, Virginia, organization that coordinates in-kind donations of all types of products to nonprofits.

Companies donate used or new equipment to GIKA, which then doles it out either to an outlet specified by the company or to some of the fifty thousand nonprofits that receive GIKA gifts regularly.

Any 501(c)(3) organization can receive free computers or software from GIKA, as long as it can demonstrate that it will use the products to pursue the organization's specific goals and objectives. To become a member and receive gift announcements regularly, groups must join GIKA's "The Agency Program" (TAP) and pay a membership fee of approximately $55. Organizations without a

501(c)(3) designation can receive donations, too, by contacting their local United Way, which is likely to be a GIKA sponsor. The United Way chapter will assess your organization's need, distribute monthly gift announcements to you, and coordinate the donation with GIKA. All donations involve a nominal shipping and handling charge that works out to about 3 percent of the goods' cost.

Keep in mind that to receive the best tax breaks, donors often earmark their products to be given only to social service groups. Environmental and cultural groups may, however, solve that problem by going through the United Way.

☞ **CHECK IT OUT:** Contact Gifts in Kind America at 700 North Fairfax Street, Suite 300, Alexandria, VA 22314, 703-836-2121.

DESKTOP PUBLISHING

The more graphically appealing your materials are, the more engaging and persuasive they'll be. You can spice up testimony, press releases, press kits, posters, fliers, and other materials by adding illustrations, compiling information into charts and graphs, and using different typefaces. Formerly, such graphic additions were the domain of specialized graphic artists. But the arrival of graphics software packages for computers has turned just about everyone into an artist.

Most word processing packages give you the capacity to construct simple charts or to put boxes around blocks of information. You'll need different software to produce fancy typefaces, snazzier bar graphs and pie charts, maps, and other illustrations.

Organizations that publish newsletters find free computerized "clip art" particularly useful. Many packages include libraries of common symbols, objects, and cartoons that can be electronically placed on a page to give it the artist's touch.

ELECTRONIC NETWORKING

Electronic networks link organizations via their computer terminals, allowing them to "log on" and send and receive messages to and from others also on the system.

Through computer networks, you can hold "conferences" with

other activists thousands of miles away; access data bases to help in your research efforts; mobilize volunteers to write letters, send telegrams, place phone calls, raise money; and inform your own targeted audiences about your important work.

Grass roots activists have their pick of a number of extremely useful national and international networks. Networks particularly helpful for nonprofits and community organizations include the "IGC" system (home to EcoNet, PeaceNet, ConflictNet, and LaborNet), the WELL (Whole Earth 'Lectronic Link), GEnie and its Public Forum* Nonprofit Connection, HandsNet, and America OnLine. Through any of these outfits you can access Internet, a "super" network that links millions of users around the world, offering access to the latest scientific or educational research results as well as ideas for organizing, fund-raising, and lobbying. Delphi, CompuServe, and Prodigy networks cater more to consumers than to activist groups, according to Tom Sherman, president of Words and Pictures, a communications consulting firm, and author of *Electronic Networking for Nonprofit Groups,* published by the Benton Foundation. But even they can help keep grass roots groups current on research, legislation, and other factors that affect their issue just by providing opportunities to "talk" (electronically) to others who may have some important knowledge to share.

For groups needing to give and receive information in their own locales, community networks and bulletin board systems (BBSs) (see Chapter 3) can be invaluable, especially for exchanging data in a specific subject area. "A grass roots organization in a smallish town, with a fairly local mission, may suffer a lot from isolation; networks enable them to act like bigger-thinking organizations and harness the expertise of others around the country," says Sherman.

The first such community network was born in Cleveland, Ohio. Dr. Tom Grundner of Case Western Reserve University's Department of Family Medicine set up "St. Silicon's Hospital and Information Dispensary" to allow people to electronically ask the university's medical staff health-related questions. The system's enormous popularity soon led to electronic networking in other information areas, such as the arts and sciences, law, and government.

Today, the Cleveland Free-Net, as it came to be known, has been duplicated in fourteen communities around the country, and another sixty-two are in the planning stages. Almost all the Free-Nets are linked to Internet and can interact with each other.

In Portland, Oregon, the Office of Neighborhood Associations

(ONA) was stymied by the need to exchange information and co-ordinate a newsletter among several neighborhood groups without blowing its meager budget on word processing, faxing, copying, and mailing expenses. ONA set up a BBS to link the community groups so they could contact each other, as well as city planning and community development offices around the city with which they traded information regularly. Through the Neighborhood Information and Communication Cooperative, the groups now share a newsletter template on-line, send mailing lists back and forth, compare notes on neighborhood revitalization, and keep posted on a community crime prevention program.

☞ **CHECK IT OUT:** The Institute for Global Communications provides computer networking tools for international communications and information exchange. It also offers capability for electronic mail, conferences, and information resources that give you access to the media, grantmaking foundations, or bibliographies. For more information, write 18 De Boom Street, San Francisco, CA 94107, 415-442-0220, fax 415-546-1794, telex 154205417, E-Mail: Support@igc.org.

The Technology Resource Consortium helps nonprofit organizations improve their abilities to use computer technology. Contact the Consortium at 666 Pennsylvania Avenue SE, Suite 303, Washington, DC 20003, 202-544-9234. Check the appendix of this book for names of local organizations might also be able to help you.

To set up an electronic network in your community, contact the National Public Telecomputing Network at 216-247-5800, or write 34555 Chagrine Boulevard, Moreland Hills, OH 44022. NPTN sends information to existing Free-Nets and helps new communities set up their own systems.

Here's one more example of how a local group tapped into an electronic network to boost its organizing efforts:

LOCAL HELL-RAISER
Plugged In

The National Committee for Prevention of Child Abuse (NCPCA) has at least one chapter in every state. Such a broad-based network may be terrific for reaching out to all sectors of the country, but it can be a nightmare when it comes to keeping chapters apprised of important information and aware of what various parts of the organization are up to.

In early 1990, to ease communications among its chapters and with other supporting organizations, the NCPCA began to tap into HandsNet, a nonprofit information and communications network that connects a broad range of human service organizations.

A year later, while organizing a symposium for teenage African Americans on male responsibility, the Greater Chicago chapter logged onto HandsNet during a brainstorming session and searched for the words "male responsibility."

To their surprise, they discovered a social services agency in South Bend, Indiana, that had produced a poster on male responsibility and advertised it on the electronic network. A subsequent phone call revealed that the agency had conducted a similar project on male responsibility, and the two groups began to share their ideas and successes. Each came away with a valuable new resource and greater insights for dealing with common concerns.

"The value of electronic networking is the range of people you can talk to," Philippa Gamse, a senior specialist at the NCPCA's headquarters in Chicago, said. Neither the committee nor the agency it contacted had known the other existed before they hooked up electronically.

The HandsNet network is growing rapidly; currently about two thousand "members" take advantage of this low-cost networking tool, exchanging articles and information with groups as far away as Australia.

In 1992, monthly maintenance costs were only a $25

membership fee, with a mere $.04 charge per thousand characters to upload (send) or download (receive) information, and a usage fee of $12 per hour during prime time. Monthly on-line costs average $30 to $40.

Today, the NCPCA uses electronic mail to publicize events and offer public awareness information and research to other human service providers and agencies that are connected, making good use of HandsNet's Children, Youth and Families bulletin board. The group also accesses the technology to inexpensively transmit proposals, training material, surveys, documents for editing, and even faxes among its chapters. Given the varying technological capabilities of NCPCA's chapters, almost half have currently committed to installing HandsNet; those who have not installed the program but have fax machines may still receive faxes from chapters with electronic mail.

"In the human service area, people are not used to working this way, using electronics," said Gamse. "It's quite exciting to see the evolution into a technological frame of mind."

☞ **CHECK IT OUT:** HandsNet is a national nonprofit information and communications network connecting a broad range of human service organizations. Members include national research centers, community-based service providers, local and state agencies, public policy advocates, legal services programs, and grass roots health, hunger, housing, and community development coalitions. The network offers daily summaries of newspaper and wire stories, public policy analysis, poverty statistics, abstracts of key studies and reports, and *Federal Register* notices. Contact HandsNet at 20195 Stevens Creek Boulevard, Suite 120, Cuperton, CA 95014, 408-257-4500.

Telecommunications Cooperative Network is a not-for-profit cooperative that offers cost-effective communications services to nonprofits such as World Wide Access, 800 numbers, fax broadcasting services, and more. For more information write to 1333 H Street, NW, Suite 700, Washington, DC 20001, or call 202-682-0949.

COMMUNICATIONS
DO's and DON'Ts

- **Do** make communications an integral part of your organizing strategy right from the beginning.

- **Do** know what you want to say, then frame your issue in terms people can understand. Be positive.

- **Do** know who you want to say it to.

- **Do** adjust your material to the medium you use; a highly visual spot that grabs TV viewers won't do much for radio listeners who need more exciting sound.

- **Do** match your medium to your audience (e.g., younger people will probably ignore direct mail but respond to a music video asking them to phone an action line).

- **Do** tie into national news stories to give your local efforts a boost.

- **Do** rework complicated or old data into a new and interesting format to help the public perceive your issue in a fresh light.

- **Don't** wait for the media to come to you.

- **Don't** expect someone else to generate your story.

- **Don't** blame the media for bad stories.

- **Don't** use jargon and rhetoric people don't understand.

- **Don't** ignore unexpected opportunities for media coverage.

- **Don't** expect a "free ride." If you don't have news, don't expect a story.

- **Don't** be hostile toward reporters or editors.

- **Don't** make media coverage the sole focus of your organizing strategy. Keep your sights focused on your organizational goals and objectives and choose media tactics accordingly.

COMMUNICATIONS
DO's and DON'Ts (cont.)

• **Do** localize and personalize your story; make it relate to the people you're trying to reach.

• **Do** seize unexpected opportunities to generate positive media coverage for your issue.

• **Don't** expect good press to happen overnight. Take time to cultivate solid relationships with reporters and editors based on your reliability and credibility.

• **Don't** overlook stories that contain incorrect information about your group. If the story is wrong, correct it.

7.
LOBBYING

WILL ROGERS, THE great American humorist, once wryly remarked, "A president only tells Congress what they should do. A lobbyist tells 'em what they will do."

What makes lobbyists so powerful? Sometimes, it's simply a matter of money. Lobbyists (particularly those who represent corporations, trade associations, and well-heeled special interest groups) buy influence through campaign contributions; given the thousands, hundreds of thousands, and maybe even millions of dollars it takes to get elected to public office today, those contributions count for a lot.

Grass roots activists earn their clout "the old-fashioned way": through hard work, commitment, creativity, and the ability to convince legislators that their constituents care enough about an issue to keep it in mind when they vote. Nancy Ricci turned to these "old-fashioned" tactics when she began her campaign to reduce drunk driving in the state of Connecticut.

ACTIVISTS in ACTION ▸ ▸ ▸ ▸

Remove Intoxicated Drivers

In an age when word processors and faxes have become household items, Nancy Ricci, founder of the Wallingford, Connecticut, chapter of Remove Intoxicated Drivers, lobbies her state legislature from her kitchen table, writing letter after letter in longhand. She capitalizes words and underlines entire sentences when she wants to grab her reader's attention, and doesn't bother consulting a dictionary. Her letters might make an English teacher shudder, but they can move the stubbornnest lawmaker to action. "That's just Nancy's style," her husband, Jim, says. "Everyone is so hepped up about being businesslike. Legislators get thousands of letters—and they pay more attention to the handwritten ones. People advised her to type her letters . . . she wouldn't do it."

Ricci's pen and the words that flow from it convey the passion and earnestness that drive her fight to free the roads of drunk and drugged drivers. Activism

against those who drive while intoxicated (DWI) is fueled by anger over countless senseless deaths, but Ricci's story is particularly heart-wrenching: Both her brother, Ray, and her sister, Anne Marie, were killed in independent automobile crashes caused by drunk drivers in 1965 and 1980.

For four years after Anne Marie's death Ricci lived in a vacuum, not believing that this tragedy had struck her family twice, that they had not learned from Ray's death to save her sister. In 1984, Ricci saw an ad in the paper for a meeting of the activist group Remove Intoxicated Drivers (RID) in Southington, Connecticut. It was, she decided, time to attend.

"All I Knew How to Do Was Keep House and Raise Kids"

The Connecticut chapters of RID are branches of RID-USA, Inc., a 501(c)(3) organization and "the only totally volunteer grass roots organization working nationally and effectively to reduce drunk driving," according to founder and president Doris Aiken. Aiken launched her ambitious campaign to make the roads safe in 1978 when two students—the same ages as her own kids—in her hometown of Schenectady, New York, were killed by a drunk driver. "We realized how quickly life can be obliterated for no reason," she said. Aiken started the original RID chapter and teamed up with several other anti-DWI organizations in New York to create the founding chapters of RID. Though novices all, the activists succeeded in passing, for the first time, "simple, reasonable laws" combatting drunk driving at the state level. At a lobbying cost of under $200, thirteen New York State laws were passed over two years, reducing deaths by drunk drivers 10 percent in the state in 1981.

Motivated by the increased awareness and lobbying successes in New York State and elsewhere, Ricci jumped into the fray in her own state. Her background was actually about as grass roots as they come: "All I knew how to do was keep house and raise kids!" she once remarked.

Undaunted, she and Jim held their first meeting for

the Wallingford chapter at the town library. She recruited volunteers by writing a press release, which got local coverage, and a column in the local paper urging people to come, learn, and help solve the problem. The Riccis also posted flyers all over town and talked among their friends.

The public support they received was astounding—the Riccis described it as a "time bomb" that went off in their hands. "Most people who knew Nancy wanted to help her out," Jim Ricci said.

Still, they had their work cut out for them. "It's amazing," Ricci said. "No one is for drunk drivers, but we sure are having a hard time getting rid of them."

The First Victory

Members of the state RID chapter had been pushing for drunk driving legislation since 1982, and Ricci and the new chapter from Wallingford were quick to contribute to the lobbying efforts in the capital. In 1984, RID landed its first major legislative victory when the activists passed a bill raising the Connecticut drinking age to 21. A year later RID was instrumental in setting the blood alcohol content (BAC) limit for legal drunkenness at .10. Both successes were secured by writing and calling legislators, attending hearings on the bills, and educating the public about the laws and the issue.

In 1986, when the president of RID Connecticut resigned, Ricci was appointed her successor and RID Wallingford merged with the state group. A third challenge sprang up shortly thereafter. In the course of meeting with a Liquor Control Board agent to prepare for a hearing, RID discovered that these agents worked only during the day, not at night, when the liquor laws most needed to be enforced. By meeting with and garnering the support of the board and its union, and by writing more letters to legislators, the activists convinced lawmakers to draft a proposal allowing liquor control agents to work flexible hours, then lobbied for its passage, which occurred in 1987.

It seemed that each victory led to another battle. The longest, most difficult, and most significant effort waged by the RID crusaders against drunk driving in

Connecticut was the fight to pass the administrative license suspension law, also known as the administrative per se law. This law, proposed in 1986 by State Representative Edith Prague, immediately suspends the license of those arrested for drunk driving for twenty-four hours, whether or not they submit to a blood alcohol test. The offender then receives a temporary driver's permit good for thirty-five days, after which he or she has seven days to request a hearing. Offenders lose their

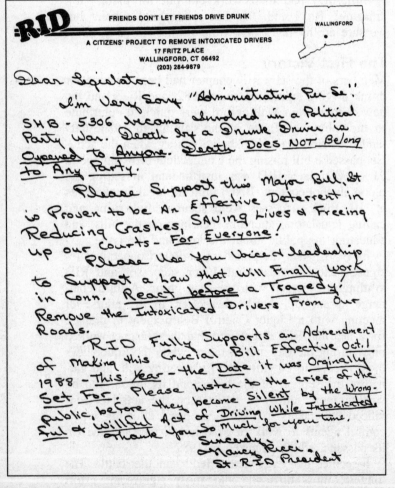

This letter to Connecticut legislators, penned by Nancy Ricci, speaks for itself.

license for ninety days if the result of the BAC test is
.10 or greater, and for six months if they refuse to submit to a blood, breath, or urine test.

What makes this law effective is that the proceeding
is an administrative hearing, overseen by the Department of Motor Vehicles, separate from any court hearings related to the offense and therefore much more
expeditious. Delays caused by court schedules and lawyers often allowed drunk drivers to keep their licenses
while awaiting trial. This bill would remove dangerous
drivers from the streets while expediting the legal process. The administrative per se law had significantly reduced drunk driving deaths elsewhere—why not in
Connecticut?

Writer's Cramp

Twenty-two states and the District of Columbia already
had the administrative per se law on their books when
RID Connecticut began fighting to pass it, but that
didn't make RID's job any easier. Ricci got a boost in
1986, when RID-USA held a workshop on administrative per se that described the benefits of the law, and
handed out a booklet written by the National Highway
Traffic Safety Administration (NHTSA) explaining the
law's technicalities and suggesting tactics to help get it
passed.

One of those tactics was to write letters, and write
letters they did. Ricci encouraged everyone to write by
hand—to every state legislator (even the opposition),
state agents, the governor, the attorney general, the chief
state attorney, judges, prosecuting lawyers, and anyone
else who might be remotely interested in the bill and its
progress, urging them to support the passage of administrative per se. The activists cited success statistics
from states such as Minnesota, which has seen a 60 percent reduction in traffic/alcohol fatalities that occur after
midnight since its version of the law was implemented;
they included newspaper articles when possible. She
photocopied editorials and sent them to legislators, always scrawling an impassioned note on the side, compelling the recipient to read and appreciate facts
attesting to the law's effectiveness.

In 1987, as RID was gaining momentum, the General Federation of Women's Clubs of Connecticut, Inc., adopted RID and the anti-DWI cause, offering them eighty-four chapters around the state from which to organize. Suddenly RID had contacts all over Connecticut who could write or call their respective representatives right in their home districts. With Ricci serving as liaison between RID and the federation, the two groups worked cooperatively—a "united front," she called it—for two of the three years it took to pass administrative per se.

Reaching Out

RID helped mold public attitudes about DWI through a broad-based educational effort aimed at getting people to "pass the keys" when they have been drinking. "Education and awareness go hand in hand," Ricci said. RID volunteers researched and compiled fliers, booklets, and brochures containing information about blood alcohol levels and DWI, which they distributed at fairs and conferences. The organization contacted the Board of Education and set up programs in local schools, even at the elementary level, where volunteers taught youths about the dangers of drunk driving, passed out literature, and encouraged kids to tell their parents to pass the keys. Ricci, as state RID president, traveled throughout Connecticut, sometimes speaking six out of seven days a week.

A "pink elephant" named Tipsy became RID's mascot and helped the chapter communicate with youths. Tipsy (a volunteer dressed in a specially made costume paid for by a car dealership) was the featured guest at RID's Drunk Driving Awareness Night, where community members and kids got together at a local nightclub to dance, take advantage of nonalcoholic drink specials, and consider the risks stemming from drinking and driving. The night had a lobbying purpose, too: a state senator, local council members, and media celebrities drank "real" drinks to show how many—or few—it took to reach .10 percent BAC.

RID also offered victim support services, accompanying citizens who had been accosted by drunk drivers to trial after trial until their cases were settled, a process that frequently dragged on a year or more after the crash occurred.

Tipsy the elephant (embodied by Jim Ricci inside) stands with Janet Peckinpaugh of WTNH-TV in New Haven at the General Federation of Women's Clubs of Connecticut Convention in May 1987.

How the Media Helped

In the meantime, RID's lobbying activities for administrative per se did not let up. The local media helped draw attention to RID's campaign by covering the organization's activities liberally, accompanying articles with boxed quotes such as " 'We have no other choice. Our lives are just too precious to be playing political games.'—Nancy Ricci" in large, bold type. One state senator's joking proposal that drunk drivers be put in the stocks on the Green in New Haven, as they would have been in seventeenth-century Connecticut, generated an editorial uproar. Such a cavalier attitude masked his indifference and even opposition to the administrative per se bill, charged one columnist, an allegation that proved to be prescient when the legislator later proposed a damaging amendment that almost upheaved the bill. More editorials ran when seventy RID members showed up to testify before the state legislature and were sent home after four hours without being given a chance to speak.

By this point, reporters were calling Ricci when they needed statistics or an opinion on any topic related to drunk driving; the activist continued to contact the press each time the bill stalled in the legislature. The first time Ricci called a local radio station, she was asked to make a statement on the air. She told listeners to telephone their legislators and urge them to pass stricter anti-DWI laws. Thereafter, she became a frequent guest on local radio talk shows, and appeared as a guest on several TV programs as well, including the nationally syndicated "Sally Jessy Raphael Show." When a local TV station called her one New Year's Eve, Ricci got up from dinner with her family to meet with the reporter, who filmed her telling listeners to drink responsibly and "pass the keys" to a sober friend.

Again, it was Ricci's style to do extra research and legwork for reporters. She kept in constant contact with them, provided them with quick information, and thanked them for stories—not to butter them up, Jim Ricci said, but because she considered so many of them her friends. Yet while Ricci maintained the friendly rapport with them that the fight sorely needed, she did not mince her words when speaking with reporters, and could always be counted on for a heartfelt, frank, and often heated comment. In addition to frequent articles, newspapers ran pointed political cartoons focused on the problem and surveyed townsfolk as to whether people arrested for DWI should have their licenses taken away.

Though tightly budgeted, RID members found they still had to watch their spending carefully. The group paid just over $500 for an ad in the local paper promoting administrative per se and attracted the attention of the state ethics commission. RID soon learned that any expenditure over $500 related to a bill constitutes formal lobbying. The organization had to register as a lobbying body with the commission—and resigned the same day. "We are legislators' conscience," Ricci said. "We don't wine and dine legislators, and we don't get paid." She advised community groups of any size to know their state lobby regulations and think twice before spending large sums on tactics that could be considered lobbying.

A Champion in the House

RID members showed up at the state capitol each time the administrative per se bill was debated to speak their case or simply make their presence known, telegraphing unequivocal signals about their dedication to the fight. Ricci had organized a loose coalition of organizations touched by the issue over the years, and she called on that network regularly when she needed support, in word or deed, in fighting for or against legislation. Some weeks she placed a hundred calls to alert RID and coalition members to the legislative sessions and urge them to attend. More calls were necessary when RID had to scramble to adjust to last-minute legislative agenda changes.

Their lobbying trips hinged on information provided by the bill's diligent sponsor, Representative Prague. The main proponent in Connecticut for not only this but all drunk driving legislation, Edith was Nancy's equal in energy, stubbornness, and commitment to combat DWI, for her twenty-one-year-old niece had been killed in 1980 by a drunk driver who went free. "You have to have one contact that you can really count on, who will stick their neck out," observed Jim Ricci. "Other legislators promised, but didn't call. Edith is one of the few I have total respect for."

The feisty legislator "grabbed on like a bulldog and wouldn't let go," remembered Jim Ricci. She lobbied for the administrative per se bill among her colleagues, called the Riccis whenever the legislation was slated for discussion at a hearing or committee meeting, and kept them generally informed about its progress behind closed doors. Sometimes, the going got pretty rough. One state representative stood up during a debate and gave Edith (who is Jewish) the "Heil, Hitler" salute; a

state senator rudely declared that administrative per se
"would never be passed as long as Nancy Ricci is fighting for it!"

Ricci built a valuable resource group upon whom she
called for information to bolster her lobbying efforts. A
local doctor provided RID with health care information
and spoke out forcefully in favor of RID's positions;
NHTSA and the National Transportation Safety Board
offered a steady flow of facts and figures. Even sympathetic workers in the mailrooms of legislative offices
helped, allowing Ricci to drop off her letters with names
only, no addresses or stamps, and then placing them in
the correct boxes.

They Won!

Eventually, the activists' perseverance, hard work, and
the enormous public support they had drummed up paid
off. Final approval for the administrative per se bill was
won with a unanimous vote in June of 1989, three years
after its initial proposal. In a state where the average
blood alcohol content of convicted drunk drivers was
.18, the equivalent of ten to fifteen drinks, a major ordinance was finally put in place to try to deter people
pulled over for DWI from drinking and driving again.

The law's success in removing drunk drivers from
Connecticut's roads has been substantial. In 1989, almost no drivers lost their licenses for a DWI offense; in
1990, almost all those accused of DWI did—a total of
1,657 drivers relinquished their right to drive for ninety
days and 622 had to settle for the passenger's side for
six months.

Ricci's eight-year reign as president of RID Connecticut was dotted by additional legislative triumphs, as
RID continued to eliminate opportunities for citizens to
drink, drive, and get away with it. The group helped
pass a law that closed all bars at the same hour and defeated an effort to push back that hour. They also fought
to change the labeling of a deceivingly potent wine
cooler whose alcohol content was 20 percent; proposed
an anti-DWI postage stamp; helped persuade the state to
open a prison in Windsor, Connecticut, exclusively for
repeat drunk driving offenders; and, in November 1992,

convinced the nationally syndicated Maury Povich show to feature a couple whose sixteen-year-old daughter died from binge drinking.

Don't Be Afraid

RID's story offers a remarkable example of how people who are initially outside the system can get involved, influence others, and effect change. Jim Ricci attributes RID's effectiveness largely to Nancy's instinct: She was not afraid of doing "the wrong thing"; all that mattered to her was fighting to save lives. The Riccis faced deep-seated opposition from their state government: Family members of the chairman of the ordinance committee owned a restaurant in Wallingford; an assistant attorney who was found drunk and slumped over his steering wheel in a running parked car was not suspended; the governor was a former bar owner. Still, the couple pressed on. The lack of bureaucracy in the RID network attracted and kept advocates like Edith Prague motivated. And the organization's success in remaining true to its grass roots effectively inculcated in policymakers a sense of responsibility for helping to stop drunk drivers.

"Nine years ago Nancy probably couldn't have told you the name of her governor," her husband Jim said, "let alone how to push a bill through the state legislature." But lack of organizing and political experience clearly did not stop Nancy and her band of activists. In fact, says Jim Ricci: "One of the reasons we were so successful is that we didn't ask anyone what to do or whether we should do it, we just did it."

"People can't be afraid to speak up and get things accomplished," said Nancy Ricci. "Anyone could do what we did. You just have to believe in what you are doing."

WHAT IS LOBBYING?

As *New York Times* columnist William Safire has observed, the practice of "lobbying"—in which citizens, interest groups, or their representatives seek to influence the action on proposed laws by meeting directly with legislators—probably is as old as lawmaking itself. According to Safire's *Political Dictionary,* however, the term "lobbying" did not enter our language until the mid-1600s, when a large anteroom near the floor of the British House of Commons in London became known as "the lobby," a place where legislators would gather when not debating measures before the House. People who wanted a word with members of Parliament could meet them there.

In this country, the posh and venerable Willard Hotel in Washington, DC, just a few blocks from the White House, likes to say it gave a boost to the term "lobbying" during the administration of Ulysses S. Grant, the Union Army general and Civil War hero who was president from 1869 to 1877. President Grant was known to relax from the rigors of office by strolling over to the Willard and sitting in its lobby to have a brandy or two and smoke cigars. "Many would-be power brokers approached him on individual causes and he called these people 'lobbyists,' " according to the official history of the Willard, which traces its own lineage to hotels that have occupied its same site since 1816.

After Daniel Willard purchased the property in 1850, he quickly turned his hotel into an important social and political location—so much so that author Nathaniel Hawthorne, visiting the Willard during the Civil War, wrote that the hotel "may be much more justly called the center of Washington and the Union than either the Capitol, the White House, or the State Department."

Today, any communication with an elected federal, state, or local representative, or his or her staff, for the purpose of influencing legislation, an appointment, a resolution, a referendum, a constitutional amendment, or an initiative is considered lobbying. Personal contact, a telephone call, a letter, a telegram, or a fax all may be considered lobbying if they are sent to legislators (or their staffs) with the intention of influencing the way they vote. And while these actions can certainly take place in the lobby of a swanky hotel, they're just as likely to occur from an activist's home, a church basement, or in the legislator's own office.

For many Americans, lobbying has an unsavory sound to it,

connoting an underhanded way of gaining special or even excessive favors from the government. But citizen lobbyists who participate in the political process through lobbying can play a critical role in protecting the public interest. Indeed, lobbying is frequently the only course open to activists committed to obtaining just laws and fair treatment on their issues. As one citizen lobbyist once put it, lobbying can be a weapon as well as a shield for both sides of any issue.

WHY LOBBY?

When you lobby, you take advantage of one of the most democratic tools available to you to bring about change.

You also provide a much-needed service to legislators who don't have the time (or perhaps the inclination) to become experts on every issue that concerns you and the people you represent. Citizen lobbyists offer valuable assistance to legislators, public officials, the media, and the general public by compiling accurate information on complex or provocative subjects while supplying the ammunition needed to defeat counterproductive proposals. By communicating with your elected officials throughout the process of drafting, debating, and enacting legislation on your issue, you greatly increase the chances that laws and regulations will meet your needs.

Lobbying is a tough, highly competitive business, and it's certain that you will win only some of the fights you undertake. But whether you win or lose, the experience should allow you to participate more effectively in the legislative process when the next opportunity arises. And there will be other opportunities, for as certain as it is that the gavel will drop to convene a new legislative session every year, there will be a next time for you to urge consideration of issues important to you.

CAN YOU LEGALLY LOBBY?

Two forms of lobbying are commonly recognized: *direct lobbying* and *indirect lobbying*. Direct lobbying involves personal contact with the members or staff of a legislative body, or a government employee who participates in the legislative process, in an effort to encourage them to introduce, support, oppose, repeal, or otherwise influence legislation. Indirect, or grass roots, lobbying encourages individual members of the public to write, telephone, or otherwise contact legislators, their staffs, or particular government employees and persuade them to take action with respect to specific legislation. A letter you write to your senator urging support for the Family and Medical Leave Act constitutes direct lobbying. A mailing you send to members of your organization, asking them to write to their senators about the act, qualifies as grass roots lobbying. Under the provisions of the 1976 Tax Reform Act, nonprofit groups are allowed to engage in both kinds of lobbying, with certain restrictions.

A 501(c)(3) organization (one that is classified by the Internal Revenue Service as both nonprofit and capable of receiving tax-deductible contributions) is permitted to lobby without endangering its tax status so long as the lobbying effort does not make up a "substantial part" of the organization's total activity. In order to stay within this limit, an organization must determine whether a particular activity qualifies as "lobbying," and how much of the organization's total activity may be directed toward lobbying. Courts have ruled that a "substantial part" of lobbying expenditures must amount to less than 5 percent of the organization's total annual expenditures.

However, a 501(c)(3) organization may elect to spend more of its budget on lobbying activities by filing its intention to do so with the Internal Revenue Service. The amount that an organization may spend on lobbying without incurring a penalty is called the "Lobbying Nontaxable Amount."

The IRS measures an organization's Lobbying Nontaxable Amount by adding together 20 percent of an organization's first $500,000 of total operating expenditures; 15 percent of its second $500,000 of expenditures; 10 percent of its third $500,000 of expenditures; and 5 percent of any remaining expenditures, as long as total lobbying expenditures do not exceed $1,000,000 in any one year. No more than 25 percent of total lobbying expenditures may be spent on grass roots lobbying.

What does all this legalese mean? Consider this example: If your organization has a budget of $100,000 a year, you may elect to spend up to $20,000 a year on salaries, postage, printing, and other expenses incurred in the course of direct lobbying. You may not spend more than 25 percent of that $20,000—or $5,000—on grass roots lobbying.

Failure to observe these guidelines may cause the IRS to levy fines or taxes, revoke your tax-exempt status, or both.

The Federal Regulation of Lobbying Act (FRLA) requires individuals and organizations lobbying the U.S. Congress on legislation to register and report these activities. An organization that plans to lobby the Congress should ask the Secretary of the Senate or Clerk of the House about registration and reporting requirements. Many states have similar registration and reporting requirements for lobbying activities in the state legislature. Information regarding these requirements generally is available from the office of your state's Secretary of State.

If you have created a 501(c)(4) organization specifically to lobby or otherwise engage in political activities (remember, a 501(c)(4) is exempt from paying federal income tax but contributions and membership dues are not tax deductible), your group may conduct unlimited lobbying activities without jeopardizing your tax-exempt status, provided the legislation that you are attempting to influence pertains to the purpose for which your group was formed. For example, an organization devoted to reforming health care may spend unlimited funds to promote legislation to nationalize health insurance, but not to advance a bill that would increase the budget for the Environmental Protection Agency.

 ACTIVIST'S NOTEBOOK

"Safe" Legislative Activities

If you are a 501(c)(3) tax-exempt organization, here is a list of the kinds of legislative activities in which you may engage without endangering your tax status:

- **Provide technical advice.** You may respond to *written* requests from a legislative committee or body to testify or provide advice on legislation. Requests you solicit, and appearances you make without a written request, will be considered direct lobbying.

- **Distribute nonpartisan informational materials to all legislators as well as to the general public.**

- **Invite public officials to attend meetings or otherwise become informed about your issue without attempting to influence votes** (as long as you invite officials of all political parties).

- **Lobby on your own behalf.** This so-called "self-defense" lobbying must deal with issues that may affect the existence of your organization or your tax-exempt status.

- **Communicate with your own members.** You may tell your members about legislation provided you don't ask them to do something specifically about it.

- **Work on regulations.** Once a bill has been passed, you may work to help draft regulations to implement and enforce the legislation.

- **Lobby on a volunteer basis.** You may lobby an elected official as an individual, rather than as a representative of your group, provided you are not reimbursed for expenses incurred in doing so.

THE ELEMENTS OF AN EFFECTIVE LOBBYING CAMPAIGN

Michael Pertschuk, who was appointed chairman of the Federal Trade Commission by President Carter in 1977 and now works for the Advocacy Institute in Washington, DC, as a citizen lobbyist, noted in his book *The Giant Killers:* "The basic ground rule in public interest lobbying is to find something that is not just winnable on its own terms but has the capacity to have a rippling effect . . . we have to choose our battles."

In other words, since we do not have the resources to fight every fight, we must pick our confrontations carefully, and hope that the ones we pick—and win—will lead to subsequent victories on other issues.

An effective lobbying campaign takes many twists during the legislative process. It can influence the choices legislators and their committees make about the issues they champion, shape bills before they are introduced, affect the way hearings are held, influence the "mark up" sessions in which amendments are considered and the content of bills evolves. Your lobbying efforts can also affect the timing of votes by the full legislative body, sway the conference committees in which senators and representatives meet to resolve differences between their two houses, and influence whether a mayor, governor, or the president signs the legislation—or vetoes it.

The lobbying process can take months, years, even decades. Rarely does a law get conceived, written, amended, and passed in the same legislative session—sometimes the session is just too short (especially in those states where the legislature meets for only ninety days each year!); frequently, it takes days and even months for competing interests to work out their differences, or for one side to muster the resources necessary to emerge victorious. And legislators themselves have elevated the science of procrastination to an art. Sometimes it seems that the procedural rules governing the way a bill becomes law are designed to actually prevent the bill from becoming law. The reluctance of many legislators to want to be pinned down to a position or a vote, particularly on a controversial topic, combined with the sizable investment required to build a coalition, mobilize the media, recruit co-sponsors, and take other essential steps, all conspire to make lawmaking a very time-consuming process.

Where Should You Start?

Though the point of a lobbying campaign is to influence legislators, most lobbying efforts begin long before a vote is cast. Like Connecticut's RID, this Indiana group began working years in advance to build coalitions, prepare position papers, and generate widespread public support for its legislation. Its strategy paid off in victory.

Local Hell-Raiser
The Indiana Home Care Task Force

If you were a senior citizen, what concern would be utmost in your mind? When the Indiana Health Care Campaign in the summer of 1985 asked senior Hoosiers that very question, they were stunned by the replies. An overwhelming 90 percent of the state's elderly responded that their greatest fear was institutionalization—being put in a nursing home. Campaign members were floored; many had aging parents themselves who would soon be needing extra health care. What could they do to convince the state legislature to pass laws securing a healthy future for Indiana's elderly?

The Campaign, a coalition of thirty senior citizen, religious, labor, and human service organizations, had been working since 1984 toward a more progressive health care program for the state. The activists embarked on an ambitious effort to draft appropriate legislation and search out key lawmakers to sponsor it.

The Citizen's Action Coalition of Indiana (CAC) and United Senior Action (USA), advocates well-known for other health-related lobbying battles they had waged, helped research and write the bill. CAC legislative director John Cardwell coordinated the lobbying effort on their behalf.

Soon the people's support for home care laws began turning into grass roots manpower. By April 1986, forty organizations, including the Campaign, the American Association of Retired Persons (AARP), Citizens Action Coalition of Indiana, United Cerebral Palsy, United Senior Action, and the state Nurses Association had formed a larger coalition—the Indiana Home Care Task Force.

During the 1986 General Assembly, this first home care bill was amended to provide for a commission to investigate the problem. Task force members lobbied hard to get "their people" seated on the commission: Experts from different affiliated agencies testified at hearings, while volunteers met with officials face-to-face and wrote to the governor, who was in charge of appointing citizens to the commission. Such pressure tactics paid off; many individuals who were asked to serve on the commission were friendly to consumer-oriented health care programs.

Indiana legislators found themselves learning about home care for the elderly whether they liked it or not: The forty well-established organizations, many of which were branches of national networks, were hard to ignore, especially when their members kept calling and sending letters! The task force had shaped the composition of the commission, but still had to lobby to steer its recommendations in the right direction. To the group's satisfaction, the commission proposed a bill that was modeled after a Wisconsin home care program.

Indiana's program, dubbed Community and Home Options to Institutional Care for the Elderly and Disabled, or CHOICE, would utilize case managers from area agencies on aging who could work with elderly clients to design a package of at-home care services to suit their needs, and a cost-share plan to make care affordable. The services could include the provision of health aides, transportation, companions, and other activities, all designed to keep the individual at home, comfortable, and independent. Home care saves seniors the inflated costs of nursing homes, for which Medicare and private insurance payments are usually inadequate.

Opposition to the home care bill swelled up from the state's Budget Agency, but the increasingly visible grass roots supporters stood fast. The task force had become a favorite of the media and enjoyed substantial state and local coverage of the press conferences it held at the statehouse and of the rallies it organized. The senior citizens proved to be enthusiastic demonstrators whose events made for great photo opportunities. In one particularly eye-catching action, 150 seniors leaned over the balcony on the third floor of the statehouse and unrolled a scroll containing the signatures of 9,000 older Hoosiers who supported the CHOICE bill. The scroll reached the floor of the statehouse rotunda—a full two stories long! Legislators gaped at the sheer numbers who were demanding a home care program, while press photographers snapped pictures furiously.

Despite officials who protested spending state money on any sort of human service, the task force was able to engage sympathizers from many different camps. "Because the response from the seniors was so overwhelming, we decided early on that it would be hard to find an adult who didn't have to care for an aging relative," Cardwell said. Party affiliation was not an issue; the goal was to build a cadre of supporters so numerous and diverse that they'd be impossible to ignore when they arrived in the capitol to lobby legislators.

The activists approached leaders of agencies such as the Chamber of Commerce who historically had opposed this type of government expenditure. "Have you ever had a family member in need of extra care?" Chamber representatives were asked. The task force urged those who had experienced the challenge of caring for an elderly relative to pledge their neutrality in the battle.

The Indiana Home Care Task Force taped together a two-story curtain of petitions, full of signatures from advocates of the CHOICE bill, and draped it over the railing at the state capitol. A banner at the top read "Home Care or Nursing Home Care—*Will you have a choice?*"

Some legislators proved unlikely proponents: One of their biggest champions was a conservative state representative whose father was caring for his mother, stricken with Alzheimer's disease, in Michigan. According to Cardwell, that representative confessed that his parents couldn't have survived in Indiana. Cardwell, who currently is looking after his own parents, could relate. "It's a very personal process," he said. "You try to keep it in the family. The CHOICE bill helps families stay involved, but not wear out."

The CHOICE legislation survived a determined attempt by another Indiana legislator to scuttle the bill. "People saw on TV and in the papers that it was close to being killed, and they went crazy," Cardwell chuckled. An intense public uproar swamped the governor's office, tying up phone lines for two days. Eventually, the opposition capitulated and the General Assembly unanimously passed the CHOICE bill at the end of April, 1987. From the beginning, Cardwell had advocated building grass roots support so diverse that no one could label it a "special interest" effort. "The grass roots was the ultimate factor that passed the legislation. Smart lobbying alone would not have passed it; intense public pressure was necessary."

The battle continues to secure adequate funding for the CHOICE program, but the task force—which still meets monthly to work on home care—succeeded in getting $62.5 million allocated for CHOICE in 1993, up from $1.5 million in 1987.

HOW TO LOBBY, STEP BY STEP

As the activists' experience in passing the home care bill showed, lobbying involves more than personal contact with legislators and showy media events at the capitol. Persuading influential people in relevant public and private agencies to back your efforts is vital. John Cardwell met with hospitals, nurses' associations, and senior citizens' groups to explain the thrust of the bill and how it would benefit the organizations' members, to ask that each group add its name to the list of those supporting the bill, and then to encourage those members to testify before the legislature themselves.

In the long run, of course, the target for your influential campaign is the legislators themselves: the people who have the power to pass, defeat, or amend your law. When your organization is ready to embark on a grass roots lobbying campaign, take the following steps.

1. Form a Lobbying Committee

Organizations with larger budgets may be able to hire professional lobbyists (either on-staff or as consultants) to represent them in their statehouse or before Congress. But volunteer-based groups probably will have to form a committee of people who either have some experience talking to legislators or who are willing to learn how to do so despite their lack of expertise.

2. Learn About Your Legislator

The more familiar you are with each of your target legislators, the more effectively you can lobby. What are their educational, professional, family, and religious backgrounds? Do they have any particular interests or hobbies? What professions did they pursue before their election to office? Are they new to elective office or "seasoned pols"? What are their voting records? How does your issue "fit" into their districts? Are they mirror images of their constituents, or mavericks? Where did these legislators receive their greatest financial and voting support? When must these legislators seek re-election? How tough is the opposition? Knowing the answers to these questions will help you make convincing arguments when you ask for a legislator's support.

☞ **CHECK IT OUT:** For personal and political profiles of each member of Congress, see *Congressional Yellow Book* (a quarterly by Monitor Leadership Directories, NY, NY, $225 per year), or *Almanac of American Politics,* by Michael Barone (published every two years by The National Journals, Inc., Washington, DC). To begin compiling information on your state legislator, contact your local Common Cause chapter. Check the telephone directory for assistance. For help in identifying your federal, state, and local representatives, contact your local League of Women Voters. You can also check with your city hall or the library.

While it's important to know who your legislator is, it's probably even more valuable to understand what he or she does.

Legislators introduce or co-sponsor legislation, hold hearings on legislation, lobby other legislators to get them to adopt their position, testify themselves, make statements on the floor of their legislative body, and help organize or lead coalitions of legislators to help pass a bill or raise the profile of an issue.

Most legislators are elected from among the public at large; occasionally a candidate will be appointed to fill a vacant House or Senate seat, but these politicians, too, will have to seek reelection if they hope to maintain their position.

Legislators are motivated by two frequently competing factors: votes and money. Ostensibly, legislators are elected to represent the voters who live in their districts, and their reelection depends to a great degree on their ability to meet the needs of those voters over time. But in today's political environment, elections are getting exorbitantly expensive. According to the public interest lobbying group Common Cause, the 435 victors in the races for the U.S. House of Representatives in 1992 spent an average of $495,834 over a two-year period to win their seats. The 36 winners of 1992's U.S. Senate races spent an average of $3,647,218 to win election to their six-year terms. Many politicians feel they must rely on political action committees and campaign contributions from special interests to enable them to pay for their election or reelection campaigns. But don't be discouraged if you have to rely on your ability to deliver votes, rather than money, to help sway legislators in your direction. Said Vin Weber, a former Republican congressman who represented southwest Minnesota for twelve years, "The most powerful lobbyists are the ones that can mobilize voters—it's really the ability to move voters, more than the ability to give money, that influences the political process."

THE LEGISLATIVE STAFF

Though legislators are the ones who actually cast the votes, their staff are often the ones who tell them how to vote. Most legislators rely on assistants to help draft legislation, research issues, and define positions. The number of staffers varies according to the size of the legislative body and the importance of the legislative office being occupied. For example, in the U.S. Congress, a representative will probably have a staff of around twenty, while committees are served by staffs of around fifty to sixty. (The number of committee staff members varies greatly: The Senate Ethics Committee has thirteen while the Appropriations Committee has 206.)

Regardless of how many people are on it, a legislator's staff is important. For one thing, the staff help sway votes, not only those of their bosses, but of other legislators as well. In fact, staffers usually know more about a particular piece of legislation than the legislator for whom they work; they are certainly more accessible to members of the public.

In a large office, staff members perform a variety of functions: An administrative assistant, usually the legislator's closest political aide, runs the office and may exert the greatest influence over how the legislator votes. A legislative assistant reads and analyzes bills and often advises the legislator on what position to take on a specific piece of legislation. In an office with several legislative assistants, each one will specialize in one or two issue areas. The press secretary handles media requests and generates press coverage; an appointments secretary controls access to the office; clerks help answer the mail.

Though you may not always be able to meet with a legislator, you can usually get in to see staff. Staff will help you refine your lobbying strategy by telling you how a legislator is thinking of voting and what problems the legislator has with your bill, and by suggesting solutions.

COMMITTEES

A committee is a group of legislators who consider specific issues, bills, and amendments before they are presented to the entire legislative body. Among the most common committees are those that work on the budget, health and human services, environmental protection, and education.

A committee may be the most important forum in which you operate. Committees refine the contents of a proposed bill or amendment, convene hearings to get input from the public, work out compromises among committee members who disagree on the legislation, and vote to determine whether to send the bill on to the entire legislature. If you lose in committee, it's harder to win on the floor; victory in committee creates a precedent and momentum that place a heavier burden on those seeking to halt the progress of a bill at the next stage of the process. Understanding the membership and dynamics of the committees that have jurisdiction over your issues is as important as knowing individual legislators. Consult with other lobbyists and review any available news stories to better understand the chairperson of the committee that has jurisdiction over your bill. Is he or she strong and authoritative, weak and passive, confrontational or a consensus builder? What about the subcommittee chairmen? Are they forceful and independent or dominated by the chairman? How strong is the staff at all levels? To whom do they answer? Do members of the committee get along well or split along party lines?

ACTIVIST'S NOTEBOOK

About the U.S. Congress

Legislators elected to the U.S. Congress may be either senators or representatives.

The U.S. House of Representatives has 435 members. The number of representatives from a state depends on that state's population. Thus, California, with a population of about 30 million, sends fifty-four representatives to Washington. Little Vermont, whose population totals only 560,000, has but one representative to look out for its interests.

Representatives must run for reelection every two years. All representatives maintain offices in their home districts to deal with what is called "case work": handling requests from constituents to find lost Social Security checks, help get a new post office, secure funding for a new park or wastewater treatment plant, or otherwise go to bat for citizens against the federal bureaucracy.

In addition to the representatives it sends to Congress, each state elects two senators who are supposed to represent the entire state. Each of the one hundred members of the Senate maintains several offices scattered around the state to help stay in contact with constituents. Since each senator is elected to a six-year term, one third of the entire U.S. Senate seeks reelection every two years.

Staff who work in the local offices of your senators and representatives can tell you how they voted on specific issues, provide details about a piece of legislation in which you're interested, and direct you to other agencies or offices whose help you may need.

Most legislators divide their time between Washington, DC, and their home district or state. If you are unable to visit the Capitol to lobby your representative or senator, try to meet with him or her at home. If you don't know where the district office is located, check the telephone number in the phone book under the legislator's name or call Congress at 202-224-3121.

For information on legislation pending in the U.S. House or Senate, call the Office of Legislative Information and Bill Status at 202-225-1772 between 8:30 A.M. and 5:30 P.M. EST. You will reach a knowledgeable researcher who will search for pending bills and their status using a data base that is updated daily and dates back four or five congressional sessions. The office does not have the text of the legislation and will not interpret it, but will answer questions about the legislative process in general and where in that process your bill stands.

The *Congressional Record* (U.S. Government Printing Office, Washington, DC 20402) offers a daily verbatim transcript of everything that happened on the floors of the House and Senate that day, and includes a summary of committee activity as well. The *Federal Register,* also a government-published daily, offers printed text of all laws, regulations, and executive orders that are passed. Congressional Quarterly publishes the *Congressional Weekly Monitor,* which focuses on daily and monthly committee hearing schedules and news of committee activity, including schedules for bills that are up for a vote in the House and the Senate. The *Congressional Daily Monitor* carries the same information but in a daily format. The publications cost $598 and $1,299 per year, respectively; contact Congressional Quarterly at 1414 22nd Street, N.W., Washington, DC 20037, to order.

You can access the *Congressional Record* and the *Federal Register* on-line via Legi-slate, a *Washington Post*–owned public computer network that covers Congress and federal regulations. The network provides an update of the day's activities on the floor, daily congressional and committee hearing schedules, and the text of all bills and committee reports. Researchers looking for legislation on a particular issue can search by subject. Annual subscriptions to Legi-slate allow you unlimited use, but cost a minimum of $1,900—and increase depending on the services you request.

State Net, based in Sacramento, California, provides a similar service but for both Congress and all fifty state legislatures—and at a much higher cost.

Most of the issue-oriented E-mail networks include updates on floor activity, schedules, and analysis of legislation in their particular subject area—often on both the state and local levels. For example, dial into SeniorNet if you want information on legislation affecting senior citizens.

The Global Action and Information Network (GAIN) provides (via EcoNet) regular news, data, and analysis of pending congressional and state legislation that has to do with the environment. GAIN also distributes information through the mail. Contact GAIN at 575 Soquel Avenue, Santa Cruz, CA 95062.

> ☞ **CHECK IT OUT:** *You and Your National Government,*
> published by the League of Women Voters Education Fund,
> explains the three branches of government (legislative, judi-
> cial, and executive) and how they work. To order a copy,
> send $1.75 to 1730 M Street, NW, Washington, DC 20036.
> Also available from the League is *Impact on Congress: A
> Grassroots Lobbyist Handbook for Local League Activists,*
> $2.25.

Some state legislatures are divided into two sections: a House of
Representatives (also known in some states as a House of Delegates or State Assembly), as well as a State Senate. Other states
are served by a "unicameral" legislature, which means only one
legislative body makes their laws. If you don't know how your
government runs, contact your local League of Women Voters,
your public library, or the Secretary of State's office.

Most of the lobbying examples in this book have focused on efforts by activists to influence their state legislators. Hell-raiser
John Randolph of Birmingham, Alabama, put many of the same
skills to work when he successfully convinced the U.S. Congress
to pass legislation protecting a pristine wilderness area.

Local Hell-Raiser
Wilderness Winner

John Randolph is a real estate attorney in Birmingham,
Alabama, with what some might view as an anomalous
yet fervent appreciation for wide open spaces and unde-
veloped land—especially for the wilderness sections of
his own state.

The activist had a particular fondness for the Sipsey
Wilderness Area, located about one hundred miles
northwest of Birmingham, a pristine world of rare
plants, wildflowers, and trees, interlaced with streams,
surrounded by high cliffs, and containing hundreds of
waterfalls.

Randolph seized an opportunity to push for expansion of the 12,700-acre Sipsey Wilderness Area in 1977, when Jimmy Carter, newly elected president of the United States, ordered a review of all potential wilderness areas in the country. Randolph believed the existing boundaries of the Sipsey area would not protect the watershed of the region's lovely West Fork of the Sipsey River.

Initially, along with organizing the Alabama Wilderness Coalition—the core members of which were the Alabama Conservancy, the Birmingham Audubon Society, and the Alabama Chapter of the Sierra Club—Randolph persuaded Representative Ronnie Flippo, a Democratic member of the House of Representatives from the region in which Sipsey is located, to sponsor federal legislation protecting the entire 30,000 acres of the upper watershed of the Sipsey's West Fork.

Representative Flippo's bill easily passed in the House but was killed in the Senate by Alabama's senior U.S. Senator, Howell Heflin, a Democrat who bitterly opposed the measure on the grounds that it threatened Alabama's timber business, the state's largest industry. Though the 30,000 acres that Randolph wanted to encompass in an expanded Sipsey Wilderness Area represented only .2 percent of all the forest lands in Alabama, the issue was an emotional one for the state's loggers, particularly a group of small timber firms that dubbed themselves SWUFFL, the Society for the Wise Use of Federal Forest Lands. The veteran senator was their powerful advocate.

Randolph and his coalition might have lacked Heflin's congressional muscle, but they marshaled what would prove to be a more potent force: a grass roots movement that persuaded city councils, Alabama state legislators, every major newspaper in the state, and two Alabama governors to endorse the move to expand Sipsey. Under Randolph's leadership, the Alabama Wilderness Coalition grew, burgeoning with diverse members that included wildflower societies, scouting groups, even an organization of hunters who use only bows and arrows.

"We wouldn't have gotten all those politicians without the support of the local people," Randolph recalls. "We spent a lot of time talking with these officials, corresponding with them—we had a thick file of endorsements from public officials throughout the state." And after persuading the conservation aide to then-governor Fob James that more of the Sipsey needed to be protected, the Alabama Wilderness Coalition won not only the endorsement of James but that of his predecessor and eventual successor, George Wallace, both Democrats.

Randolph organized press tours of the wilderness area and oversaw creation of films and slide presentations about it to keep the controversy in the public eye. His list of formal endorsements for expanding the region's protection grew to include the state's conservation and tourism departments, as well as the five municipal water districts—Birmingham and Jasper among them—which drew water from the West Fork.

Senator Heflin nevertheless remained adamantly opposed to the Sipsey legislation and killed two more proposals aimed at expanding the wilderness region. When it became clear that he would never support the Sipsey bill as it was written, Randolph and the other coalition leaders sought a compromise by scaling back their expansion request to 13,000 acres.

Another 5,000 acres of the West Fork were added through the use of a National Wild and Scenic River designation. The U.S. Forest Service redesignated the remaining acreage a "semiprimitive" region, putting limits on the logging that could be done there. The area thus received "a mix of protection," as Randolph put it, and even the industry-backed Alabama Forestry Association agreed to this proposal, persuading Senator Heflin to drop his opposition to the Sipsey expansion.

Finally passed in 1988, the Sipsey Wilderness Area law doubled the size of the protected region and added fifty-two miles of the West Fork and its tributaries to its National Wild and Scenic River designation.

For his eleven-year achievement, Randolph received an Award of Recognition from the Alabama Chapter of the Sierra Club and was named 1988 Conservationist of the Year by the Alabama Conservancy.

Word of his accomplishment spread far, and in 1989

the State University of New York gave the activist its Sol Feinstone Award, a $1,000 prize bestowed for achievements in regional environmental causes.

Randolph later told the national Sierra Club that he was most pleased to be able to show "that one person with right on his side can make a difference in the political arena, despite the most powerful opposition, if he will just stand his ground."

3. Develop a Lobbying Strategy

When he was lobbying Congress to defeat the MX Missile, Fred Wertheimer, president of Common Cause, made some remarks that seem to apply to all lobbying campaigns. He noted: "You cannot win these fights without intensity, you cannot win these fights without focus. You cannot win without persistence, you can't win without a long-term willingness to lose. You can't win a fundamental battle easily or quickly." But ultimately, Wertheimer concluded, you can win—with luck, fortitude, and the right strategy.

Whether you're writing a new law or amending an old one, you follow the same essential steps in developing your lobbying strategy. (These steps should look familiar by now; they're the same ones you followed to devise your general organizing and communications strategies!)

- **Identify your goal.** Is it to pass or defeat a bill? Strengthen or weaken amendments to a bill? Oppose or favor a nomination for a cabinet position or court nominee? Endorse a resolution? Each lobbying campaign should revolve around a specific desirable result.

- **Take stock of your organizational resources.** Is this campaign going to require a lot of travel, the printing and distribution of a variety of publications, intense long-distance telephoning or faxing, and other capital expenditures? Do you have enough staff or volunteers to prepare materials, meet with legislators, and keep on top of the process? Who is going to educate the media, write letters to the editor, and meet with editorial writers? Do your staff and volunteers have the credibility to wage the campaign? Answer these questions honestly, then look to ways you can build coalitions and stretch your own resources to get the job done.

Richard Lombardi, a highly effective grass roots lobbyist

based in Lincoln, Nebraska, advises the groups he works for to inventory their members frequently as to possible connections they might have with the public officials the group is targeting in its lobbying campaign. "One reality that never ceases to amaze me in this work is the endless intriguing networks and connections of people that take place," he says. "Targeting the people your members know reaps bountiful dividends and helps build relationships that endure."

- **Pick your targets.** Identify legislators who can provide leadership on the bill and help guide it through the legislative process. Get one or more legislators to introduce the proposed measure, and recruit as many co-sponsors as possible. Focus on legislators who have not yet made up their minds about how they will vote. While you will also want to shore up your supporters, don't waste precious time or resources on opponents whose minds you know you can't sway. Target the swing voters for lobby visits, letters, and telegrams, and aim your publicity about the bill to press outlets in their home districts.

  Compile a list of elected officials with their telephone numbers and addresses. Keep track of those who support your position, those who oppose it, and those who have yet to make up their minds. Identify the most influential people in the legislative body; pinpoint whether anyone in your organization knows them personally. Rank elected officials according to your ability to influence them on the issue.

- **Choose your tactics.** Hold media events like the one the seniors held in Indianapolis to give your issue credibility and generate public support. Offer testimony at hearings about your proposed bill to help legislators understand your position. Visit a legislator's office in person to convey the importance of the issue. (You may have to do this several times: initially, to present your position and ask for support; later, to get a commitment if you haven't done so already; and finally, right before a vote.) Generate letters, telegrams, faxes, and telephone calls to legislators to turn the heat up before a vote. Meet with legislators in their home districts.

- **Know your time frame.** How much time do you actually have before your bill faces crucial votes? If a vote is imminent, you'll have to be able to mobilize volunteers quickly and generate substantial visibility to help advocate your position.

More time will allow you to reassess your goals regularly while winning interim victories that keep activists and supporters motivated, sustain interest from the media, and create opportunities for your leaders and activists to develop their skills through training workshops or other activities.

Investigate the legislative schedule to determine when hearings, debate, and votes are slated on your issue—then expect it all to change. Delays on other legislation, a local or national crisis, or political maneuvering can throw the schedule off by weeks or even months.

As the League of Women Voters advises its citizen lobbyists: "Anticipate the rhythms of the political calendar, then choose the moment when your message can be heard with minimal distraction and maximum sympathy . . . timing is everything." Take cues from legislators who support you and staff of those who still haven't made up their minds. If you've been able to amass adequate legislative support, arouse the passions of the public, and mobilize your members, it's probably time to push for a vote. Otherwise, postpone action, particularly if you believe a losing vote now will sound the death knell for future efforts to enact the legislation.

Nancy Ricci's Lobbying Strategy

GOAL

- Pass the administrative per se bill.

ORGANIZATIONAL RESOURCES

- Stretch dollars by getting services and supplies donated; enhance expertise by recruiting doctors and government experts to testify and provide information; augment outreach capabilities by working in coalitions.

TARGETS

- State legislators.

TACTICS

- Generate substantial publicity, hold media events with "Tipsy the Elephant," testify, write letters, meet with legislators in person, buy an ad in local paper, distribute information at fairs and conferences.

TIME FRAME

- Three years.

4. Build a Coalition

It is not unusual to have to contact personally a hundred or more people during a legislative campaign, considering how many lawmakers and staff you need to reach. In fact, you'll probably want to communicate in some way with every member of the legislature before your lobby campaign is finally put to rest.

In addition to the strictly legislative activities you undertake, you may also need to generate editorials that support your position (and send an unmistakable message to an elected official); organize media events; produce materials for distribution to legislators, the media, and the public; and raise money to keep the whole affair going.

Few groups can mount such an effort alone. And even if they could, broad-based coalitions that represent a diverse array of constituents and interests are much harder to ignore than a single-issue interest group who can't convincingly claim to speak for "the people."

Many tips for building a successful coalition were outlined Chapter 3. But remember these key points:

- **Begin by working with your natural allies.** These may include religious, parent, social service, women's, labor, minority, senior citizen, and peace groups.

- **Explore the possibility of working in coalition with groups or individuals with whom you may not have agreed in the past, but who may share your interests on this issue.** Remember how powerful the Mothers of East L.A. became when they teamed up with local businesses to defeat the prison plans?

- **Even if other groups won't join your coalition, they may agree to help lobby a legislator on the issue.** Ask them to write or sign postcards that you can collect and submit for them, circulate petitions for support among their members, offer the names of other people and organizations that can be contacted, write letters to the editor, or print an article about the situation in their newsletters.

- **Give all coalition members an important job to do.** Some may meet with legislators while others write press releases, organize media events, prepare brochures and fact sheets, and mail or fax information.

Remember, you do not need to have a name, set up a steering committee, write bylaws, or staff an office for the coalition to be effective. It is not unusual for ad hoc committees to come together around a particular piece of legislation—as did the Indiana Home Care Task Force—and then to disband until another important issue calls.

5. Formulate Your Legislative Arguments

Organizations take various approaches in presenting their position to legislators. Some groups favor a "go for broke" approach, crafting firm positions that represent their entire legislative wish list. Others work incrementally: Though they keep their sights fixed on their long-term goals, they focus on short-term objectives to move their agenda along bit by bit. Whatever the strategy, it is usually determined by the political realities of the moment: whether a majority of legislators supports your opinion, whether a group has the resources to sustain a campaign over time, or whether the issue is so immediate that regardless of available resources or legislative support it must be brought to a vote.

To determine what your position should be, consult with other members of your coalition as well as with legislators and their aides who are friendly to your cause. Brainstorm several legislative options you may have, and explore the positive and negative ramifications of each one.

Once you have opted for a particular position, develop a series of arguments that clearly and concisely support your point of view. It is critical that these arguments be as persuasive and compelling as possible: They will become the backbone of any testimony you give on the legislation, will be presented in meetings you have with individual legislators, and will probably become the topics of news releases to promote your work on the legislation as well as fund-raising appeals to raise money to support it.

Don't assume that all legislators will be interested in your legislation for the same reasons. Some lawmakers support legislation to protect the environment because they value the intrinsic nature of the ecosystem; others believe new investments in environmental protection will create jobs.

Develop a backgrounder or fact sheet that clearly lays out your legislative rationale. In simple, nontechnical language, offer five to ten reasons why legislators should support your approach. Present your information in short paragraphs to make the document easy to read for the legislators, staffers, reporters, and other activists to whom you will send it.

6. Lobby!

Once you've formed your committee, learned about your legislators, developed your lobbying strategy, built your coalition, and formulated your legislative arguments, you'll be ready to begin talking to legislators and their staffs. Don't treat the occasion lightly. Prepare for your lobbying meeting as thoroughly as you would anticipate an important interview with a reporter.

PLAN THE LOBBYING VISIT

Develop a strategy in advance. What's the purpose of the meeting? What do you want to accomplish? Do you need to give information or get information? Do you want to elicit support for a particular position or generate opposition to a developing situation?

Review the facts supporting your position so that everyone understands the issues and you can discuss your presentation and develop your ideas together. Consider arguments for and against your position. Research past election data, as well as personal background and voting behavior of the official so you know something about him or her personally. Pay particular attention to the legislator's voting record on issues similar to the one you will be discussing.

MEET WITH THE LEGISLATOR

"The best lobbying is done back in your congressional district by a group of organized citizens that makes an appointment to see you in your office or stands up at a town meeting and asks you a tough question . . . they have the most influence—certainly more than any high-priced lobbyist," Tom Downey, former Democratic member of the U.S. Congress from Long Island, New York, told National Public Radio during a special segment on lobbying shortly after the Clinton administration took office. Take the congressman's advice. Visit your legislators in person, either in the district they represent or in their offices (this could mean in the mayor's chambers, in a senator's or congressman's Washington, DC, office, or at a state capitol). It may be difficult to secure an appointment, especially with a member of Congress, but persevere. A staff member will probably be available to meet with you if the legislator is not.

Having a delegation of others accompany you to the meeting will help convey the impression that a broad base of support exists for your issue in the community. When deciding who should go to the meetings, consider who would be most influential. Obviously, someone from your organization who knows the issue should be present. But the elected official may be more influenced by family members, local politicians, business leaders, contributors, church officials, synagogue representatives, or representatives of other organizations perceived to have more clout in the community. Five to seven people make for a good-sized delegation.

ASK FOR SUPPORT

Arrive at the meeting early so that you don't feel rushed and can collect your thoughts. Don't get upset if you have to wait; elected officials often run late.

Introduce yourself and begin to find out where the legislator stands on your issue. Has he reviewed the bill you're concerned about? Is there other legislation he favors? By looking around his office, can you find personal mementos—such as photographs, trophies, or plaques—that provide insights about the legislator? Can you pick up any background information about his interests or hobbies from the conversation that you can use later in building a personal relationship?

When presenting your position, always use facts; don't exaggerate. Never threaten. Be polite and to the point. Thank the official for meeting with you and for any support you have received in the past. As you present your case, personalize it as much as possible. Explain your agenda in very simple terms and ask for support. If the legislator agrees to help you, find out if he will play a leadership role in getting more support for the bill. Ask the legislator to make a public commitment to your bill so that the story gets picked up in the media. If the legislator hasn't made up his mind, additional lobbying may be called for. If the legislator is opposed to your point of view but still has reservations about the legislative package that will be offered, perhaps you can persuade him or her to abstain or be absent on the day of the vote.

Even if the legislator doesn't agree with you, don't get angry. Leave the meeting on a pleasant note—then regroup and plan your next steps.

Remember that legislators and their staffs are focused on the "here and now," so don't talk to them about a floor vote that's two months away when the committee hasn't even debated the bill yet.

<u>FOLLOW UP</u>

After the meeting, determine how to proceed. Does the legislator need more information? Do more people need to lobby the official? Keep a written record of your meeting by putting a short report about it in your files. Understand that there's a world of difference between a legislator's remarking "I understand what you're saying," and "I support your legislation." Only the latter is a commitment. If a legislator is considered a "swing" vote and hasn't made up his mind, ask others to help lobby through phone calls, letters, telegrams, and personal visits.

Send a thank-you note to the official raising your position again and requesting that you be kept informed.

7. Monitor the Entire Process: How a Bill Becomes a Law

Meeting with your legislator is only one of many actions required to convert an idea into legislation. The process is a complicated one that you may have to monitor every inch of the way.

The following chart shows, in general, the steps involved in creating legislation.

How a Bill Is Passed	Your Role
• A bill is prepared.	• Suggest provisions to be contained in the bill.
• The bill is introduced in one house of the legislature by one or two sponsors and many co-sponsors.	• Help line up legislative co-sponsors for the bill.
• The bill is examined by a committee according to its subject area (e.g., a bill to control drug traffic would be assigned to the judiciary committee).	• Work with the committee to help its members understand the bill and prevent any tinkering that could weaken it.
• Hearings are held to discuss the bill.	• Testify at the hearings.
• The committee votes on the bill.	• Lobby committee members to support the bill.

How a Bill Is Passed	Your Role
• If the bill passes through the committee, it is sent along to the entire legislative body to be debated further and voted upon.	• Lobby swing voters in the full house to support the bill, and generate letters, phone calls, lobby visits, and media from your supporters to do the same.
• If it passes, it is sent to the other house in the legislature (unless the other house is working on similar legislation already.)	• Line up co-sponsors for the bill in the other house.
• The bill is examined by the relevant committee in that branch of the legislature.	• Work with members of that committee to strengthen support for the bill.
• Hearings are held again.	• Testify.
• The bill is debated on the house floor.	• Lobby swing voters.
• If the bill passes, it probably does so with amendments that now need to be negotiated in a "conference committee" made up of representatives of both houses before it is finally voted on.	• Lobby members of the conference committee to protect the bill from weakening amendments.
• If it passes, it is sent to the executive (the governor or president) for signature.	• Publicize the bill's passage and generate support for signing it.
• If the executive signs the bill, it becomes law.	• Celebrate!

LOBBYING TOOLS

Meetings with legislators and staff offer one way to influence legislation. But such meetings are infrequent at best, and often you can't even arrange a meeting if you haven't already demonstrated that there is public concern about the issue on which you're working. Your campaign will need to develop and maintain momentum to succeed. The following materials will help get your legislator's attention.

Common Cause of Rhode Island
FACT SHEET

**SYNOPSIS OF 1992 REFORMS
RESEARCHED, DRAFTED AND SUPPORTED BY COMMON CAUSE,
BACKED BY THE RIGHT NOW! COALITION,
ENACTED IN 1992 BY THE RHODE ISLAND GENERAL ASSEMBLY
SIGNED INTO LAW BY GOVERNOR BRUCE SUNDLUN
AND/OR APPROVED BY REFERENDUM ON NOVEMBER 3, 1992**

ETHICS:
• Prohibited political insiders from "seeking or accepting" judgeships and other lucrative, permanent state jobs for one year after they leave office. Without question this is the strongest "revolving door" bill in the nation.
• Banned nepotism, by making it illegal for public officials--whether elected or appointed--to use their post to benefit their parents, siblings, children, in-laws and a large circle of other relatives by blood, adoption or marriage.
• Streamlined the Ethics Commission to ensure prompt, vigorous, efficient prosecution of officials who violate the public trust.
• Opened the Commission's adjudicative hearings--which follow a finding of probable cause--to the press and public.
• Increased maximum fines for ethics violations from $10,000 to $25,000 per violation.

CAMPAIGN FINANCE:
• Banned all personal use of campaign contributions and testimonial proceeds.
• Outlawed corporate contributions to Rhode Island campaigns and the laundering of otherwise illegal contributions through political parties.
• Required greater disclosure by political action committees (PACs) and established effective controls on their operations.
• Capped spending, slashed contribution limits and lowered the threshold of reportable contributions.
• Closed loopholes which previously enabled many politicians to escape reporting how much they got and where they got it.
• Encouraged small contributors through a double match for donations under $250.
• Expanded matching funds program to all five general officers for 1994 election.

CONSTITUTIONAL CHANGES:
• Amended the state constitution to provide four-year terms, with recall and two-term limit, for state general officers. After three previous defeats, the measure passed with a 60% to 40% margin.
• Laid the groundwork for reform in the General Assembly through creation of a "blue ribbon panel" to study the size, compensation and appropriate functions of the state legislature. The thirteen-member commission will report comprehensive recommendations in April, then conduct statewide hearings. Since Rhode Island has no initiative process, a draft of constitutional amendments will be presented for approval by the General Assembly in the spring of 1994 and by the voters that fall of that year.

Common Cause is PEOPLE RIghting Rhode Island

**Much accomplished in 1992. Much more remains.
To learn more or help, call 861-2322.**

Fact sheets like this helped Common Cause of Rhode Island members and the public at large grasp the volume and thrust of the many bills the group proposed and lobbied to pass.

The Issues Packet

Develop an issues packet that you can give to a legislator when you meet and that you can leave with the staff. The packet should include:

- A cover letter urging the legislator to adopt your position on the bill, appointment, or amendment in question.

- Fact sheets arguing your position.

- Endorsements from prominent constituents and experts in the field.

- Copies of news stories or editorials that favor your position.

- A list of supporting organizations, interest groups, national figures, and local leaders who support your position.

Some groups also prepare a detailed description of the objections to their argument that are likely to be raised by their opponents. Take the offensive. Identify your adversaries, by group and by name if possible. Describe their arguments and counter with your own. Such a tactic will also help prepare legislators for arguments they'll have to make on behalf of your position.

Video

More and more organizations are turning to video as an effective way to convince their legislators to take appropriate action. Many legislators are frequently too busy to get out of their offices and observe problems firsthand; others simply do not have the budget to travel to a distant destination. Offering legislators a short videotape that captures the essence of your position can help focus attention on your issue and persuade legislators of the importance of your campaign, as long as the tape provides crisp images that clearly convey exactly what you say they do.

While well-heeled corporations may spend tens or even hundreds of thousands of dollars producing a glitzy video package, citizens groups appear to be having just as much success pulling together video testimony on their home camcorders. As the activists in the following story discovered, your footage doesn't have to be of broadcast quality or narrated by a Hollywood celebrity to be effective. In fact, sometimes it's even more dramatic—and effective—because it seems so real.

LOCAL HELL-RAISER
Eco Spy

Most people buy camcorders to chronicle family vacations or send tapes of their kids to out-of-town grandparents. Terri Moore used hers to lobby the Indiana state legislature.

When an existing landfill expanded to engulf 90 acres just 1,700 feet from her home, it didn't take long for Terri and other Center Point, Indiana, residents to notice the lines of idling semis waiting to drop their foul cargos basically in their backyards. Apart from the sheer volume of garbage that was being dumped, Terri was bothered by the fact that most of the trucks sported license plates from the East Coast. Why were places like Pennsylvania, New York, and New Jersey trashing Center Point? Terri, a mother and former lab technician who works for her family's petroleum marketing business, began to investigate, purchasing a Sears camcorder to help her out.

"I'm just a local citizen exercising my right to document what's going on in my landfill!" she said as she narrated one of the videos. A few preliminary inquiries revealed that neither the state of Indiana nor the federal government required any documentation or certification for out-of-state garbage dumpers. Terri's backyard could be used to store trash from anywhere in the United States. With landfills on the East Coast and elsewhere rapidly filling up, Center Point's facility was turning out to be everyone's dump.

Terri was already a member of HOPE (Hoosiers Opposed to Pollution of the Environment) of Clay County, which worked to promote recycling in the area. To document the dumpings, she organized fellow HOPE members and other townspeople into a Dump Patrol. Six days a week, seventy-five people—a significant percentage of the town's population of 250—split up into four-hour shifts and, ten hours a day, manned patrol posts just outside the landfill gates.

The activists recorded every truck that entered the dump. They wrote down license plate numbers and focused their lenses on truck accidents, which were frequent. Ten percent of the trucks, they found, were lost or off route, and it was commonplace to see trucks run over signs and front yards. The videocamera captured the disgustingly dirty condition of the trucks and their vile contents, and filmed any suspected violations of waste rules, creating a powerful body of evidence.

In addition to the videotape, the Dump Patrol took still photographs of the violations, which proved a handy tool to combat industry arguments. The patrol mailed the photos to the companies whose advertisements appeared on the sides of the trucks. When they saw the abhorrent trucking procedures, the advertisers were quick to pull their ads. The threat of losing ad dollars sent some companies spinning into action to improve their policies. These photos would become part of the activists' testimony before Congress—both state and federal!

The Dump Patrol soon learned of a practice that was not only disgusting, but endangering the health of many people beyond Center Point: To the activists' horror, truck drivers reported that they usually hauled meat, seafood, and produce back to their home states in the same maggot-infested trucks in which they had transported the garbage. Vowing to stop the revolting practice, Terri and the Dump Patrol turned to their state legislature.

In the summer of 1989, the group brought its most graphic photographs to a meeting of the legislature's Interim Study Committee on the Environment. Many committee members had heard about the landfill and the Dump Patrol via the local and state media, which had begun to cover the growing piles of garbage and the activists' mission. Still, the legislators were shocked by what they saw and heard. For five minutes, Terri recounted what the group had discovered, displaying her photographs for all to see. The packed room fell totally silent. "I have never felt so much power. I had everyone's ear," she said. "I handed the pictures to the legislators, and they just shook their heads. It was the most

This trailer served as the Dump Patrol's headquarters, as activists monitored the trucks that hauled garbage into their community landfill.   *Terri Moore*

effective five-minute lobby I have ever done!" The next speaker, one of the many professional lobbyists from the waste industry who were present, opened her address by stating that she didn't know how to follow what Terri had just said.

HOPE members attended every Interim Study Committee meeting after that, fueled by the media coverage they were now receiving on the state's major TV stations. Legislators' eyes really opened when the group stumbled into lobbying on the federal level. When *USA Today* ran a story about a backhauling bill proposed by Representative Chris Smith of New Jersey, HOPE contacted his office and alerted him to the backhauling going on in Center Point, offering him the still photos that had shaken up the Indiana legislature. Smith, thrilled to receive the Dump Patrol's graphic snapshots, distributed them to other members of Congress; suddenly HOPE's efforts on the state level were receiving federal attention as well.

Meanwhile, HOPE members were busy attending

state hearings and writing to their state legislators demanding change. They taught themselves to be more effective lobbyists by bringing home letters and speeches that professional lobbyists had submitted, and copying their language and format. They also became acquainted with the waste industry lobbyists. "We did a learn-as-you-go lobby, but it worked! Legislation was the only way we could right these wrongs," Terri said. The activists attended every "crackerbarrel," or county meeting, with kids in tow to impress upon lawmakers why the stakes were so high. Now very aware of the power of their video evidence, they created a "newscast" in which Terri as reporter showcased the dumping and the condition of the trucks and described the sheer number of out-of-state vehicles they had documented: over 2,500 filthy semis in twenty-three weeks.

Terri was asked to testify before the House Subcommittee on Transportation and Hazardous Materials on behalf of another federal bill, and again brought pictures, this time enlarged to allow the entire room to see. The national lawmakers were engrossed in her "homemade" visual proof of exploitative dumping on a "defenseless" community! ABC-TV not only covered the hearing but aired her video on "Nightline," showing the country the inside of a refrigerator truck—before it was loaded with food—crawling with maggots from its previous load of garbage.

The ABC feature became the turning point in HOPE's fight. Within a month of the broadcast, Terri was asked to appear on "Donahue," on an episode concerning food safety. Through the compelling pictures and startling videos, viewers connected with the people of Center Point, Indiana. "Our video got a lot of attention because it was something people could relate to. And with the news media, who have only a short period of time, just a small amount of footage can say a lot," Terri said. "Without documentation and media exposure, we could not have been empowered enough to accomplish our goals."

When Terri returned to lobby her state legislature, even waste industry representatives supported HOPE's battle. They communicated ideas for bills to the governor's office, which then wrote the legislation and gave it to key lawmakers to push through the state legislature. The most important bills HOPE helped propose passed unanimously. Backhauling became outlawed in the state of Indiana, withstanding constitutional challenges by the waste industry. And the state now requires documentation of every out-of-state truck that crosses its borders to dump a load of garbage.

No matter how rudimentary it is, a picture is still worth a thousand words.

Here's what the Dump Patrol was trying to stop: The use of filthy trucks like these to haul produce back east from Indiana. *Terri Moore*

One production tip: Much home video is difficult to watch because it is produced by camera operators who insecurely hold their cameras on their shoulders, making for a very wobbly picture. You'll get a more stable visual image by basing your camera on a tripod or resting it on some other solid foundation as you shoot. You can buy a basic tripod at most photography supply shops for less than $100.

Personal Letters

Video has not yet supplanted the written word as a means of personal contact and expression. Planned Parenthood used to tell its members "letters are the barometers that measure political pressure back home," and that's truly the case. An average member of Congress receives more than one hundred pieces of mail a day; prominent legislators receive as many of five thousand a week. And when an issue is controversial, tens of thousands or even hundreds of thousands of letters may pour into Washington on behalf of or in opposition to a bill or an appointment.

Letters are equally effective on the state level—Nancy Ricci sustained her organizing efforts for years through the momentum generated by letters from her and many other RID supporters.

Each letter is estimated to represent at least twenty to thirty other people who did not write. Thus even ten or fifteen letters on one topic will get the attention of a staff person and perhaps the legislator himself.

••STEP BY STEP••

How to Write a Letter to Your Legislator

- **Include your name and address** on the letter itself (you'll want a reply, and your envelope may get lost).
- **Write neatly by hand,** or use a typewriter or word processor.
- **Keep it short,** one page or less, and focused on only one subject.
- **Ask for a specific action,** either to support or oppose a bill, amendment, appointment, etc. Be as specific as possible, referring to the bill by name and number, stating who introduced it, and summarizing what it will do.

How to Write a Letter to Your Legislator (cont.)

- **Demonstrate your knowledge** of the legislator's record on the issue.
- **Make it clear** why your letter counts: For example, are you a constituent or do you have some particular knowledge of how the bill will affect voters in a legislator's district or state?
- **Decide** whether you should be writing as an individual or as a member of a group. If the group to which you belong is considered particularly influential in the legislator's district, it will help to let the legislator know who you represent.
- **Make a concise argument** for your point of view, using meaningful examples and easy-to-understand statistics that underscore your position.
- **Thank the legislator** for his or her time.
- **Ask for a reply.**

Address the letter in the following way:
U.S. Senators:
The Honorable (Name)
United States Senate
Washington, DC 20510

Dear Senator (Name):

Members of the U.S. House of Representatives:
The Honorable (Name)
U.S. House of Representatives
Washington, DC 20515

Dear Representative (Name):

State Senators:
The Honorable (Name)
(Name of State) State Senate
City (Capital), State Zip

Dear Senator (Name):

State Representatives (or Assemblymen):
The Honorable (Name):
(State) House of Representatives, House of Delegates
or Assembly
City (Capital), State Zip

Dear Representative (Name):

Individual letters pack a lot more clout than form letters or post-card campaigns. Still, there are times when hundreds or even thousands of letters are needed to generate a response. Large, better funded organizations mobilize letter writers through direct mail; smaller groups frequently hold a letter-writing party.

Invite activists to gather at someone's home, office, or church for an hour or so. Bring some light refreshments. Distribute blank postcards, sheets of paper, envelopes, and pens, but also ask participants to bring their own personal stationery and/or a typewriter if they have one.

Letters have more impact if they are individually written and reflect the writer's own concerns. But people may need assistance in remembering the bill name and number and the most important issues to raise. Distribute a one-page sheet containing essential information, such as the proper title and number of the bill involved, and a summary of your reasons for supporting or opposing it. Make sure your letters tell legislators how they should vote on the issue.

When the letters are written, provide stamps and take them to the post office or mailbox. If a large number is to be mailed, stagger the mailing over several days so that they are not received all at one time (unless the measure you're supporting or opposing is coming up for a vote soon and timing dictates immediate delivery).

Phone Calls

If a vote has been scheduled unexpectedly, or a candidate you support or oppose has just been appointed to an important position, you may not have enough time to write a letter. In that case, pick up the phone and register your opinion.

If you don't know your representative's or senator's phone number, dial the general switchboard for the U.S. Capitol (202-224-3121), and ask to be connected to the member's office. (You can get the direct number from the member's district office.) To contact state legislators, get a telephone directory to their offices from the state capital, or call directory assistance for the state capitol and give the name of the legislator with whom you wish to speak.

Even though you can ask to speak to the legislator directly, unless you are personally well acquainted, you will probably end up speaking to a staffer. On some issues, where large numbers of constituents are calling primarily to weigh in on a controversial issue, all the staff will want to know is whether you favor or oppose the issue at hand.

Telegrams

Through Western Union, you can send your legislator a public opinion message (POM) on any issue. These messages can be up to twenty words, excluding your name and address, and the price is only $9.95. POMs are delivered within two hours of the time they are sent. Western Union will charge the cost to your telephone bill or a major credit card, or you may prepay at any Western Union office.

Call 800-325-6000 to get the number of the Western Union office nearest you.

For a slightly higher cost, you can send a Mailgram, also via Western Union. A mailgram allows a fifty-word message, including the sender's and receiver's names and addresses. The mailgram will be delivered with the next day's mail. As in the case of a public opinion message, the cost of the mailgram can be charged to your telephone bill.

Don't be afraid to get creative with the messages you send to legislators. Here's another example of a clever tactic that sent a message to lawmakers—over and over and over again.

LOCAL HELL-RAISER
Lobbying with Post-Its™

The American Arts Alliance (AAA), based in Washington, DC, is a lobbying group whose members include 2,600 art museums; dance, theater, and opera companies; symphony orchestras; and performing arts presenters from around the country. The Metropolitan Opera in New York City, the Guthrie Theater in Minneapolis, and the Hubbard Street Dance Company in Chicago all belong to the alliance, as do many smaller local arts organizations.

AAA is dedicated to charting a comprehensive federal arts policy; it serves as a resource on arts matters for federal agencies, Congress, and the White House, and lobbies frequently on issues that affect nonprofits in general, such as taxes and policy, as well as to increase funding for the National Endowment for the Arts (NEA).

The NEA, which provides public money to support artists and exhibitions, is constantly being battered by conservative lobbyists who tout family values and effectively discourage the federal funding of "controversial" art projects. When, in response to a barrage of negative lobbying, the House and Senate Appropriations Committees threatened to cut off NEA's funding altogether, Anne G. Murphy, former executive director of the alliance, had a brainstorm. First, remind legislators that only a small minority of the works NEA has ever funded have been controversial. Second, show legislators how NEA's relatively minor budget compared to the huge amounts of money Americans spend on other pastimes. Discarding the "mundane" practice of writing letters, Murphy opted to deliver her message via blue Post-Its, the small rectangular sticky notes found in virtually every office on Capitol Hill.

The Alliance came up with sixteen catchy "Artfacts" that demonstrated the worth of the arts and how little funding they receive. Each page in a 3-by-4-inch Post-It pad was printed with one of these surprising messages. A faint AAA logo in the center allowed for note writ-

ing, but the writer couldn't help but read the Artfact
first, tuning in to such tidbits as "More tickets are sold
to dance performances each year than to National
Football League games" and "Annually, each American pays taxes of $1,137.28 for the military, $201.00
for education and 68 cents for the arts." Some notes
contained quotes supporting artistic freedom from
Franklin Roosevelt, Dwight Eisenhower, and Ronald
Reagan.

*More tickets are sold to dance performances each year than to
National Football League games.*

*Annually, each American pays taxes of $1,137.28 for the
military, $201.00 for education and 68 cents for the arts.*

"What we call liberty in politics results in freedom of the arts."
Franklin D. Roosevelt

In the spring of 1992, the American Arts Alliance
used clever sticky notes to lobby members of Congress
to increase funding for the arts.

Four pads—each sporting a different set of Artfacts—
were shrink-wrapped together, so that every legislator received each of the sixteen facts. The alliance produced
around a thousand of these packages and distributed
them in the spring of 1992, before the Appropriations
Committee funding vote in July.

The campaign was an immediate hit. Several lawmakers wrote to the alliance to express their apprecia-

tion for the AAA's sense of humor and brevity, and legislative staffers requested more pads for their desks.

In conjunction with the Post-It effort, the Alliance set up a series of meetings with "swing" members of Congress—those who had not yet firmly supported or opposed the endowment—to persuade them to join the ranks of NEA advocates. When all was said and done, AAA had secured funding for the NEA for another year.

The Post-Its offered a refreshingly different approach from conventional lobbying tactics. Beyond being useful, they repeated the Alliance's message over and over again. 3-M, Post-It's manufacturer, claims that each sticky note is read three times before it is thrown out.

Isn't repetition the key to understanding?

Faxes

Most, but not all, members of Congress make their fax numbers available to the public, and letters can certainly be faxed to those who do. Call your state representative or congressperson's office to get his or her number. Fax numbers are also published in *The U.S. Congress Handbook,* written and published by Barbara Pullen (call her at 703-356-3572 to order).

E-Mail

The U.S. House of Representatives is currently experimenting with hooking its members up to E-mail, but not all representatives have E-mail addresses. You can send postcards to your legislator's office requesting his or her E-mail address, and the office will contact you with the information. Or call your state and federal representatives and ask if they have an E-mail address or plan to get one.

The federal governments's executive offices are already plugged into the network. In its September 6, 1993, issue, *Newsweek* reported that President Clinton and Vice President Gore receive up to 4,000 E-mail messages a week (as compared to 60,000 to 80,000 letters).

☞ **CHECK IT OUT:** President Clinton's E-mail address: president@whitehouse.gov; Vice President Gore's: vice.president@whitehouse.gov

---

Lobbying Days

To demonstrate widespread support for an issue and focus attention on an upcoming vote, many organizations recruit volunteers to assemble in the state capitol or halls of Congress for a specific day of lobbying. (Sometimes, volunteers are asked to participate for two days: the first day, for a training session in lobby techniques; the second, to do the actual lobbying.)

To maximize publicity, you can hold a news conference before the lobbying begins and inform reporters who cover legislation for a newspaper, radio station, or television station not based in Washington or your state capital that someone from their community is in town lobbying.

Give citizen lobbyists persuasive materials to deliver to each legislator they visit. In addition to written materials such as those described above, they may want to distribute some item that would provide a good "photo opportunity" for the media (like the seat cushions Jan Garton and the Cheyenne Bottoms task force distributed to their Kansas legislators).

Oral Testimony

At some point in the legislative process, hearings will be convened to discuss a bill on which you are lobbying. (The only way to know precisely when this will occur is to keep in constant contact with the committee convening the hearings.) Someone from your organization may be invited to testify; you can also request the opportunity to testify. That testimony will become a permanent part of the public record, where it may be used by other legislators, advocates or opponents of your position, researchers, journalists, and interested citizens.

Your oral testimony should be brief (ten to fifteen minutes), consisting of a summary of your most pressing concerns. But you should also submit written testimony for the record that can be much more detailed. In addition, you can offer eye-catching charts or exhibits that might result in newspaper photos or television coverage of your testimony.

Expect tough questions during your testimony. Rehearse your statement beforehand, with other members of your group acting as "hostile" questioners to prepare yourself for the experience.

Publicity

Lobbyist Michael Pertschuk wrote in *The Giant Killers:* "Ink is the public interest lobbyist's holy grail. Almost any media coverage of a campaign is panted after. All public interest advocates are confident that right would prevail if the people only knew."

There are several steps you can take to ensure that the people do know. The following actions are appropriate to take whether you're lobbying Congress, the state legislator, or your city council:

- **Develop a comprehensive media strategy to complement your lobbying strategy.** Organize committees that can be mobilized to write letters to the editor. Begin planning media activities that will bolster the efforts of your lobbyists. Time the staging of press conferences and media events and the release of reports to have maximum impact on the legislative process (remember to "leave room for escalation"; build up your press coverage over time and save events and activities that pack the biggest punch for the time in the campaign when you really need them). Target individual legislators for media events, ad campaigns, and letters in their districts when you need to affect the votes of specific lawmakers.

- **Prepare a list of reporters and editors whom you can educate about the issue and rely upon over time to write stories.** The list should contain fax and phone numbers as well as addresses of reporters who cover your issue at radio and television stations, daily and weekly newspapers, newsletters and magazines, and other media outlets. The list should also contain the contact information for the capitol press corps, those reporters who are based in the capitol building (this is true whether you're focused on your state capital or on Washington, DC) and who specifically cover the activities of the legislature.

- **Build relationships with reporters.** The capitol press corps cover what the legislator from their district does, when the legislature is scheduled to consider a measure or vote, and other legislative issues. The capitol press corps is probably the easiest group of reporters to track down, because they almost always work out of a press room that's located in the capitol building. Drop by their desks, introduce yourself, and leave behind some background information about the legislation you're working on. Try to set up a one-on-one lunch or meeting with individual reporters to talk to them about your agenda for the current legislative session. Keep in touch through mail or faxes regularly. Invite them to attend press events, urge them to cover any testimony you present, and alert them to pending votes. When your measure is being debated in a committee or by the full legislature, seek reporters out so you can put your spin on the story and give them a quotable quote.

- **Meet with the editorial board of your local newspaper to explain your position and ask for support.** Follow up with a thank-you note, and continue to send news releases and other press materials.

- **Mount a letters-to-the-editor campaign prior to an important vote.** Also send out letters to editorial writers apprising them of the situation and asking for a supportive editorial.

- **Draft an opinion editorial that concisely explains your point of view, and try to get it placed a couple of days before the vote to help convince legislators to support you.** Send legislators a copy as soon as it's printed.

- **Stage a media event either in the capitol or in the legislator's home district.** Make sure the event is well covered by television, radio, and newspapers.

- **Organize a petition drive.** Deliver the petitions en masse to the legislator's office, accompanied by reporters who will give the story good play in the local news.

- **Hold a press conference on the steps of the capitol to demand action.** Distribute press releases and other materials to reporters in attendance, and fax copies to those who don't show.

- **Take out newspaper ads (weekly papers are cheaper) or advertise on the radio (in twenty- or thirty-second sound bites) to let your legislator know you mean business.** Send copies of the ads to reporters who might do stories about the ad campaign.

- **Release a report that substantiates your concerns about the issue, and hold a news conference to distribute copies and answer questions.** Excerpt a portion of the report into an op-ed piece that you ask your newspaper to print.

As with other media efforts you make, your chances of generating publicity for your lobbying campaign will increase exponentially if the story you're pitching is "newsworthy": if it's new, timely, controversial and colorful; if it revolves around crisis or conflict; and if it will affect large numbers of people in a significant way (see Chapter 6 for a lengthier discussion on what the media consider to be newsworthy).

Here are some specific examples of when you have lobbying news:

- When you petition the legislature (or city or county council) for a change in the law.
- When supportive legislators introduce a proposal.
- When opponents lobby against your proposals.
- When public debate and controversy over the issue develops (through talk shows, interviews, debates, etc.).
- When a group of your advocates congregates for a day of lobbying.
- When editorial writers and columnists endorse your position.
- When hearings on the bill are held.
- When the legislature debates the measure.
- When the legislature votes on the measure.
- When the bill is enacted by the legislative body and signed by the executive.
- When plans for its implementation are developed.
- When activities commemorate the date of implementation.

When you generate news that endorses or substantiates your position, reproduce the news clips and send them to legislators. The stories will shore up your supporters, help sway those who haven't yet made up their minds, and perhaps neutralize your opponents. In general, when it comes to influencing legislators, print media—especially news stories and editorials in the daily newspapers most important to the politician—appear to have more influence on a legislator than either television or radio coverage. Some groups amass all of their press clips into a single folder, then distribute the folders to legislators they're lobbying as a powerful reminder of how much support their issue enjoys among the legislator's constituents.

LOBBYING
DO's and DON'Ts

- **Do** develop a lobbying strategy focused on your short- and long-term goals that takes into account your organizational resources, the legislators you need to mobilize or sway, tactics available to you, and the amount of time you have.

- **Do** get to know legislators and their staffs to understand what messages and appeals to use in presenting your issue.

- **Do** familiarize yourself with the legislative process in your city, county, state, or at the federal level.

- **Don't** lobby only legislators who already agree with you. Target as well those who are uncertain or whose minds can be changed if you present your information clearly and persuasively.

- **Don't** try to lobby a bill through the entire legislature alone. The effort will put a greater drain on your resources if you can't share the burden. A solo effort is more likely to fail because it can be perceived—and dismissed—as a "special interest."

8.
ACTION AT THE POLLS

EVERY VOTER EXERCISES a public trust, remarked Grover Cleveland in his inaugural address on March 4, 1885, and that is as true today as it was a hundred years ago.

How voters exercise that trust is another matter. What do they vote for? Who do they vote for? These are the essential questions activists grapple with when they decide to try to motivate voters to cast their ballots for or against specific candidates or issues.

Mounting a political campaign to affect the actions voters take at the polls is complicated, expensive and intense. For these reasons, taking the steps necessary to put an issue directly to voters is a tactic usually considered only as a last resort. Sometimes you don't have a choice, as these Maryland gun control activists discovered when a law they had fought hard to pass was challenged at the polls.

ACTIVISTS in ACTION ▸ ▸ ▸ ▸

The NRA Shot Down

Dr. Samuel Johnson, the eighteenth-century British essayist, lexicographer, and raconteur, said that the prospect of being hanged in a couple of weeks concentrated a person's mind wonderfully. Today, getting shot does the same thing.

Victims who haven't given much previous thought to handgun control suddenly realize how important an issue it is—and they become determined activists on its behalf. In March 1981, Olen J. Kelley, an assistant manager of a suburban supermarket, was shot in the chest by a robber wielding a short-barreled .38-caliber revolver, the kind of cheap handgun known as a "Saturday Night Special." That same month, John W. Hinkley had used a Saturday Night Special manufactured by the same company to shoot President Ronald Reagan. Olen Kelley decided to do what no one had ever done before: sue Florida-based R. G. Industries, the manufacturer of that pistol, for damages.

Doing the Unprecedented

After three years of litigation, Kelley wound up with a landmark decision by the Maryland Court of Appeals, which ruled in October 1985 that the makers and sellers of low-quality Saturday Night Specials indeed could be held liable for the damages suffered when shooting victims were wounded by one of their weapons.

The Maryland court's precedent-setting decision created the first legal definition of what a Saturday Night Special handgun is. But it also set in motion a chain of events involving legislators, high-paid political consultants, grass roots organizers and hundreds of volunteers in a multimillion-dollar battle between people like Mr. Kelley, who were seeking curbs on handgun sales, and the powerful National Rifle Association (NRA) and its allies in the retailing and insurance business.

The Battle Lines Are Drawn

The face-off began as soon as Maryland's Court of Appeals issued its *Kelley* ruling. Handgun control advocates in the state knew the NRA would fight tenaciously to overturn the finding. A member of the Maryland Attorney General's Office, Vincent DeMarco, and one of his former college roommates at Baltimore's Johns Hopkins University, Bernard Horn, correctly suspected that the NRA would try to ambush the *Kelley* ruling in the halls of the state legislature during its 1986 session by seeking passage of a law to invalidate the court's handgun decision.

DeMarco and Horn, who was a private citizen and attorney, formed a citizens group, Marylanders for Victims' Rights, to lobby against any NRA effort to void *Kelley* and began working in coalition with another one of their former college roommates, Len Lucchi, a private attorney representing the Maryland Fraternal Order of Police, to lobby hard against the NRA-backed bill. Intense lobbying against hardball progun lobbyists during both the 1986 and 1987 legislative sessions helped the ruling survive those sessions intact, though only by a thread. "The full force of the House leadership was on the NRA's side, but you need seventy-one votes to enact a bill," DeMarco recalls about the bill proposed in 1987 to overturn *Kelley*. "The progun side only got sixty-seven."

More Legislative Wrangling

Despite their victories, DeMarco and the other gun control advocates realized that they could not defend the court's ruling in *Kelley* forever. During the summer of 1987, DeMarco began drafting legislation that would "codify" the court decision—and create a handgun review process that would list handguns that could be banned in Maryland, using the criteria the appellate court cited in the *Kelley* ruling when it defined a Saturday Night Special: concealability, barrel length, quality of materials, accuracy, reliability, and other factors.

A New Law Is Born

In 1988, weary gun control advocates found themselves mobilizing for yet a third round of lobbying to defend the *Kelley* ruling. But this time, Sarah Brady, the wife of presidential press secretary James Brady, who was severely wounded during Hinckley's attempted assassination of President Reagan, stepped into the ring. She and her organization, Handgun Control, Inc., held an emotional press conference to urge passage of bill 1131. Progun lobbyists responded with relentless counterattacks, until a compromise was reached: The *Kelley* decision would be overturned and in its place would be created a nine-member handgun review board, appointed by the governor, that would decide, based on whether the weapons were destined for self-protection, law enforcement, or sporting purposes, which ones could be sold and which should be banned because they were of use only to criminals.

Gun control forces put one radio advertisement on the air, urging citizens to call their state senators and delegates in support of the Saturday Night Special law, and legislators were flooded with phone calls. On the last day of the legislative session, the handgun control bill passed the General Assembly overwhelmingly and was signed into law on May 23, 1988.

Putting It to the Voters

Though angry and embarrassed, the progun forces did not lick their wounds for long. A local citizens group, Gunowners of America, immediately launched a petition drive to have the new law placed on the Maryland ballot in the upcoming November general election, enabling citizens to vote it up or down. By the July 1 deadline, the progun forces had collected the necessary 33,000 signatures to place the new law before the voters.

The NRA, initially reluctant to join the battle, decided that once the local group had succeeded in placing the law before the voters, the progun forces must give it all they had.

And all they had was quite a lot—$6.6 million, in fact, the largest amount of money ever spent in an election in Maryland and more than ten times the amount the gun control forces managed to raise. With the stakes thus boosted to a new level, at least in the state of Maryland, the fight between the progun and antigun forces came to overshadow even the 1988 presidential contest in Maryland.

The Battle for Ballots

Calling on its 49,000 Maryland members, the NRA helped create its own local citizens action group, the Maryland Committee Against the Gun Ban. Although the progun forces received some 3,400 small contributions and sold thousands of $1 raffle tickets, they only raised $100,000 this way. The bulk of their money came from the NRA.

The largest single contributor of the gun control forces was Handgun Control, Inc., which donated $95,000. The gun control advocates raised the rest of their money locally, obtaining much of it from the deep-pocketed supporters of Maryland governor William Donald Schaefer. Angered by the NRA's heavy-handed entry into Maryland politics, the governor strongly backed the gun control law.

Vincent DeMarco, the intense, energetic native of Italy who was brought by his parents to the United States as a child, took an unpaid leave of absence from the Maryland Attorney General's Office to serve as director of a new group, Citizens for Eliminating Saturday Night Specials (CESNS). At thirty-one, he had been a grass roots organizer for various causes, but he had never run a political campaign. Nevertheless, he knew the issues and the law intimately. The lawyer assembled a small paid staff and a huge number of volunteers to fight on the gun control bill's behalf. Pictures of Robert F. Kennedy and John Lennon graced his small, cluttered office as "inspiration," he told reporter Jef Feeley of the *Daily Record,* Baltimore's legal newspaper. Both Kennedy and Lennon had been killed with Saturday Night Specials.

Early on, DeMarco formed grass roots organizations in Baltimore City and each of Maryland's twenty-two other political subdivisions. CESNS evolved into a diverse coalition whose members included the Maryland

State Teachers Association and Baltimore's Interdenominational Ministerial Alliance, a group of 150 black pastors who pledged to lend support from their pulpits. Having only a bare-bones budget, DeMarco's campaign strategy focused heavily on obtaining free publicity via a well-orchestrated series of carefully timed endorsements of the law by top politicians, law enforcement officials, business executives, and religious leaders.

Going Public

DeMarco aimed to have most of the gun control endorsements issued after Labor Day, timing them one a week, if possible, to build momentum. Morning press conferences were held around the state so local police chiefs and key elected officials could endorse the law and make it onto the noon TV news. Even the Reverend Jesse L. Jackson, a national "celebrity," came to Baltimore to tape a radio ad in favor of the law.

DeMarco's tactics for gaining free publicity, along with financial contributions, sometimes were ingenious—and stunning. At a fund-raising breakfast hosted by Governor Schaefer for top business leaders in one of Baltimore's swankiest private clubs, the state Secretary of Public Safety and Correctional Services, Bishop L. Robinson, slyly educated the unknowing guests about how easy it is to conceal a small, cheap handgun. Calmly, he pulled two such weapons, one after the other, from his business suit and brandished them at the startled breakfasters, who nearly choked on their scrambled eggs. "My God, it was so dramatic," one of the guests told James Bock, a reporter for the *Baltimore Sun.* It also was effective. The business executives promptly made on-the-spot pledges estimated at between $75,000 to more than $100,000 to back the efforts of Citizens for Eliminating Saturday Night Specials.

DeMarco's coalition needed every penny it raised that morning, given what the NRA-backed progun group was pouring into Maryland. Its $6.6 million was being used to pursue every avenue—purchase TV time, buy radio ads, print a comic book, launch a free newspaper—that might persuade Marylanders to vote against the gun control law. Hiring a savvy Los

Angeles–based political consultant, George Young, to spearhead its effort, the NRA wasted no time issuing its first barrage against the gun control law. Young, who said he was pleased to be called the NRA's "hired gun," was best known for his work in defeating a 1982 handgun initiative in California. He knew the turf and had well-honed tactics ready to deploy—among them, massive radio, television, and print advertising campaigns aimed at convincing Maryland's voters that the gun control measure was "a bad law" that ultimately would lead to banning the sale of all guns.

Taking Shots

The NRA's ads hit upon themes that were cagey and almost—but not entirely—deceptive. They delivered their messages against the referendum—that the bill's wording was vague and could lead to the banning of other guns as well, and that the handgun review board would become politically corrupt—with impressive and well-funded force.

Their opening salvo—a full-page newspaper advertisement in every major paper in the state—appeared on July 4, just three days after their referendum petitions were declared valid. DeMarco and his group replied with what he now describes as "a Clintonesque quick response," calling a press conference at which Baltimore County Police Chief Cornelius Behan angrily waved the NRA's newspaper advertisement for the TV cameras and branded it untrue.

The NRA fusillade continued unabated—and intensified. Well-produced radio spots saturated voters as they were driving to and from work. Animated and live-action television commercials propagandized during prime time—one spot was broadcast during a break in ABC's "Monday Night Football" game, one of the most expensive time slots on the air. Ten thousand copies of a twenty-eight-minute videotaped "infomercial" protesting the law were distributed around the state, as were 100,000 copies of a comic book that ridiculed the gun control law and its supporters, and likened the measure to the efforts of foreign dictators to disarm their people.

One dramatic TV ad portrayed an elderly woman

reading in bed as a burglar noisily jimmies open the door to her home. She frantically dials her telephone for the police—but as the sound of a busy signal is heard, the doorknob to her room is shown slowly turning. The implication, of course, was that had "Granny" been able to reach for a handgun, she could have defended herself.

The gun control forces did not have the money for such slick advertising, but they had a more powerful weapon now at their disposal: the fabled temper of Governor William Donald Schaefer. Outraged at what he considered the progun force's distortions and lies about the new law, he lent his considerable prestige to backing the law. He attacked the out-of-state consultants and money brought to bear against the statute and taped two thirty-second television ads urging voters to "win one for Maryland" and support the new law.

The NRA continued its attacks, publishing a newspaper, the *Free State Press* (co-opting Maryland's old anti-Prohibition nickname as "The Free State"), and mailing 2 million copies of it to voters. This further enraged the governor, whose well-connected former campaign associates began raising more money to counter the NRA's efforts.

Some of the NRA TV advertisements ran into trouble. In mid-September, several stations declined to air the ads, deeming them distorted. The gun control forces got additional free publicity in October when two televised debates were held between progun and antigun representatives. When the NRA paid $13,000 to place seventy-six large signs and thirty-seven smaller advertising messages on Mass Transit Administration buses in mid-October, the MTA, a state agency, agreed to donate space on as many as two hundred buses for "public service announcements" prepared by the groups supporting the gun control law.

Getting Out That Vote

Both progun and gun control forces worked hard to personally contact as many voters as possible—and to convince them to take their side at the polls. With the progun forces employing a twenty-five-line phone bank each night to telephone voters and urge them to defeat the new statute, the gun control side had to install its own fifteen-line phone bank in response, supplying volunteers with a "voter mobilization fact sheet" to use in making their pitch. Citizens groups on both sides distributed fliers, dispatched speakers to public forums, and went door-to-door to argue their cause. The progun forces hired a public relations firm in Washington, D.C., owned by African Americans, to target black residents of Baltimore's inner city, and paid telephone and door-to-door canvassers well to inflame voters living in urban areas with false reports that the new law would make it impossible for poor people to purchase weapons to defend themselves against crime.

A few days before the November 8 balloting, a story in the *Evening Sun* reported that the progun committee was offering poll workers $10 apiece in illegal election day "walk-around money" in return for passing out literature in the vicinity of polling sites and getting out the vote—a dated campaign tactic that had been banned in the state some years earlier. The night before the election, the city's top prosecutor obtained a subpoena for the financial records of the Maryland Committee Against the Gun Ban and sent police to the group's city office to look for evidence that any leafletters were being paid. Simultaneously, DeMarco obtained a court order against the progun committee, restraining it from paying any poll workers, and showed up at its Baltimore headquarters to personally serve it—in front of local TV crews, whom he had called in advance.

Victory

Preelection polls had shown the progun and antigun votes running neck-and-neck, but when the ballots were counted the gun control side won by 58 percent to 42 percent. The $6.6 million spent by the NRA and its local associates in their intense, three-month effort had shattered all previous records for election spending in Maryland, with the votes cast against the law costing them roughly $10 apiece, according to a postelection story by the *Sun*'s James Bock. The Maryland Committee Against the Gun Ban ended the campaign more than $1.3 million in debt—$951,000 of it owed to the NRA.

Here's Your Opportunity To Help Stop Crime

SATURDAY NIGHT SPECIAL FOR SALE

VOTE FOR QUESTION #3
ON TUESDAY, NOVEMBER 8th

Let's stop Saturday Night Specials and Undetectable handguns.

This law **will not** affect law-abiding citizens who keep handguns in their homes for self defense or anyone who wishes to own a shotgun or rifle.

Question #3 seeks to end the sale of Saturday Night Specials, which are made especially for criminals but which are too unsafe and unreliable for use by law-abiding citizens, and plastic guns which are invisible to protective screening devices at airports and courtrooms.

If you wish to volunteer, call 889-1477
If you wish to make a contribution, make checks payable to:

Citizens for Eliminating Saturday Night Specials
2530 N. Calvert Street, Suite #10*
Baltimore, Maryland 21218

Authority Honorable Ralph M. Hughes, Chairman
Citizens for Eliminating Saturday Night Specials

This flier urged Marylanders to use their voting power to join forces with Maryland handgun control activists.

By contrast, the gun control forces spent $752,107—or less than the progun groups spent on salaries alone. Of that sum, $554,000, or about 70 percent, was spent on advertising. The gun control forces ended the campaign with a $17,000 surplus.

Since Maryland's voters approved the 1988 Handgun Roster Law, the gubernatorial appointed board, which began operation in 1990, has banned over a hundred low-quality Saturday Night Specials from sale in Maryland and the avalanche of state and local gun control legislation that the NRA feared did indeed follow rapidly in the wake of the group's defeat in Maryland.

Bernard Horn, the onetime college roommate of Vincent DeMarco, went on to head Handgun Control, Inc., in Washington. He reports that the years since the 1988 Maryland referendum have been the best ever for gun control efforts in state legislatures. Significant antigun laws now have been passed in California, Connecticut, Florida, Virginia, Oregon, Massachusetts, New Jersey, Rhode Island, Delaware, Iowa, Nebraska, Wisconsin, Missouri, Indiana, Louisiana, Minnesota, and Hawaii, as well as in more than thirty cities and counties nationwide. And in 1993, President Bill Clinton signed into law the Brady Bill, imposing new federal regulations on the sale of handguns. The law was named in honor of James Brady, the press secretary to President Ronald Reagan, who was severely wounded in the 1981 assassination attempt on Reagan.

"Four years ago, the NRA had a virtual stranglehold on state legislatures," Horn says. "We were able to pass little or no significant gun restrictions until the Maryland Saturday Night Special law was enacted in 1988." Today, he says, "on the state and local level, we are beating the NRA across the nation."

For his work as director of the Citizens Committee for Eliminating Saturday Night Specials, Vincent DeMarco was named 1988's "Marylander of the Year" by the *Baltimore Sun*. He returned to the Maryland Attorney General's Office but left in 1992 to head Marylanders Against Handgun Abuse, which continues to oppose weapons whose only purpose is to kill the victims of violent crime.

Not only Marylanders but gun control forces throughout the nation owe a lot to Olen Kelley. His lawyer, Howard Siegel, told the *Sun:* "The beauty of the system is that when every legislator was terrified at going up against the NRA, they couldn't keep Olen Kelley out of the courtroom. He was the one who got it started."

Senate President Thomas V. (Mike) Miller, Maryland Governor William Donald Schaefer, and Speaker R. Clayton Mitchell gather in 1988 to sign the bill that effectively banned Saturday Night Specials in the state. Sarah and Jim Brady look on.

MAKING THE DIFFERENCE

As Maryland's battle for gun control showed only too well, participating in the political process is a "hardball game" that requires commitment, tenacity, and—perhaps most of all—a thorough understanding of the myriad complex elements that lead to success—or failure.

Will your organization launch—or defend itself against—a referendum campaign? Attempt to register voters who support your position? Or even field your own candidates for political office to insure that your issues receive proper attention?

There's no way to succeed at any of these ventures without a thorough understanding of what it takes to turn a dubious scheme into a smashing success.

REFERENDA, INITIATIVES, AND OTHER "DO-IT-YOURSELF" STRATEGIES

Three election devices enable organizers to become "citizen legislators" by taking their issues directly to the public for approval.

As is evident from the Maryland handgun battle, a **referendum** is a process by which voters approve or reject a law the state legislature already has passed. Referenda can either be put on the ballot via a citizens' petition drive or by the state legislature itself. Twenty-four states out of fifty allow citizens to petition a law to referendum; twenty-five of the fifty states also give the legislature the power to put a law before the public for a vote.

An **initiative** is a proposed *new* law that is put directly before voters for approval or rejection after organizers have collected a required number of signatures in order to place the issue on the ballot. Only twenty-three of fifty states allow the initiative process: Arizona, Arkansas, Alaska, California, Colorado, Florida, Idaho, Illinois, Maine, Massachusetts, Michigan, Missouri, Montana, Nebraska, Nevada, North Dakota, Ohio, Oklahoma, Oregon, South Dakota, Utah, Washington, and Wyoming. (Though it is technically inaccurate to say so, once an initiative effort makes it to the ballot, it is generally referred to as a referendum.) Most frequent in placing initiative questions on the ballot are California, Oregon, Colorado, Arizona, and North Dakota.

Most organizations mount initiative and referendum campaigns to demonstrate popular support for an issue in order to persuade the legislature to act upon it, to curtail the efforts of their opponents (which is what both the progun and the gun control forces in Maryland were doing), to ratify legislation that has already been passed, or to create laws when the legislature is unwilling or unable to do so.

A **recall** allows voters to petition for a special election to remove an elected official from office.

The laws governing initiative, referendum, and recall efforts vary widely from state to state. Check with the office of your state's attorney general or secretary of state to determine what options you have.

Going to the Voters: Should You or Shouldn't You?

Whether you're mounting a referendum, an initiative, or a recall campaign, all three actions involve essentially the same steps. You must describe the action you are proposing in writing that is legally accurate, collect signatures (unless the issue has been put on the ballot by legislators) on petitions that describe your proposed action, and mount an intense publicity and grass roots organizing campaign. What are the pros and cons of engaging in this process?

Taking an issue directly to voters via the polls is a very difficult, expensive, and time-consuming way to advance your political agenda. Before going ahead with such a campaign, groups should make sure that neither an effective lobbying campaign nor pressure on the executive branch of government to pass an executive or administrative order would work to instate their law. The effort to collect signatures and mount a campaign can take a year or more, detracting from your other organizational business, exhausting your staff, and, should you fail, undermining the credibility of your group.

On the other hand, a successful campaign can work wonders for your organization and its goals by allowing you to take charge of the political process and achieve results that might otherwise be impossible. Managed properly, such a campaign can boost fundraising, raise the visibility of your group and its beliefs, and build a membership base that endures long after the last campaign button is unpinned.

Though the Maryland gun campaign describes a successful effort by a community group and its allies to defeat a negative ballot measure, activists around the country have used the initiative or referendum process in an affirmative way to pass legislation on a number of topics. (We'll take a closer look at recall later in this chapter.) Local Common Cause groups in California, Idaho, Florida, Washington, and Massachusetts have petitioned to pass campaign and lobbying finance reform measures. Environmentalists in Maine and Michigan used initiatives to enact "bottle bill" legislation; citizens of Washington, DC, passed a 1992 initiative that limited the amount individuals and political action committees (PACs) may contribute to support local politicians; and consumer and environmental activists in California mounted a successful initiative campaign to protect themselves against toxic hazards.

According to the Northern Rockies Action Group, which provides organizational and management consulting services to social change groups in Montana, Idaho, and Wyoming, more than 90 percent of the attempts to place issues on the ballot through an initiative or referendum fail. However, half of those issues that do make it onto the ballot win voter approval.

Given those odds, how can you insure that your effort will be a success?

First, before going ahead, determine whether a popular vote is the best way to achieve the policy change you desire. Ask yourself the following questions:

- **Can you win approval for your proposals by waging an effective lobbying campaign instead?** Don't launch an initiative crusade if the issue is so narrow that the general public will have no interest in it. Fight those campaigns in the state legislature, where you'll have greater opportunity to present the minute details involving your issue in detail to an audience of lawmakers who will have more time to listen to your arguments.

- **Can the change you're seeking be achieved through an executive or administrative order instead of by legislation or at the polls?** Working through existing channels may be quicker—and much cheaper—than going to the polls.

- **How strong is public support for your initiative or referendum?** Your base of support should be solid enough to sustain the withering attack you'll probably receive from your opponents. If at least 50 percent of the public does not endorse the ideas behind your initiative when you launch it, you may want to postpone the campaign and spend a year cultivating more favorable public opinion.

- **Do you have the financial and organizational resources needed to mount a successful initiative or referendum**

campaign? While, as the Maryland handgun campaign showed, you don't have to match your opponent's war chest dollar for dollar, you do need enough capital to pay for essential salaries, office expenses, printing, and advertising costs.

- **What allies do you have?** You can't win an initiative campaign working all by yourself. The broader your coalition is based, the more likely it is you'll succeed.

- **How powerful is your opposition?** Answer this question not only in terms of money, but in terms of clout. The Maryland chapter of Gun Owners of America was intimidating because it was backed by the National Rifle Association, one of the country's most powerful political lobbying groups.

- **What will be the repercussions if you lose?** Will defeat devastate your organization and render your issue irrelevant? Or will it raise needed visibility for your fight that will help you win next time around?

Sometimes, as in the case of Maryland's handgun campaign, you have no choice but to engage in the process to protect gains you have made in the courts or in the legislature. In other cases, you may feel like you have to do what the legislature either can't or won't do. Regardless of what motivates you to go directly to the voters, you've got to do your best to win.

A STEP-BY-STEP GUIDE TO INITIATIVE AND REFERENDUM CAMPAIGNS

The steps involved in running an initiative or referendum campaign are essentially the same. The following guide refers to both initiatives and referenda as "initiatives" unless otherwise noted.

1. Set Up a Timetable

If you can, give yourself a full year to get organized, raise money, recruit volunteers, collect signatures, and mount the campaign.

2. Plan the Campaign

Timing is critical to the success of your campaign; so is long-range planning. Work back from election day to develop a publicity, advertising, and fund-raising strategy. If your election is in

November, plan to have your petitions signed by the beginning of August so that you have a full three months to deploy your strategy.

3. Draft the Measure

Assemble a small committee to write an initial version or draft of your law. Use language that is not so radical it will offend a majority of voters or so balanced that it fails to inspire activists to participate in the campaign. In *Be It Enacted By the People; A Citizen's Guide to Initiatives,* published in 1981 by the Northern Rockies Action Group, author and grass roots organizer Mike A. Males urges activists to keep in mind the charges your opponents will inevitably throw against you: that your proposed law— whatever it is—is "vague, unconstitutional, will raise your taxes, will create more government bureaucracy, will fuel inflation, will increase unemployment, is backed by special interests, and may even kill you. Vote no."

Make a radical plan more appealing by having it enacted in stages and allowing a period of voluntary compliance with the law before it becomes mandatory.

Make the law "readable" by using short words and sentences and clumping information into subheadings, almost in outline form. Ask many people to review what you've written: Lawyers, who can check it for constitutionality; other constituency groups that will be affected by the bill if it becomes law; even your opponents, who may identify nuances in the language that could become troublesome during the initiative campaign later on. Perhaps most important, meet with professional bill drafters at the legislature (you may need a legislator to formally request their help), who can tell you whether your law is too broad or too narrow to accomplish what it intends.

When filing a referendum measure, before signatures are gathered, you generally must include the text of the law to be referred. The petition form then includes an official summary of the measure and a ballot title written by the government—a short description of the referendum that will appear on the ballot. The signature sheet may also have a spot for the subject of the referendum.

In Washington State, for example, after receiving the referendum measure, the state attorney general prepares a short official description and a longer summary of the referendum. In Oregon, the state voters' booklet, which is sent to all registered voters prior to the election, contains an explanation of the initiative or referendum and any arguments that are officially filed by citizens against it or on its behalf.

Once you incorporate all the comments you receive into a final draft, proofread the document several times. Check the accuracy of all names, dates, and numbers. Be absolutely certain that the law says exactly what you want it to say, for once you file your draft with the secretary of state, you cannot change a word.

4. Design Your Petition Form

The petition is the piece of paper that you will use to collect the signatures you need to put your proposed law on the ballot. It should contain the law as you have written it, plus signature lines allowing people to write and print their names and addresses. Check with the secretary of state to make certain that your petition form uses the exact headings, warnings, and signature format prescribed in the laws governing your state's initiative and referendum procedure.

5. File Your Petition

File the final draft of the law and the petition form with your secretary of state. Propose at least two titles for the petition that explain the issue the way you want it described. Then contact your state's attorney general, who writes the title, and closely monitor that titling process. A measure to regulate construction of nuclear power plants could be opposed on the grounds that it constituted a "ban" on nuclear power as an energy source or supported because it gives voters "control" over energy growth, depending on how the title is worded.

6. Organize Your Signature-Gathering Campaign

While you are awaiting approval of your petition, identify coordinators for each area of the state who will be responsible for lining up volunteers, distributing petitions, getting petitions notarized, and returning them to the proper location. Hold an organizing meeting to motivate volunteers, build "team spirit," and answer questions.

Put together an organizing kit for each coordinator. It should include:

- The campaign timetable.
- A "fill-in-the-blanks" news release to help kick off the campaign in each city.

- Instructions (gleaned from state regulations) on how to collect signatures properly.

- A fact sheet on the petition and the reasons for circulating it.

- Quotas for the number of signatures you want each coordinator to collect.

- Tips on how to recruit and manage volunteers.

Once your petition is approved and titled, proofread it again for typographical errors. Depending on how many signatures you need, you'll probably have to print a few thousand petitions. Distribute them to volunteers and hold a news conference (or at least issue a news release) announcing that your signature drive has begun.

Your group may need to register as a political action committee when you file your petitions. Check with your state's campaign practice office so you operate within the law. Also, make sure you designate or elect officers of the group, especially a treasurer, who can establish a bank account and keep track of all financial transactions in which the group engages.

7. Gather the Signatures

Collecting thousands of signatures is a grueling task. Some states impose limits on the maximum amount of time you may have to amass the names you need. All states have rules governing the "correct" way a signature must be written, which usually ends up disqualifying a significant percentage of those a group collects. To be safe, a group often must collect a larger number of signatures than the law requires.

Recognize that you'll be able to collect more signatures in large urban areas than in less populated rural ones, and set up quotas for signatures accordingly. Focus on collecting signatures at shopping malls, on university campuses, outside transportation hubs, at county fairs, at sporting events—anywhere it will be easy to approach a lot of people in a short period of time and in a concentrated space. (In the recall example that follows, activists amassed their signatures from voters on election day.) You can also try canvassing door to door for signatures, but this time-consuming tactic is not really effective if you're facing a tight deadline.

Some states will ask that you submit your petitions to the secretary of state; in other states, you may have to turn the petitions over to the clerk and recorder in the county where the signatures were collected. Determine the correct procedure for your state, and submit the documents accordingly.

Eventually, your petitions will be tallied. If you have collected enough good signatures to qualify for the ballot, trumpet the news in a press release—and keep going! Even if you fall short of your goal, you can draw some positive conclusions from the experience and regroup for a future effort.

ACTIVIST'S NOTEBOOK

Use Computers

Because of the help they provide in managing large lists of names and addresses, word processing, streamlining accounting, interfacing with other computers, and data processing, computers can be an invaluable campaign asset. On the other hand, computers can cost campaigns a lot of money, especially when the purchase of an expensive computer leads to much greater expense in software, gadgets, and maintenance. The Republican National Committee's *State and Local Campaign Manual* advises workers to clearly define their campaign goals and then determine if a computer can help you reach those goals more cheaply, quickly, and efficiently. If you do buy a computer, look for specialized software packages that generate ready-to-submit campaign finance reports, create lists for mailing and phoning, list donors, and perform other useful functions (see Chapter 6).

8. Develop a Persuasive Press Kit That Can Be Sent to the Press As Well As to "Opinion Leaders"

The kit should contain:

- News releases, updated as needed.
- A backgrounder or fact sheet about the bill.
- Lists of endorsements from other community groups and "celebrities" (the mayor, local sports heroes, business leaders, etc.).
- Polling data supporting your initiative, if it's available.
- Background information about your organization or coalition and why it is mounting the campaign.

9. Raise Money

Once you collect your signatures and qualify for the ballot, your fund-raising efforts will probably pick up considerably. Nevertheless, you're always going to have to raise more money than you have, and you'll almost never be able to match your opponents' war chest. Read the tips provided in the fund-raising chapter of this book, Chapter 5, for more suggestions on how to generate the resources you need to keep your campaign going. And by the way, don't forget to comply with your state's campaign finance reporting requirements; you'll have to begin reporting as soon as you start collecting any money.

10. Conduct a Media Campaign to Convince People to Vote Your Way

To a large degree, initiative campaigns are media campaigns. A well-conceived publicity strategy will heighten the profile of your organization across the state, recruit new members and contributors, build media relations, and form new coalitions. These benefits will accrue to your group even if you lose the campaign.

As with any media campaign, the most successful strategy in an initiative effort is one that takes full advantage of the many publicity tools available, targets the right message at the right audience, and builds momentum throughout the campaign. You also need to expect to engage in a substantial advertising effort, so that you can counter attacks from your opponents and present your own message to voters in an undiluted way.

You can review your publicity options from Chapter 6. As for targeting your audience, you'll probably want to focus on what is known as the "swing vote," that block of citizens who haven't made up their mind about the issue and could either vote for you or against you. Try to identify the reasons voters may swing the other way, then present arguments that shore up your position.

Generate momentum throughout the campaign by lining up a string of endorsements, scheduling debates, and organizing media events every few weeks to keep your side of the story in the papers. As the Maryland organizers of Citizens for Eliminating Saturday Night Specials showed, you can time your endorsements from prominent supporters at regular intervals—say, one a week—to build momentum; also schedule your press conferences before noon to get on midday TV news, "drive-time" radio later that day, and on the early evening TV news. Hold press conferences at different locations around the state to capitalize on the support of prominent local citizens and organizations and to get your message on the smaller radio and TV outlets, as well as in the weekly papers that serve individual communities—and are well-read. Don't be shy about using gimmicks and eye-catching tactics that will grab the attention of newspaper photographers and TV cameras. Remember the surprise "handguns for breakfast" stunt that got the antigun forces in Maryland a lot of publicity, as well as contributions.

As for the advertising side of it—well, resign yourself now to the inevitable truth: Chances are, you could be outspent by five or ten to one—and a lot of that money will be spent on a radio, television, and newspaper "blitz" intended to bury your issue once and for all.

You can counterattack by being prepared. Line up endorsers to come to your defense about the misleading content of the opposition's materials, and paint your group as the underdog. (If you believe the opposition's ads are out-and-out distortions, file a formal complaint with the attorney general and the TV or radio stations airing the ads, a move that will at least get you some media attention.) Always try to keep your issue before the voters.

Conduct one or more public opinion polls to determine what voters think of your measure and to track the impact of the opposition's campaign on your supporters. If you can, hire professional pollsters to keep tabs on voters' attitudes; you should also pay attention to the polls your local newspapers conduct on the issue.

If you have built up any kind of an advertising war chest, use it judiciously. Consult with professional "time buyers" about the best periods of the day and programs for which you should buy time to air your radio spots, and perhaps television ads; compare costs and advantages of purchasing space in newspapers, on billboards, and on the backs of buses. Explore cable television for advertising time, where, compared to its commercial counterpart, air time usually sells for a fraction of the cost.

And don't ignore the "free" forms of advertising: bumper stickers, lawn and yard signs, posters, leaflets, door hangers, buttons, and banners. Though the design and printing of these items could be expensive (unless you can get those services donated), placing them is free. By the way, don't overlook unexpected opportunities for help; one of the last places the Maryland handgun control forces expected to secure PSA placements was on the sides of two hundred city buses. But secure them they did.

Evaluate the Results

If you have won, uncork the champagne—but keep a watchful eye over your opponents and any tactics they may use to try to circumvent you (such as challenging the constitutionality of your law in court). If you have lost, take stock of the mistakes you've made, and identify what other means exist to help you achieve your goals. Thank volunteers for all their hard work.

--- **The Initiative Checklist** ---

___ Set up a timetable.

___ Assemble a draftwriting committee.

___ Write and refine your law.

___ Create your petition form.

___ File your petition.

___ Organize your signature-gathering campaign.

___ Gather the signatures.

___ Celebrate if you qualify for the ballot!

___ Develop a press kit.

___ Raise money.

___ Conduct a media campaign to convince voters to vote your way.

___ Evaluate the results.

LOCAL PROPOSALS

Initiative and referendum campaigns can be conducted on the city and county level, as well as statewide. While you may face many of the same tasks—collecting signatures, mounting a publicity campaign, raising money—the magnitude of the task usually is much more manageable.

In 1987, for example, two hundred residents of Jacksonville, Florida, banded together to reduce visual pollution in their city through a voter initiative on billboard control. Coalescing as Citizens Against Proliferation of Signs (CAP Signs), the group decided to put a proposal on the ballot to prohibit the construction of new billboards in Jacksonville. The measure would also have removed all billboards on non–federally funded highways over a period of five years.

In one day, CAP collected more than fifteen thousand signatures, handily surpassing the required 5 percent of the city's registered voters required by law. After a fierce campaign, CAP prevailed, passing the initiative by a margin of fifty-nine to forty-one.

CAP has gone on to form an effective network of civic groups and environmental organizations across the state that continues to rebuff the billboard industry and restore Florida's natural scenic character.

☞ **CHECK IT OUT:** For a detailed step-by-step explanation of the initiative process, see *Be It Enacted By the People; A Citizen's Guide to Initiatives,* by Mike A. Males, available for $10.00 from the Northern Rockies Action Group, 9 Placer Street, Helena, MT 59601.

And to learn more about your state's initiative, referendum, or recall process, call your secretary of state's office in the state capital.

You can also contact the Center for Policy Alternatives, a nonprofit organization that helps involve citizens in their government. Contact them at 1875 Connecticut Avenue, NW, Suite 710, Washington, DC 20009, 202-387-6030.

RECALLS

Just as activists can help elect a candidate, in seventeen states they can mount a recall campaign to remove an official from office. (Those states include Alaska, Arizona, California, Colorado, Georgia, Idaho, Kansas, Louisiana, Michigan, Montana, New Jersey, Nevada, North Dakota, Oregon, South Dakota, Washington, and Wisconsin.) City council members, mayors, state and federal representatives, and even governors may be vulnerable to this form of citizen retribution if they fail to execute the duties of their office satisfactorily; in some states, only local officials may be subject to recalls.

Specific recall procedures differ from state to state, but the general steps involved in mounting a recall campaign are similar to those taken during an initiative or referendum effort. Concerned activists notify the city, county, or state government (depending on whom they're trying to recall) of their intent to recall; gather a certain number of signatures on petitions to put the question before voters; have the signatures verified; and mount the recall campaign.

In most states, in order to spend money on a recall effort, your group must register as a political action committee (PAC) with your state elections office. You must also appoint a treasurer, and record and report to the state elections office any campaign expenditures you make.

Both the organization initiating the campaign (the "initiator") and the voters who sign the petitions must be registered in the same electoral district as the official they are trying to oust. When filing an intent to recall an official with the state elections office, the initiator often has to submit a statement explaining why the recall demand is being made. The initiator then has anywhere from three to nine months, depending on the state, to collect signatures showing that a substantial number of people in the official's district support the recall effort.

Recallers print up their own petitions, but the format and design are strictly regulated and must be approved before the petition is circulated. The number of signatures required to place the official's name on the ballot for a recall vote may be based either on a percentage of the total number of eligible voters in the district or on the number of votes the official received when he or she was elected.

After the signatures have been collected and the petitions have been filed, a special election is scheduled for the recall vote. Then both the recall forces as well as those who support the candidate begin to battle, developing and implementing the same kind of media strategies and get-out-the-vote tactics to recall an official as they would use in waging an initiative or referendum campaign.

In some states, voters must choose a new official during the same election in which they recall the existing one. New Jersey is one such state, and that requirement could have been the undoing of a group of Atlantic City residents seeking to oust their corrupt mayor. Fortunately, it wasn't. Read on.

LocaL hELL-RaiSER
Recall Roulette

It was in many ways a "numbers game," observed a reporter for the *New York Times,* employing an appropriate metaphor for the ultimately successful effort by a group of citizens in the beleaguered gambling resort of Atlantic City, New Jersey, to obtain enough signatures on petitions to force a victorious 1984 recall vote against the mayor.

Once a sparkling seaside vacation favorite of well-to-do and middle-class families, Atlantic City had fallen on hard times by the middle 1970s. Its formerly grand hotels had turned dowdy; its fabled Boardwalk attractions seemed tame; visitors stopped coming. With a population of just forty thousand permanent residents, it had become, as the *Times* put it, "an urban disaster" with rampant unemployment and deteriorating housing conditions. The Casino Control Act was passed in 1977 to revitalize the city, but legalized gambling only served to increase crime and, many poor residents believed, worsen the housing situation.

Attempts to improve the efficiency of local government also initially backfired. Voters in 1982 approved a switch from the city's old-time commission form of government to a mayor-council system and held their first direct election for a mayor in seventy years. By just 359 votes they picked Michael J. Matthews, a white Democratic state assemblyman, over James L. Usry, a black Republican assistant superintendent of city schools, in balloting that reflected the city's racial division.

A group calling itself the Citizens Committee to Make Mayor-Council Government Work, formed largely by Usry supporters, believed Matthews was completely uninterested in their problems and doing nothing to help them. (Events would show he was corrupt as well as inattentive.) In March 1983, the citizens group decided to launch a petition drive for a recall vote on Matthews in order to make Usry mayor.

Under New Jersey's complicated law governing recall elections, at least 25 percent of a city's registered voters are required to sign petitions in order to force a vote to recall a mayor. After one unsuccessful attempt at collecting signatures, the Citizens Committee decided to start over. On election day, November 9, they launched a second petition drive, concentrating their efforts on polling places in the black areas of the city, where registered voters who were Usry supporters would likely turn out in large numbers. The organizers distributed fliers to the homes in these areas and bought advertising time on local radio stations two days before the November 9 election, urging voters to sign the petitions. The petitioners contended that Matthews, then sixteen months into his four-year term, not only ignored Atlantic City's problems of crime, poverty, and inadequate public transportation, but spent too much money promoting casino interests and had created municipal jobs for political supporters. What the recall petitioners did not know then was that Mayor Matthews also was under investigation by the U.S. Department of Justice, which suspected him of having contacts with organized crime and of accepting bribes.

On January 3, 1984, the city clerk of Atlantic City ruled that the signature drive had fallen 257 signatures short of the number needed to force a recall vote. But after some legal wrangling, Superior Court Judge Philip Gruccio ruled on January 24, 1984, that the recall petitions were valid and ordered the city clerk to certify them and set a date for the recall vote.

Scheduling such an election became a bureaucratic hassle. The committee had to work in an impossible time frame: According to New Jersey law at that time, the recall vote had to be taken within ninety days of the first filing of valid petitions, which would be March 2, 1984. But the mayor's attorneys argued that the committee would miss this deadline because the simultaneous election of a new mayor couldn't be held until April 10, forty-seven days after new candidates' petitions were submitted, and those petitions had not even been printed yet.

The "numbers game," and election timing, seemed to become even more convoluted, with the Citizens Com-

James Masland, spokesman for Citizens to Make Mayor-Council Government Work, hands completed petitions to the Atlantic City Clerk.

Walter O'Brien, The Press of Atlantic City

mittee contending that Mayor Matthews and his allies simply were playing for time, trying to delay a vote until a May election for ward council members. The embattled mayor, they said, might fare better in such a balloting, running on a ticket with candidates from each ward, rather than alone on a recall ballot that concentrated only on his record as mayor.

Judge Gruccio agreed. On February 1, he ordered a recall election for March 13, 1984. Mayor Matthews's attorneys appealed the judge's finding, but on March 6, the New Jersey State Supreme Court refused to block the recall election. Once again, Matthews's chief opponent would be James L. Usry.

In what state law required to be a nonpartisan election, with no political party affiliations listed on the ballot, voters actually had to decide two issues: whether to recall or keep Matthews as mayor and, at the same time, select who they wanted as the city's next mayor. The winner of the separate mayoral race would replace Matthews immediately if the voters decided to recall him—but, ironically, the possibility existed that Matthews could succeed himself. He might lose the recall vote, which would be decided by a simple majority, and yet simultaneously be reelected as mayor, since that balloting could be won by a plurality among Matthews, Usry, and a third candidate, John Pollilo, an activist who was a laborer in the city's Public Works Department.

But this irony never came to pass. After just twenty months in office, Mayor Matthews lost the recall vote by a three to two margin and was replaced by Usry. Again the voting was along racial lines, but this time Usry had significantly more campaign money to spend, while Matthews, hobbled by the corruption probe, had far fewer resources and campaigned little.

Two weeks after the recall election, Matthews was indicted on federal bribery, extortion, and conspiracy charges, and on December 31, 1984, a federal district court judge sentenced him to fifteen years in prison for extortion, calling him "a crook . . . who betrayed the people of Atlantic City."

Unfortunately for the people of Atlantic City, honest government was not in the cards for them with James L. Usry, either. In 1989 he was arrested and later indicted for taking a $6,000 bribe. He was defeated for reelection in June 1990, and in December 1991 he pleaded guilty to violating campaign contribution laws in return for dismissal of the remaining corruption charges.

To paraphrase the old song, luck has not been a lady to Atlantic City.

GETTING INVOLVED IN A CANDIDATE'S CAMPAIGN

Becoming a "citizen legislator" through the initiative, referendum, or recall process is one way to affect public policies and achieve social change. Electing candidates who support your position is another.

Activist groups have a lot to offer a candidate. Your membership provides a ready pool of volunteers who can man telephone banks, pass out literature, raise money, display bumper stickers and lawn signs, and write supportive letters to the editor, all ingredients most candidates need to get elected.

What's in it for you? If your group participates in a campaign, you may be rewarded with easy access to the candidate once he or she gets elected. And if your group plays a decisive role in elections, you will earn additional clout in your community as an organization that makes a difference in the political process. Your proposals will be taken more seriously, and you'll be more effective holding elected officials accountable for their campaign promises.

How do you decide whether or not to get involved in a candidate's electoral campaign? Answer the following questions:

- **Can you make a difference?** Or is the race so marginal that no matter what you do, your candidate will lose? (Some groups have a "sure loser" policy—they won't endorse candidates they've decided have no chance of winning.)

- **Do you have adequate resources to get involved?** Can you afford to divert money, staff, and resources away from your organizing agenda and into a political program? (We'll discuss whether that's legal below.)

- **How strong is the candidate on your issues?** Will helping to elect the candidate get you a real friend in office—or will your issues still be relegated to the candidate's back burner?

Though you may feel as if you should participate only in campaigns where the candidate you favor is an underdog, incumbents and challengers who have an easy race should not be ignored, either. Most candidates "run scared," even supposed shoo-ins. Working to elect sure winners still builds up your image in their eyes—and helps keep your issue on the candidate's political agenda.

What Your Nonprofit Group Can and Can't Do Under the Law

Before planning your political program, develop a clear understanding of what activities you can conduct legally. Be aware that if your organization is incorporated as a nonprofit 501(c)(3) under the Internal Revenue Code, it is prohibited from engaging in *any* partisan political campaign activity, whether on behalf of or in opposition to a candidate for public office.

If you want to engage in activities that directly or indirectly support a candidate, consult applicable federal and state law. It is possible to set up organizations separate from your 501(c)(3) to endorse candidates, make campaign contributions, and provide other political services, but only if your organization conforms to state and federal tax and electoral laws. In general, if your group is classified as a 501(c)(3), follow these basic rules:

- **Don't endorse a candidate for public office**; in fact, don't even recruit a candidate for public office. (Of course, you as an *individual* can favor one candidate over another.)

- **Don't publish or distribute statements** that either favor or oppose a candidate for public office.

- **Don't give money** to a candidate, political party, or political action committee (PAC); a contribution to a political party constitutes indirect support for candidates and an expenditure to influence voter preference. (You may contribute your own money.)

- **Don't give "in-kind" contributions** to a candidate, political party, or PAC, such as mailing, membership or donor lists, office space, staff time, volunteers, or research on the candidate's opponents. (Again, you may volunteer, but acting as an individual, not as a representative of your organization.)

Violation of these rules may cause the IRS to revoke your tax-exempt status or impose a tax on your organization and its managers. Your 501(c)(3) can, however, provide the following services:

- **Publish voting records** of incumbent members of the U.S. Congress, state legislature, or town council. (A voting record lists selected pieces of legislation, a brief description of each, and an indication of how members voted.) The voting record should not identify which candidates are up for reelection or relate the voting record to a political campaign. Neither should it rate candidates or directly compare the voting record of a candidate you favor to one you oppose. Voting records that indicate whether the incumbent voted in favor of or against positions taken by your group can be distributed only to your members. Voting records that include no editorial opinion or an indication of approval or disapproval of votes may be distributed to the general public and media.

- **Prepare a questionnaire** and circulate it among all candidates running for public office. The questionnaire should cover a broad range of issues and avoid all appearance of bias. Results of the questionnaire may be distributed to the general public.

- **Educate the public** about your issues—as long as you do so in a way that is strictly nonpartisan and educational.

- **Conduct and publicize the results** of an issue survey or poll. However, no questions on such an issue survey may relate to candidate or political party preference.

- **Organize a public forum** to allow all candidates to present their platforms and answer questions from the audience. Make sure you invite all legally qualified candidates from the voting district in which a public forum is focused (see the step-by-step checklist on page 373).

- **Conduct training workshops** on the electoral process. Workshop sessions may focus on giving participants tips on mounting a voter registration drive or get-out-the-vote effort, fund-raising and media training, how the political process works, or how to get elected as a delegate. No workshops should attempt to involve participants in a specific campaign.

- **Brief candidates** on issues and encourage candidates to adopt a similar position—as long as you brief *all* the candidates.

- **Help register people to vote**—as long as you avoid all references to a candidate or political party.

- **Volunteer on your own time**. However, you must not speak or act in the name of your organization while engaging in political activity. And an organization may not organize or direct its volunteers to work for a candidate's campaign or political committee.

••STEP BY STEP••

How to Organize Nonpartisan Candidate Debates

A 501(c)(3) organization may invite candidates to a regularly scheduled meeting or hold a special public forum for candidates to discuss their views and answer questions on issues of interest to the organization. If your group decides to hold a candidate forum, follow these guidelines, prepared by the law firm of Perkins Coie in Washington, DC:

• **Invite all legally qualified candidates in your voting district.** At least two candidates must appear at the forum. If the debate is held during a primary election, the sponsoring group is not required to invite candidates of both parties to the debate. A separate Democratic and Republican party debate may be organized.

• **Address a wide variety of issues.** Include issues considered to be of important educational interest to the members of your organization.

• **Convene a nonpartisan, independent panel of individuals to ask the candidates questions.** Ask local newscasters, heads of organizations other than your own, the school principal, or the president of a local company to participate.

• **Select a nonpartisan moderator to preside over the debate.** The moderator will make sure that the ground rules you've established for the debate are observed, and make it clear that the views expressed are those of the candidate and not of the sponsoring organization.

• **Give each candidate an equal opportunity to speak and answer questions on his or her views.** The candidates must be treated fairly; the moderator should disavow any preference or endorsement by the sponsoring organization.

If your organization sponsors a candidate forum, treat it like any other media event. Post fliers and posters several weeks in advance to encourage as many people from the community to attend as possible; run an article about the debate in your group's newsletter and an announcement on your local cable TV channel, and activate your telephone tree if you fear turnout will be low. Prepare a news advisory about the debate and send it to your list of reporters and editors; call reporters a couple of days before the event to remind them it is happening and to urge them to attend.

Hold the forum several weeks before the election to give those who attend an opportunity to digest what they have heard and come to some conclusions about the candidates.

 CHECK IT OUT: To help you decide what you can and cannot do under the law, get *Nonprofit Organizations, Public Policy and the Political Process: A Guide to the Internal Revenue Code and Federal Election Campaign Act,* prepared by Perkins Coie, a law firm based in Washington, DC that specializes in election activities. IT'S FREE. Write 607 Fourteenth Street, NW, Washington, DC 20005–2011.

Local Hell-Raiser
Just a Bunch of "Boschwitz"

Beating the incumbent in an election year is no easy task, especially when the challenger is an unknown. But Ginny Yingling and a band of environmental activists in Minnesota were able to do just that by finding the right candidate, raising the visibility of her issue, and using the media to keep the public informed.

In 1990, Yingling, the chair of the Sierra Club's North Star chapter in Minnesota, observed, "We had some of the best and some of the worst on environmental issues in our congressional delegation." Among those she considered to be the worst was Republican

Rudy Boschwitz, a two-term senator whose environmental sleights-of-hand angered many Minnesota environmentalists.

A particular bone of contention was Boschwitz's support for damaging amendments to the federal Clean Air Act that would have allowed three fourths of the worst air polluters in the country to avoid their obligations to meet clean air standards. But environmentalists weren't surprised: Boschwitz had a habit of supporting every amendment that would weaken an environmental bill—then later voting for the bill as amended and getting credit for being an environmentalist.

Yingling and her Sierra Club chapter determined to replace Boschwitz with someone they could trust. They ended up throwing their support behind an extremely unlikely candidate: a Minnesota college professor who was both a champion for social justice and a budding environmentalist.

"We liked Paul Wellstone from the very beginning; even though he was not well known and had little campaign money, we believed in him," Ginny remembers.

To win, the activists knew they would have to convince the media that the environment was a critical issue. The group sponsored a series of five press conferences to outline environmental concerns that were important to Minnesota and the nation, among them clean air.

Because hitting the incumbent's weak points was also critical to winning the election, Yingling prepped Wellstone for three debates, again focusing on the environment, an issue polling data showed was a major concern for 80 percent of Minnesota's voters.

Interestingly, Yingling had a very difficult time convincing the national office of the Sierra Club that the organization should endorse the local affiliate's candidate. Early on in the campaign, Boschwitz had been considered a "shoo-in"; what would be the ramifications for the Sierra Club if it attacked an incumbent senator—and then he won?

But Yingling prevailed. Upon determining that 80 percent of Sierra Club members in the Minnesota/St. Paul area vote in national elections, Yingling focused on

SIERRA CLUB
North Star Chapter

RUDY'S SUPPORT FOR THE CLEAN AIR ACT IS JUST A BUNCH OF "BOSCHWITZ"

Senator Rudy Boschwitz has been getting a lot of press lately about his "hard work" on the Clean Air Act that recently passed in the U.S. Senate. The truth of the matter is that Sen. Boschwitz did his best to sabotage the bill and now is trying to earn political points by painting himself as an environmentalist. Below is a comparison of what Sen. Boschwitz says and what he actually did, as recorded in the Congressional Record:

What he said: "Some in the Senate (including me) tried to act as honest brokers." (St. Paul Pioneer Press - March 28) and "I was a leader on the Clean Air Act." (Almanac, KTCA - April 6)

What he did: Sen. Boschwitz introduced a late-night amendment that would have allowed two-thirds of all air toxics emitters to avoid regulation because it would have mandated special emissions permit reviews for them. The resulting paperwork, as he well knows, would have taken decades for environmental agencies to catch up with. In the meantime, the polluters would be free to evade the Clean Air standards.

How did Boschwitz act as an "honest broker" on this amendment? By clothing it in the guise of a small business technical assistance program -- the catch is he defined small businesses as emitting up to 100 TONS of air toxics per year (200,000 pounds). This would have exempted such "small businesses" nationally as Bethlehem Steel, emits 82 tons per year, Uniroyal Chemical 73 tons, and DuPont, 61 tons; and locally Champion International in Sartell, 60 tons, Control Data Corporation in St. Louis Park, 68 tons, Honeywell Inc. in Hopkins, 40 tons and in St. Louis Park, 78 tons, Potlach Corp. in Bemidji, 79 tons and 3M in Maplewood, 86 tons. (By no small coincidence, many of his PAC contributors would have also been exempt). The amendment was later changed by other legislators to be a true small business assistance program. The "honest broker" now claims that was what his amendment said all along!

1313 Fifth Street SE, Suite #323 • Minneapolis, MN 55414 • (612) 379-3853

This fact sheet offered straight talk about Wellstone's opponent to draw voters to the Wellstone camp.

What he said: "...we're getting a dam good bill -- a bill that would accomplish 96 percent of the cleanup suggested by the Senate Environment Committee." (Minneapolis Star-Tribune, March 28).

What he did: The Senator knows that regardless of how much a bill could accomplish (and the figure that he quotes is highly debateable), it will only be effective if it can be enforced. Yet on March 26, Senator Boschwitz voted **twice** in support of the Heflin-Nickles amendment which would have severely weakened the ability of the EPA and states to investigate Clean Air violations, to enforce air quality standards and to prosecute violators. It also would have shut the door on citizen's rights to challenge unlawful permits and to sue sources that violate their permits -- the most effective environmental law and order provision in the Clean Air Act.

Even if the bill does achieve 96 percent of the original bill, it's no thanks to Boschwitz. When strengthening amendments were offered to reduce toxic emissions from automobiles (the source of half of all urban air-toxic related cancer cases) and to reinstate the EPA's ability to institute smog cleanup plans, he voted against them.

What he said: "...the 96 percent bill we are agreeing upon costs about $20 billion a year. The Environment Committee bill would have gotten the last few percentage points by more than doubling the cost." (Minneapolis Star-Tribune, March 28)

What he did: Senator Boschwitz is well aware (because Sierra Club has told him frequently) that an American Lung Association study determined that the cost of NOT cleaning up our polluted air is $40 to $100 billion a year in health care costs and lost productivity. The costs in terms of ruined resources, such as Lake Superior and smaller lakes in Minnesota which are being slowly poisoned by airborne toxics, are difficult to estimate.

What he said: "Over the years, I've learned this about bandwagons: They are often crowded, but frequently they are headed down the wrong road." (St. Paul Pioneer Press, March 28 - referring to the original Clean Air legislation).

What he did: After doing his best to gut the Clean Air Act of 1990, Senator Boschwitz quickly jumped on the environmental bandwagon, voted for the weakened bill and hurried home to plant a tree and blow his own horn.

Minnesota deserves much better than Sen. Boschwitz's self-serving political manuevering and election year rhetoric. We need a Senator who will represent the concerns and goals of the people of Minnesota, not rich industry PACs that often are not even located in this state.

mobilizing her own constituency as well as the general public. Of particular value, she said, were the feature articles on Wellstone the North Star chapter ran in its own newsletter.

For help "in the trenches," Yingling phoned club volunteers who had already been identified as political activists, and recruited others from membership surveys and special events.

In addition to numerous press conferences, the pro-Wellstone forces organized several masterful media events: At one point, they compiled an environmental report card—and gave Senator Boschwitz a D. They then had T-shirts printed that said, "SIERRA CLUB TRUTH SQUAD," illustrated with a picture of a report card with the grade "D" prominently marked.

At the state fair, Sierra forces distributed five hundred to seven hundred copies of fact sheets on Boschwitz's record. "Rudy's Support for the Clean Air Act Is Just a Bunch of 'Boschwitz,' " said one. "Save a Tree—Defeat Rudy Boschwitz," urged another.

Back at campaign headquarters, other club volunteers distributed literature, manned phone banks, and helped keep the office going.

Yingling later was told by Wellstone's campaign manager that "focusing on the environment was one of the main things that put him over the top." And over the top is where he went. On election night, Wellstone beat Boschwitz 50 percent to 48 percent.

While it's conceivable that, once in office, Wellstone could have "pulled a Boschwitz" and failed to live up to his conservationist campaign pledges, apparently that hasn't been the case.

"He's been an absolute champion of energy and environmental issues," crows Yingling. "This is the way it ought to work."

Efforts like Yingling's are being played out all over the country. There was a particular increase in action at the polls in 1991, when community groups in different locales around the country decided to oppose the candidacies of their city administrators and city council members who favored incineration over more environmentally benign programs to reduce pollution and solid waste. Several of their stories are told in *Facts to Act On: Stopping Waste Incinerations through Local Self-Reliance,* published by the Institute for Local Self-Reliance in Washington, DC.

For example, in Oyster Bay, New York, Angelo Delligatti, the incumbent Republican chief administrator, supported a new incinerator as a method for dealing with the garbage crisis, contending that recycling alone was not enough to cure the problem. In fact, many argued that the incinerator would have created a classic case of overkill, costing taxpayers $400 million to build and requiring the city to *import* garbage from other cities in order to operate it efficiently.

When Lewis Yevoli, a Democrat, vigorously opposed the construction of the incinerator, the New York Public Interest Research Group and the Nassau-Suffolk Neighborhood Network jumped into the fray. The groups began distributing flyers attacking the economically and environmentally costly incinerator, which they identified with Delligatti. Yevoli won, thanks in large part to the Nassau-Suffolk Neighborhood Network.

Elsewhere, articulate citizens gave voice to the same sentiments that motivated voters in Oyster Bay. An anti-incineration activist in Lowell, Massachusetts, announced to her city council representative, "If you vote for this plant, not only will we never vote for you or your party again, but neither will our children."

In explaining their collective activism, Susan Shattock of Mothers Opposed to Mass-Burn Incineration in Hempstead, New York, noted, "Whether people like it or not, if you want things to run well you have to be an involved citizen and you have to be political."

Organizations attempting to engage in political activity must satisfy different requirements set forth by the Internal Revenue Code, the Federal Election Campaign Act, and the laws of the state in which they are operating. If you're considering engaging in political activity, the best advice is to consult with a lawyer who specializes in the provisions of election statutes in your state.

Independent Expenditure Campaigns

In addition to contributing directly to a candidate or volunteering to work on a campaign, your organization can help elect a candidate by working independently of the campaign. In such an "independent expenditure campaign" the group can have no interaction with that candidate's official campaign.

Mounting an independent expenditure campaign has several benefits. It allows your organization to inject your issue into the campaign in the way you wish to have it positioned. It enables you to throw more resources behind the candidate of your choice without needing to worry about violating federal or state election laws. And it helps you build a strong relationship among your organization, your issue, and the candidate that will be particularly valuable if the candidate wins.

As with other political activities, consult a lawyer or expert in federal and state election law before mounting an independent expenditure campaign.

LOCAL ḢELL-RAiSER
A "Dark Horse" No Longer

This was their chance. Advocates of reproductive choice in Colorado eagerly eyed the U.S. Senate seat being vacated by Timothy Wirth in 1992 and figured they might just have a shot at sending Congressman Ben Nighthorse Campbell, a staunchly pro-choice candidate, back to Washington as a U.S. senator.

The Colorado chapter of the National Abortion Rights Action League (CO/NARAL) took the lead. After meeting with an outside consultant to develop a statewide campaign strategy that would include efforts to support candidates for state office who were also pro-choice, organizers decided to mount an independent expenditure campaign on behalf of Campbell.

The hard work began. In a massive effort to mobilize volunteers, CO/NARAL commenced amassing a force of committed men and women that eventually amounted to eight hundred supporters available to attend demonstrations, telephone voters, and hand out leaflets and fliers. "Our people were everywhere during this campaign," said CO/NARAL's Pat Blumenthal. "It was nuts-and-bolts organizing."

Colorado NARAL-PAC 1992
Voting Guide and Endorsements

Election Day: Tuesday, November 3, 1992

Colorado State Senate Races

Senate District 4			**Senate District 23**			
• **LINDA POWERS**	(D)	PRO	Lloyd Casey	(D)	PRO	
Harold McCormick	(R)	ANTI	Ted Strickland	(R)	R	

Senate District 8
Dave Wattenberg (R) PRO-

Senate District 25
Bob Martinez (D) NR
David Mitchell (R) NR

Senate District 10
Ray Powers (R) NR

Senate District 26
Lloyd Covens (D) PRO
Tom Blickensderfer (R) R

Senate District 12
Mary Anne Tebedo (R) ANTI

Senate District 27
Bill Owens (R) ANTI

Senate District 14
• **BILL STEFFES** (D) PRO
Bob Schaffer (R) ANTI

Senate District 28
• **BELLE MIRAN** (D) PRO
Elsie Lacy (R) PRO-

Senate District 17
• **PAUL WEISSMAN** (D) PRO
David Leeds (R) R

Senate District 29
• **STEVE RUDDICK** (D) PRO
David Rowberry (R) ANTI

Senate District 18
• **JANA MENDEZ** (D) PRO

Senate District 31
• **DON MARES** (D) PRO

Senate District 19
Evie Hudak (D) PRO
Al Meiklejohn (R) PRO

Senate District 33
• **REGIS GROFF** (D) PRO

Senate District 21
• **MICHAEL FEELEY** (D) PRO
Lynn Watwood (R) ANTI

Senate District 35
• **DOTTIE WHAM** (R) PRO
Mike Johnson (D) PRO

KEY

• **BOLD TYPE**	=	Candidates endorsed by Colorado NARAL-PAC
PRO	=	Pro-choice; wants to protect reproductive rights
PRO-	=	Generally supports pro-choice positions, but supports a restriction
R	=	Supports many restrictions and wants to impede women's choice and access to safe and legal abortion services
ANTI	=	Anti-choice; wants to overturn Roe v. Wade and take away a woman's right to choose
NR	=	No response to Colorado NARAL-PAC

Colorado State Representative Races

House District 1
Jeanne Faatz (R) PRO
Marion Thornton (D) PRO-

House District 2
• **TONY HERNANDEZ** (D) PRO
Ted Harvey (R) NR

House District 3
• **WAYNE KNOX** (D) PRO
Chuck Henning (R) PRO-

House District 4
• **ROB HERNANDEZ** (D) PRO-
Ron Vertrees (R) ANTI

House District 5
• **CELINA BENAVIDEZ** (D) PRO
Tom Knorr (R) NR

House District 6
Diana DeGette (D) PRO
Clark Houston (R) PRO

House District 7
• **GLORIA TANNER** (D) PRO
Athena Eisenman (R) ANTI

House District 8
Glenda Lyle (D) PRO
Stu McPhail (R) PRO

House District 9
• **KEN GORDON** (D) PRO
Dick Bettinger (R) R

House District 10
Doug Friednash (D) PRO
Kathy Finger (R) PRO

House District 11
• **RUTH WRIGHT** (D) PRO
Bob McDonald (R) NR

House District 12
Mary Blue (D) PRO-
Bonnie Finley (R) R

House District 13
• **PEGGY LAMM** (D) PRO

(Peggy is a write-in candidate, you will have to request a write-in ballot and put her name, spelled exactly on the ballot).
Drew Clark (R) ANTI

House District 14
• **DOROTHY RUPERT** (D) PRO

House District 15
Ron May (R) ANTI

House District 16
• **BILL MARTIN** (R) PRO
James Coakley (D) PRO

House District 17
• **DAPHNE GREENWOOD** (D) PRO
Victor Mote (R) ANTI

House District 18
Tom Ratterree (R) R
Jim Pierson (D) NR

House District 19
• **MARY ELLEN EPPS** (R) PRO
Don Davidson (D) R

House District 20
Charles Duke (R) ANTI

House District 21
• **CHUCK BERRY** (R) PRO

House District 22
Michael Duncan (D) PRO
Marcy Morrison (R) PRO

House District 23
• **LANCE WRIGHT** (D) PRO
Penn Pfiffner (R) R/ANTI

House District 24
• **ROD HATES** (R) PRO-
Moe Keller (D) R

House District 25
Don Parker (D) PRO
Tony Grampsas (R) PRO-

House District 26
Charles Randall (D) R
Shirleen Tucker (R) R

House District 27
• **JIM PIERSON** (D) PRO
Pat Miller (R) ANTI

House District 28
• **VICKI AGLER** (R) PRO

House District 29
• **MICHELLE LAWRENCE** (R) PRO
Samantha Dixon (D) PRO

House District 30
• **ALICE WHITE** (D) PRO
Norma Anderson (R) PRO

House District 31
• **FAYE FLEMING** (R) PRO
Dutch Schindler (D) PRO

House District 32
• **JEANNE REESER** (D) PRO

House District 33
• **CAROL SNYDER** (D) PRO
Carol Pool (R) R

House District 34
Alice Nichols (D) R
Tim McClung (R) NR

House District 35
• **VI JUNE** (D) PRO-
Steve Wither (R) ANTI

House District 36
• **DON ARMSTRONG** (D) PRO
Don Hamstra (R) R

House District 37
Martha Kreutz (R) PRO
Scott Levin (D) PRO

House District 38
Robert Haines (D) PRO
Phil Pankey (R) R

House District 39
Mary Gruber (D) PRO
Paul Schauer (R) NR

House District 40
Ron Anderson (D) PRO
Mike Coffman (R) R

House District 41
• **PEGGY KERNS** (D) PRO
John Fritschler (R) PRO

House District 42
• **BOB HAGERDORN** (D) PRO
Eugene Hogan (R) ANTI

House District 43
Debby Allen (D) R
Roger Henderson (D) R

House District 44
Larry Schwarz (R) R
Bob Shoemaker (D) R

House District 45
• **BILL THIEBAUT** (D) PRO
Mike Occhiato (R) NR

House District 46
Gilbert Romero (D) NR

House District 47
• **JOHN SINGLETARY** (D) PRO
Mike Salaz (R) NR

House District 48
Mel Fosdoven (D) R
•David Owen (R) R

House District 49
William Jerke (R) PRO-
David Morgan (D) R

House District 50
Sue Schultz (D) PRO
Pat Sullivan (R) R

House District 51
John Irwin (R) NR

House District 52
• **BERNIE STROM** (D) PRO
Dan Nygaard (R) ANTI

House District 53
• **PEGGY REEVES** (D) PRO-
David Groff (R) PRO-

House District 54
Bill Baird (D) R
Tim Foster (R) R

House District 55
• **DAN PRINSTER** (D) PRO

House District 56
• **JAMISON SMITH** (D) PRO
Jack Taylor (R) NR

House District 57
• **DAN ARROW** (D) PRO
Russ George (R) R

House District 58
• **STEVE ACQUAFRESCA** (R) PRO
Dave Williams (D) R

House District 59
Jim Dyer (D) PRO-

House District 60
• **LEWIS ENTZ** (R) PRO
Silver Jaramillo (D) ANTI

House District 61
• **KEN CHLOUBER** (R) PRO

House District 62
• **SAM WILLIAMS** (D) PRO
Leona Hemmerich (R) ANTI

House District 63
J.B. Smith (D) R
B.D. Moellenberg (R) ANTI

House District 64
Jeanne Adkins (R) PRO-

House District 65
• **JIM BRANDON** (D) PRO-
Robert Eisenach (D) R

Paid for by Colorado NARAL-PAC, Sue Bellman, Treasurer

A key task was to identify voters who were pro-choice. With help from the national NARAL office, CO/NARAL pinpointed over forty thousand Colorado voters who claimed to support reproductive freedom. These voters received a "Pro-Choice Voting Guide" and other convincing fliers, and were targeted for a strong get-out-the-vote effort on election day.

The Washington, DC–based NARAL office helped with media as well, holding a press conference in Colorado to announce its endorsement of Campbell that received statewide media attention, and mounting a significant advertising campaign that, in the last ten days, "turned the campaign around," claimed Blumenthal.

But "nothing was automatic" with the press, recalls Blumenthal. "We had to work hard to get reporters to show up at our press conference." CO/NARAL sent out press releases, made follow-up calls to reporters, and offered to set up one-on-one interviews to generate the successful coverage they did.

The activists also prepared persuasive fliers and leaflets that criticized Campbell's Republican opponent, Terry Considine, because he "doesn't respect my privacy," while reinforcing their pro-choice message through other campaign literature lambasting state candidates who were also opposed to choice. Proclaimed one card: "Even if You are Raped, State Representative Pat Miller Would Deny You the Right to Have an Abortion!"

On election night, Campbell handily defeated Considine, becoming the first Native American elected to serve in the U.S. Senate. In his victory speech, Campbell credited his election in part to voters who supported his pro-choice views on abortion.

"It was a victory . . . for all people who have been left out of the American dream," he cheered.

Vote for Choice: Elect Ben Nighthorse Campbell to the U.S. Senate.

Vote to protect your freedom to choose.

The Supreme Court is one justice away from overturning *Roe v. Wade*. At the same time, the U.S. House and Senate are just a few votes from passing the Freedom of Choice Act to guarantee by law our freedom to choose.

Now, more than ever, we must vote for Senators who support vital Pro-Choice legislation such as the Freedom of Choice Act — and who will confirm Supreme Court Justices who respect our right to privacy.

Ben Nighthorse Campbell is Pro-choice.

Ben Nighthorse Campbell is our pro-choice Senate candidate. Campbell will fight against those who try to undermine our freedom to choose. Campbell knows that choosing whether or not to have an abortion is a private and personal decision for a woman. It's no wonder that Campbell's position has earned him the endorsement of the National Abortion Rights Action League PAC (NARAL/PAC).

Terry Considine opposes our right to choose.

Senate candidate Terry Considine believes government has an obligation to interfere in such a personal decision as abortion. We're tired of politicians telling us what to do.

According to the *Denver Post*, "Considine said that state governments have a responsibility to interfere in personal lives when the issue is abortion" (4/1/92).

Use your voice. Vote for Choice.
Elect Ben Nighthorse Campbell to the U.S. Senate.

This election year may be our last chance to save our freedom of choice. Use your voice and vote for Choice. Colorado must have a pro-choice Senator like Ben Nighthorse Campbell. We must guarantee our freedom to choose by electing those who will stand by us and protect our rights. It's our vote, our voice, and our choice. Vote for Ben Nighthorse Campbell for the U.S. Senate on November 3rd.

For information on other pro-choice candidates call: 303/831-0369

Colorado NARAL activists sent this Campbell promotional piece and the pro-choice voting guide shown on page 381 listing their endorsements for the November 1992 election to thousands of voters from their mailing lists.

A Word About Volunteers

Volunteers are the backbone of every campaign. In fact, dedicated, enthusiastic volunteers are probably among a campaign's most important assets (and this is as true for an issue campaign as it is for one focused on politics).

Few campaign treasuries can afford to pay workers to do the hundreds of jobs that volunteers do for free. But even if they could, it probably wouldn't be necessary, as long as volunteers are treated with respect and made to feel like an integral part of the campaign. Even more than salaried workers, volunteers bring to a campaign a dedication and a willingness to work that are indispensable.

How do you get people to volunteer? Many citizens are just waiting to be asked. The key is to convince potential volunteers—people who already support your candidate or your cause—that they can make an important contribution to the campaign with their time and services.

Volunteers may be motivated by many different reasons. Some are curious about the political process (more than one person who volunteered to work on Bill Clinton's presidential campaign in

1992 did so to "be a part of history"). Others have definite personal political aspirations, so they want to make contacts now for a later effort of their own. Still others become volunteers because they care passionately about an issue and want to make it a focal point of the campaign. Staunch advocates of the candidate will want to do everything possible to get him or her elected. (I volunteered to work on a congressional race in 1974 because I had just moved to Denver, Colorado, and thought it would be a good way to make friends; I also supported the candidate. Later, I worked on an initiative effort in Colorado to ban the underground testing of nuclear weapons because I strongly believed weapons testing had to stop.)

Volunteers need good direction and supervision. It's better to give volunteers a variety of options than to ask people to do the same task day in and day out. And give them a job they can finish. "Call these fifty names" works better than "Call as many people on this list as you possibly can."

Here are some other tips for boosting volunteerism in your campaign:

- **Make the campaign headquarters as pleasant as possible to work in.** Inevitably the office will get messy and crowded, but try to keep it from getting dirty and so jammed that no one has a place to sit down.

- **Ask volunteers to commit to a certain day and time.** Let them know you are counting on them to work when they say they will.

- **Invite volunteers to regular briefings on the status of the campaign.** Whenever possible, have the candidate or the candidate's spouse personally thank the volunteers for their hard work. Make volunteers feel important; the more a part of the group they feel, the more they'll volunteer.

- **Recruit volunteers from many places:** interest groups, high schools and colleges, youth groups and civic associations, seniors organizations, political parties, labor unions, and so on. At a large event, set up a recruitment table to get names and addresses of people who want to help. You can also advertise very cheaply in weekly newspapers and "shoppers" for volunteers.

  Always recruit more volunteers than you need, since some are more dedicated and reliable than others—and all of them can be helpful. Undertaking an activity for which you need thirty volunteers requires you to recruit seventy-five.

- **Once the election is over, make sure you thank all your volunteers.** Then work to integrate the new volunteers into your organization. Many of the skills they acquired getting a candidate elected are the same skills they can use in raising awareness around your issue and advancing your issue agenda throughout the year.

    Keep track of those who volunteer and the work they're willing to do by having each recruit complete a volunteer pledge card like the one on page 386.

OTHER POLITICAL OPPORTUNITIES FOR ACTIVISM

Even if your organization cannot officially participate in a campaign, there are many other ways you can become involved in politics.

Vote

Perhaps the single most important step you can take to make a difference in the political process is to vote. Today men and women all over the globe are fighting and dying—just as our forebearers fought and died—to obtain something we take for granted or even ignore: our right to vote. That each of us can cast a ballot on behalf of what we believe is a privilege. And if you dismiss its value and diminish its importance by saying that your one vote "won't make a difference," just remember: John F. Kennedy won the presidency in 1960 by the tiny fraction of a single vote per precinct.

☞ **CHECK IT OUT:** *Pick a Candidate* is a pocket-sized voter's guide prepared by the League of Women Voters to help you follow candidates' campaigns, listen to what they say, and sort out what you need to know to pick a candidate when you get to the polls. For a copy, send $.35 to The League of Women Voters, 1730 M Street, NW, Washington, DC 20036.

Sample Volunteer Pledge Card

Elect Sharon Briggs to the State House Campaign Volunteer Pledge Card
_____

Name _____ Ward # _____
Address _____ Precinct # _____
Phone # _____ (home) _____ (work)

My job preference is:
_____ Door-to-door canvassing   _____ Sign posting
_____ General office work   _____ Computers/word processing
_____ Literature distribution   _____ Transportation
_____ Telephoning   _____ Letter writing

_____ Special skills:
          _____ Typing   _____ Graphics
          _____ Photography
          _____ Fluent in _____ (language)

Other _____

I can work regularly:

_____ Monday   _____ Friday
_____ Tuesday   _____ Saturday
_____ Wednesday   _____ Sunday
_____ Thursday

          _____ 9 A.M. to 1 P.M.   _____ 1 P.M. to 5 P.M.

On Election Day, I can:

_____ Watch the polls   _____ Telephone voters
_____ Canvass   _____ Provide transportation

Register Others to Vote

A "motor voter" law was passed by Congress in May 1993 mandating that by January 1, 1995, every state must have in place a procedure whereby each resident who is handed forms to either obtain or renew a driver's license is also given a voter registration form. Twenty-one states have already implemented such a system, and more states either ask license applicants if they would like to register or have registration forms available for the public to pick up at state offices.

Another provision of the motor voter bill is that states must provide opportunities for citizens to register by giving them forms when they sign up for services at certain federal agencies. This provision is designed to register people who do not drive; ten states already use this method. Twenty-nine states offer residents ways to register by mail. States must also implement a method for ridding the books of voters who are no longer eligible to vote without removing names simply because they haven't voted.

A few states are exempt from motor voter legislation because their registration system has already been tailored for accessibility: Minnesota, Wisconsin, and Wyoming don't have to follow the provisions because they allow their voters to register right at the polls. North Dakota doesn't require residents to register at all.

Though these laws may eventually eliminate the need for voter registration drives, until 1995 at least, voter registration may be a necessity where you organize. These tips from the League of Women Voters will help you mount an effective voter registration drive.

••STEP BY STEP••

How to Mount a
Successful Voter Registration Drive

• **Set your goals.** Set an achievable numerical goal from the beginning, to inspire volunteers and serve as a guide for evaluation. Select a time period, target a group, pick sites, and so on. Focus on groups of citizens most likely to need help with voter registration, such as those who are newly eligible to vote, senior citizens, those with low incomes, or minorities.

• **Know the law.** Become familiar with the federal, state, and local laws that govern voter registration in your community. Voter registration practices—methods, deadlines, permissible sites, the role of volunteers—differ from state to state and often vary among local jurisdictions in the same state.

Federal laws such as the Voting Rights Act and the Voting Access for the Elderly and Handicapped Act protect the rights of minorities and other citizens to vote. Seek legal assistance if you believe that state or local registration practices violate citizens' voting rights. Be sure to keep your registration drive nonpartisan.

• **Work with election officials.** Make sure local election officials know about your plans. In some places, you will need their help to get voter registration forms, to register voters at special sites, or to train and deputize volunteer registrars. In other places, only official registrars may be allowed to register voters.

If you will need large numbers of forms or deputy registrars, inform election officials as early as possible and in writing so they can make any necessary arrangements. If election officials are uncooperative or do not provide assistance allowed by state election laws, and you have made a good faith effort to work with them, seek legal assistance.

• **Form a coalition.** Coordinate your effort with other organizations planning voter registration drives, and share information about scheduled activities to avoid duplication. Take advantage of each organization's special strengths.

How to Mount a
Successful Voter Registration Drive (cont.)

• **Select your sites.** Site selection and careful timing are critical elements of an effective registration drive. Use your resources efficiently by providing voter registration opportunities where there are large numbers of unregistered citizens who have the time to sign up. Consider festivals, fairs, sporting events, schools, colleges, public transit stops, or downtowns and shopping centers. Select your sites to make the best use of volunteers. Schedule volunteers for peak attendance times.

Visit sites ahead of time, and select the most visible and accessible places to station volunteers. Consider the needs of disabled people. Get whatever permission is necessary to register at a privately owned site.

• **Recruit, train, and assign volunteers.** Impress volunteers with their responsibility to protect the right to vote of the citizens they are registering by filling out the forms correctly and returning them promptly to election officials. Include training in techniques for approaching potential registrants and in messages that will motivate citizens to register.

Check state requirements for deputizing and training volunteer registrars. In states that do not permit volunteer registration, have volunteers inform citizens on how, where, and when they can register and help them get to the registration site.

Voter registration involves large numbers of volunteers interacting directly with potential voters. Make every voter registration event an opportunity to recruit more volunteers.

• **"Sell" voter registration.** Use the media—editorials, articles, talk shows, PSAs—to promote voter registration. Provide information about registration deadlines and requirements. Advertise your voter registration drive, including the sites and times where voter registration will take place. Stage an event—a celebrity appearance, band concert, or remote broadcast—to attract a crowd. Don't be afraid to get out from behind a table and work the crowd.

• **Keep good records.** Volunteer registrars should keep track of the names, addresses, and telephone numbers of everyone they register by using tally sheets. This information will

How to Mount a
Successful Voter Registration Drive (cont.)

help if forms are lost or there are other problems. Use this information to get your new registrants out to vote.

• **Show the world that your voter registration drive made a difference.** Public officials and candidates will take note. Volunteers will feel positive about their hard work.

• **Reward volunteers.** A well-timed expression of appreciation will help motivate your volunteers to work on the get-out-the-vote effort and to volunteer for the next registration drive.

• **Follow up and evaluate.** Contact the new registrants—by mail, phone, or at their homes—and remind them to vote. Find out if they need information or assistance. Evaluate your efforts and decide what worked best. Keep this information, with the names and phone numbers of your volunteers, for your organization's next voter registration drive.

(Adapted with permission from *Ten Steps to a Successful Voter Registration Drive,* published by The League of Women Voters Education Fund.)

☞ **CHECK IT OUT:** Copies of *Ten Steps to a Successful Registration Drive* may be purchased for $.75. Write The League of Women Voters Education Fund, Publication #948, 1730 M Street, NW, Washington, DC 20036.

Volunteer on Behalf of a Candidate

There is almost no end to the ways you as an individual can help elect a candidate to office. Here are a few:

- **Staff the office.** Volunteer to stuff envelopes, answer the phone, take messages, order pizza, serve as a messenger, and do anything else anyone needs to keep the campaign moving.

- **Pass out literature.** Take your candidate's message directly to voters by handing out campaign literature at fairs, farmer's markets, shopping malls, rallies, and anywhere else where people congregate.

- **Monitor the media.** Volunteers can keep files of favorable and unfavorable news coverage the candidate receives, monitor radio and television news for stories about the candidate and his or her opponent, track local issues about which the candidate should be aware, or identify events the candidate might want to attend. If the candidate is going to be on a call-in talk show, volunteers can be ready to call in with supportive questions and comments.

- **Write letters to the editor.** Volunteers can express their support for the candidate or their distaste for his or her opponent through letters to the editor of the local daily and weekly newspapers.

- **Research.** Campaigns can always use more ammunition—about the candidate's opponent and facts and figures concerning especially important issues facing the community.

- **Recruit volunteers.** Volunteers are the lifeblood of every organization—and there are never enough of them. Ask your friends and neighbors to help out, call lists of people who belong to groups sympathetic to the candidate, and sign up people who attend events where the candidate is speaking.

- **Prepare position papers on issues you're knowledgeable about.** These papers can persuade campaigns to emphasize your issue by demonstrating the electoral value of it. Your paper will show the campaign how your issue can be linked to the most important issues of the day. And position papers will help present your issue in a nontechnical, relevant way. You'll supply the campaigns with crisp, accurate information, facts, and talking points—and show the campaign organizers that your issue can attract support from ticket splitters and members of the opponent's party. Position papers can be distributed to the media, passed out at debates, and included in the packet of information candidates send to donors and the general public.

- **Organize media events.** How about a pancake breakfast so your candidate can talk about how his opponent "flip-flops" on the issues? Or a series of town meetings so the candidate can meet the people? A month in which the candidate works at a different community-oriented job every day? Even simply appearing at an opponent's rally with signs promoting your candidate may create a stir.

- **Raise money.** Hold "small donor" fund-raisers like sock hops, spaghetti dinners and fish fries, and larger money events like cocktail parties and receptions, auctions, and marathons.

- **Host a candidate "coffee."** Invite neighbors and friends to your home to meet the candidate and get involved in the campaign.

- **Become a "surrogate" speaker.** Become so well versed on the candidate's positions you can stand in for the candidate at debates and other public gatherings.

- **Staff a phone bank.** Telephone potential voters or donors to build and maintain support for your candidate.

- **Canvass.** Go door-to-door to distribute literature, raise money, and encourage people to vote.

- **Help "get out the vote" (GOTV).** Remind registered voters to go to the polls, set up car pools to help anyone who needs a ride to his or her polling place, and knock on doors to get people out to vote on election day.

If you have any question about the effectiveness of efforts like these, particularly GOTV drives, consider the impact the following group of activists had on a congressional election in Philadelphia.

Local Hell-Raisers
Knock, Knock

As the polls closed across the country on election night, 1980, the worst fears of environmental activists became reality.

Gaylord Nelson, George McGovern, Birch Bayh, Frank Church: one by one, a whole generation of environmental leaders went down in flames.

Amid the gathering darkness, however, there was one bright spot. Representative Bob Edgar (D-Philadelphia), an ardent champion of clean air and water, had won a stunning victory.

It was an unexpected triumph. Congressman Edgar's slim victory margin in 1978 (1,300 votes) and his outspoken leadership on controversial environmental issues had put him at the top of the New Right's hit list in 1980. And Edgar's opponent was both well financed and well organized. Conventional wisdom in Washington held that Bob Edgar was headed for defeat.

But while the political pundits and armchair analysts wrote off the Philadelphian's chances for reelection, environmentalists did not.

For six months, five days a week, a dozen paid environmental canvassers systematically visited almost every home in Edgar's district. The canvassers identified thousands of potential green voters and told them of Bob Edgar's unparalleled record and leadership on environmental issues. The canvass also recruited hundreds of volunteers to work on the critical "get out the green vote" drive.

On election day, the volunteer organization the canvass had built shifted into high gear, turning out thousands of green voters for Bob Edgar.

As the dust settled on election day, the full impact of the green canvass became apparent. Reagan carried Edgar's district by a landslide—36,000 votes. But the landslide had not buried Bob Edgar. The congressman won reelection by more than 11,000 votes. The environmental door-knockers had persuaded tens of thousands of voters to split their tickets and send Bob Edgar back to Congress to fight for a better environment.

(Adapted from *The Power of the Green Vote: A Campaign Training Manual for Environmental Activists,* by Americans for the Environment, 1990.)

RUN FOR OFFICE YOURSELF

Someday you may realize that you could do a much better job representing your community before the city council, the state legislature, or Congress than any other candidate who is pursuing the post.

When that moment comes, and you decide to make the leap into politics, step back, take a deep breath, and honestly answer the following questions:

- Given your message, your competition, and the mood of the voters, can you win?
- Have you amassed the political support you need to mount a winning campaign?
- Can you raise enough money?
- Can you put together a winning campaign team?
- Do you have—or can you build—a positive relationship with the media?

On a personal level:

- Can you and your family withstand the brutal public scrutiny to which your personal life will be submitted if you become a candidate?
- Are you prepared to make professional and perhaps monetary sacrifices in order to seek public office?
- Do you really want to run?

If, after honest assessment of your chances and serious personal introspection, you decide to pursue a political career, find yourself a seasoned campaign manager, develop a campaign strategy, and start raising the money you'll need to sustain your election effort.

CHECK IT OUT: *Thinking of Running for Congress? A Guide for Democratic Women,* is published by Emily's List, 1112 Sixteenth Street, NW, Suite 750, Washington, DC 20036, $5.00.

Here's the story of one citizen activist whose battles to protect her neighborhood against a new highway project propelled her all the way to the U.S. Senate:

Local Hell-Raiser
Barbara Mikulski

Barbara A. Mikulski's road to the United States Senate began on a highway she stopped from being built.

The feisty, four-foot-eleven-inch-tall Maryland Democrat, now known as the "dean of women" in the Senate, is a former Baltimore social worker who launched her political career as a community activist of a vocal community group, the Southeast Council Against the Road (SCAR), which evolved into the even bigger and more visible South East Community Organization (SECO), now considered a national model of community activism.

Although some houses in Canton, Maryland, already had been demolished to make way for the road, an extension of Interstate 83, SECO fought against its construction so fiercely and for so long—in the process helping to get Mikulski elected to the Baltimore City Council as an antiexpressway candidate—that plans for the highway eventually were dropped. Both Fells Point and Canton, now centerpieces of Baltimore's revived "Gold Coast" harbor redevelopment area, were saved.

With the battle against the road behind her, Mikulski ran for the U.S. Senate in 1974, challenging the extremely popular liberal Republican senator Charles McC. Mathias, considered unbeatable. She lost but nevertheless got 43 percent of the vote, an unexpectedly good showing. Then in 1976, the diminutive city council member from East Baltimore successfully ran for the House of Representatives. During the ensuing decade, Mikulski established a strong congressional record as a tough prolabor liberal. In 1986, she ran for the Senate once more against stiff opposition—and won handily. Overwhelmingly reelected in 1992 with 71 percent of the vote—higher than any other Senate Democrat—she now is the first woman in her party to become a member of the previously all-male Senate Democratic leadership.

KEEP IN TOUCH WITH THE CANDIDATE AFTER THE ELECTION

As soon as possible, arrange to meet with your newly elected representative. Make sure he or she realizes how much work you did on behalf of the campaign and how important the vote of your constituents was to its success.

If you can, work on or with the official's transition staff to begin implementing the policies the official advocated during the campaign.

Arrange briefings for the official on your issue and with your issue experts. Introduce the official to other elected officials your organization has supported. If the candidate has a campaign debt, help to retire it by planning a fund-raiser or making a direct contribution.

If your candidate lost, make sure he or she knows that you're still supportive. Meet with the candidate soon after the election to determine the candidate's plans and to discuss how you can continue to work together.

Even if your candidate lost, you should try to establish a working relationship with the winner. No matter who's in office, you're going to have to work with them—at least until the next election rolls around.

ACTION AT THE POLLS
DO's and DON'Ts

- **Do** volunteer as an individual to help elect candidates you support or defeat those you oppose.

- **Do** check with an attorney who specializes in federal and state election law before committing your organization to efforts to elect or defeat a candidate or engage in a referendum or initiative campaign.

- **Do** mount a referendum, initiative, or recall campaign only if you think you have a good chance of winning, or if you can advance your position even if you lose.

- **Do** vote, and register others to vote. On election day, help get voters to the polls.

- **Do** run for office yourself.

- **Don't** officially endorse candidates for public office on behalf of your organization unless you have the proper tax status. (Check with your attorney.)

- **Don't** allow your organization to contribute to candidates unless it is permitted to do so by the Internal Revenue Service.

- **Don't** publish or distribute statements that either favor or oppose a candidate for public office unless you are certain you will not endanger your tax status.

- **Don't** ignore the system. Participate!

APPENDIX:
AMMUNITION AND ALLIES

FOR MORE INFORMATION ON ORGANIZING

Ammunition: Books and Other Resources

Civics for Democracy: A Journey for Teachers and Students by Katherine Isaac, Center for the Study of Responsive Law, Essential Books, Washington, DC, $17.50, 390 pages.

The Corporate Examiner, Interfaith Center on Corporate Responsibility (ICCR), 475 Riverside Drive, Room 566, New York, New York 10115, $35 per year for 10 newsletter issues. (ICCR also offers *Church Proxy Resolutions,* copies of resolutions submitted by churches and pension funds.)

Empowering Ourselves: Women and Toxics Organizing by Robbin Lee Zeff, Marsha Love, and Karen Stults, Citizens Clearinghouse for Hazardous Wastes, Inc., P.O. Box 926, Arlington, Virginia, 1989, $6.95, 45 pages.

Fighting Hunger in Your Community: A Guide for the Development of Community Action Projects, League of Women Voters Education Fund, 1730 M Street, NW, Washington, DC, Publication #893, $5, 15 pages.

Fighting Toxics: A Manual for Protecting Your Family, Community, and Workplace by Gary Cohen and John O'Connor, Island Press, Washington, DC, 1990, $19.95, 346 pages.

Grass Roots: Ordinary People Changing America by Tom Adams, Carol Publishing Group, New York, New York, 1991, $18.95, 357 pages.

The Kid's Guide To Social Action by Barbara A. Lewis, Free Spirit Publishing, Minneapolis, Minnesota, $14.95, 185 pages.

Membership Recruiting Manual by Bruce P. Ballenger, The Northern Rockies Action Group, 9 Placer Street, Helena, Montana 59601, $10, 96 pages.

A Nonprofit Organization Operating Manual: Planning for Survival and Growth, Foundation Center, 312 Sutter Street, San Francisco, California 94108, 1991, $29.95, 484 pages.

Nonviolent Resistance by Mohandas K. Gandhi, Random House, New York, New York, 1983. Out of print but worthwhile—check your library.

Not in Our Backyards: Community Action for Health and the Environment by Nicholas Freudenberg, Monthly Review Press, New York, New York, 1984, $10, 304 pages.

Organizing: A Guide for Grassroots Leaders by Si Kahn, McGraw-Hill Book Company, New York, New York, 1982, $7.95, 387 pages.

Organizing for Social Change: A Manual for Activists in the 1990s by Kim Bobo, Jackie Kendall, and Steve Max, Seven Locks Press, Cabin John, Maryland, 1991, $19.95, 270 pages.

The Politics of Nonviolent Action by Gene Sharp, Porter Sargent, Boston, Massachusetts, 1973, $11.95 for the three-volume set.

Public Policy Manual for Community-Based AIDS Service Providers by Karen B. Ringen, AIDS Action Council, 1875 Connecticut Avenue, NW, Suite 700, Washington, DC 20009, 1992, FREE.

Roots to Power: A Manual for Grassroots Organizing by Lee Staples, Praeger Publishing, Westport, Connecticut, 1984, $19.95.

The Shareholder Proposal Process: A Step-by-Step Guide to Shareholder Activism for Individuals and Institutions, United Shareholders Association, 1667 K Street, NW, Suite 770, Washington, DC 20006. Offered with $50 per year annual membership.

You Can Change America: How to Make a Difference Right Now in Your Community, in Congress and in the Country, Earth-Works Group, Berkeley, California, 1991, $5.95, 95 pages.

Allies: Organizations and Institutions

Action for Community Organization and Reform Now (ACORN), 739 Eighth Street, SE, Washington, DC 20003

Citizen Action, 1120 Nineteenth Street, NW, Suite 630, Washington, DC 20036

Citizens Clearinghouse for Hazardous Wastes, Inc., P.O. Box 926, Arlington, Virginia 22216

Citizens Committee for New York City, Inc., 3 West Twenty-ninth Street, New York, New York 10001

Common Cause, 2030 M Street, NW, Washington, DC 20036

Jobs With Peace, 750 North Eighteenth Street, Milwaukee, Wisconsin 53233

Midwest Academy, 225 West Ohio Street, Suite 250, Chicago, Illinois 60610

National Association for the Advancement of Colored People, 4805 Mount Hope Drive, Baltimore, Maryland 21215

Northern Rockies Action Group, 9 Placer Street, Helena, Montana 59601

Nonviolence International, P.O. Box 39127, Friendship Station, NW, Washington, DC 20016

Public Citizen, 2000 P Street, NW, Suite 700, Washington, DC 20036

Rocky Mountain Peace Center, P.O. Box 11156, Boulder, Colorado 80306

Rural Southern Voice for Peace, 91901 Hannah Branch Road, Burnsville, North Carolina 28714

U.S. Public Interest Research Group (U.S. PIRG), 215 Pennsylvania Avenue, SE, Washington, DC 20003

FOR MORE INFORMATION ON FUND-RAISING

Ammunition: Books and Other Resources

The Art of Asking: How to Solicit Philanthropic Gifts by Paul H. Schneiter, The Taft Group, Rockville, Maryland, 1985, $25, 176 pages.

Chronicle of Philanthropy, 1255 Twenty-third Street, NW, Washington, DC 20037, $67.50 per year. Biweekly newspaper.

Dear Friend: Mastering the Art of Direct Mail Fund Raising by Kay Partney Lautman and Henry Goldstein, The Taft Group, Rockville, Maryland, 1991, $65, 378 pages.

Doing Best By Doing Good: How to Use Public-Purpose Partnerships to Boost Corporate Profits and Benefit Your Community by Dr. Richard Steckel and Robin Simons, Penguin Books, New York, New York, 1992, $24, 277 pages.

Dollars and Sense: A Community Fundraising Manual for Women's Shelters and Other Nonprofit Organizations, Western States Shelter Network, 870 Market Street, Suite 1058, San Francisco, California 94102, 1982, $17 for members, $22 for nonmembers.

Filthy Rich and Other Nonprofit Fantasies by Richard Steckel, Ten Speed Press, Berkeley, California, 1989, $12.95, 223 pages.

Fund Raisers That Work: Scores of Successful Moneymakers for Youth Groups by Margaret Hinchey, Group Books, Loveland, Colorado, 1988, $8.95, 102 pages.

Fund Raising Management, Hoke Communications, Inc., 224 Seventh Street, Garden City, New York 11530, $54 per year. Monthly magazine.

Fundraising for Non-Profit Groups by Joyce Young, Self-Counsel Press, Bellingham, Washington, 1989, $8.95, 161 pages.

Fundraising for Social Change by Kim Klein, Chardon Press, Inverness, California, revised 1994, $25, 245 pages.

The Grass Roots Fundraising Book by Joan Flanagan, Contemporary Books, Inc., Chicago, Illinois, updated 1992, $14.95, 219 pages.

Grassroots Fundraising Journal, P.O. Box 11607, Berkeley, California 94701, $25 per year for six issues.

Managing for Profit in the Nonprofit World by Paul B. Firstenberg, Foundation Center, 312 Sutter Street, San Francisco, California 94108, 1986, $19.95, 253 pages.

The Membership Mystique by Richard P. Trenbeth, The Taft Group, Rockville, Maryland, 1986, $36.95, 280 pages.

Nonprofit Times, 190 Tamarack Circle, Skillman, New Jersey 08558, $59 per year. Monthly newspaper.

Program Planning and Proposal Writing by Norton J. Kiritz, Grantsmanship Center, 1125 West Sixth Street, Fifth Floor, Los Angeles, California 90017, 1980, $4, 48 pages.

Revolution in the Mailbox by Mal Warwick, Strathmoor Press, Berkeley, California, 1990, $49.95, 313 pages.

Securing Your Organization's Future: A Complete Guide to Fundraising Strategies by Michael Seltzer, Foundation Center, 312 Sutter Street, San Francisco, California 94108, 1987, $24.95, 514 pages.

Allies: Organizations and Institutions

AddVenture Network, Inc., 1350 Lawrence Street, Plaza 211, Denver, Colorado 80204

American Association of Fundraising Counsel (AAFRC), 25 West Forty-third Street, New York, New York 10036

The Foundation Center, 79 Fifth Avenue, New York, New York 10003

The Funding Exchange, 666 Broadway, New York, New York 10012

The Grantsmanship Center, 1125 West Sixth Street, Fifth Floor, Los Angeles, California 90017

National Committee for Responsive Philanthropy, 2001 S Street, NW, Suite 620, Washington, DC 20009

National Society of Fundraising Executives, 1511 K Street, NW, Washington, DC 20005

Philanthropic Advisory Service, Council of Better Business Bureaus, Inc., 1515 Wilson Boulevard, Arlington, Virginia 22209

The Taft Group, 12300 Twinbrook Parkway, Suite 450, Rockville, Maryland 20852

FOR MORE INFORMATION ON COMMUNICATIONS

Ammunition: Books and Other Resources

Guide to Public Relations for Nonprofit Organizations and Public Agencies, Grantsmanship Center, 1125 West Sixth Street, Fifth Floor, Los Angeles, California 90017, 1979, $3, 16 pages.

Making Health Communication Programs Work: A Planner's Guide, National Cancer Institute, 1992, FREE, 131 pages.

Media Skills Manual, Safe Energy Communications Council, 1717 Massachusetts Avenue, NW, Suite LL215, Washington, DC 20036, updated 1994, $17.50.

Radio and Public Relations: The Inside Story, News Broadcast Network, 1010 Curtiss Street, Downers Grove, Illinois 60515, FREE, 30 pages.

Strategic Communications for Nonprofits edited by Larry Kirkman and Karen Menichelli, Benton Foundation and Center for Strategic Communications (see *Allies: Organizations and Institutions* below for addresses). Books in the series: *Media Advocacy, Talk Radio, Op-Eds, Electronic Networking for Nonprofit Groups, Voice Programs, Cable Access,* and *Using Video,* all 1991, are $7 each; *Strategic Media,* 1991, is $10. The set of eight costs $50. *Making Video,* 1993, includes a videotape, costs $50 alone and makes the set price $85.

The Video Primer, by Richard Robinson, Putnam Publishing Group, New York, New York, 1983, $10.95, 382 pages.

We Interrupt This Program ... A Citizen's Guide to Using the Media for Social Change by Robbie Gordon, Citizens Involvement Training Program, 377 Hills South, University of Massachusetts, Amherst, Massachusetts 01003, 1978, $12.50, 117 pages.

Allies: Organizations and Institutions

American Forum, 529 Fourteenth Street, NW, Washington, DC 20045

The Benton Foundation, 1634 I Street, NW, Twelfth Floor, Washington, DC 20006

Center for Urban Studies, Portland State University, P.O. Box 751, Portland, Oregon 97207

Center for Strategic Communications, 505 Eighth Avenue, Suite 2000, New York, New York 10018

Community Media Services, 121 West Twenty-Seventh Street, Suite 1202A, New York, New York 10001

Computer Support Project, Health Services Initiatives, Inc., 1804 Highway 45 Bypass, Suite 516, Jackson, Tennessee 38305

Computer and Telecommunications Services, Management Assistance Project for Nonprofits, Inc., University Avenue West, Suite 360, St. Paul, Minnesota 55114

HandsNet, 20195 Stevens Creek Boulevard, Suite 120, Cupertino, California 95014

Information Technology Resource Center, 59 East Van Buren, Suite 2020, Chicago, Illinois 60605

Institute for Global Communications, 1800 De Boom Street, San Francisco, California 94107

Nonprofit Computer Connection, Technical Development Corporation, 30 Federal Street, Boston, Massachusetts 02110

Nonprofit Computer Exchange, Fund for the City of New York, 121 Sixth Avenue, 6th Floor, New York, New York 10013

Safe Energy Communications Council, 1717 Massachusetts Avenue, NW, Suite LL215, Washington, DC 20036

The Support Center, 70 Tenth Street, Suite 201, San Francisco, California 94103

Telecommunications Cooperative Network, 1333 H Street, NW, Suite 700, Washington, DC 20005

Vanguard Communications, 1835 K Street, NW, Suite 805, Washington, DC 20006

FOR MORE INFORMATION ON LOBBYING

Ammunition: Books and Other Resources

The Cardinals of Capitol Hill by Richard Munson, Grove Press, New York, New York, 1993, $23, 222 pages.

The Dance of Legislation by Eric Redman, Simon and Schuster, New York, New York, 1973, $9.95, 320 pages.

The Elements of a Successful Public Interest Advocacy Campaign, Advocacy Institute, 1730 Rhode Island Avenue, NW, Suite 600, Washington, DC 20003, $7.50, 45 pages.

The Giant Killers by Michael Pertschuk, W. W. Norton, New York, New York, 1986, $7.95, 252 pages.

Impact on Congress: A Grassroots Lobbying Handbook for Local League Activists, League of Women Voters, 1730 M Street, NW, Washington, DC 20036, 1987, $2.25 ($1.75 for members), 8 pages.

Lobby? You? Independent Sector, 1828 L Street, NW, Suite 1200, Washington, DC 20036, $1.25, 12 pages.

The Lobbying Handbook, Professional Lobbying and Consulting Center, 111 Fourteenth Street NW, Suite 1001, Washington, DC 20005, $99, 118 pages.

Lobbying Strategies: A Basic Guide for Grassroots Environmental Groups, Americans for the Environment, 1400 Sixteenth Street, NW, Washington, DC 20036, $3, 25 pages.

To PAC or Not to PAC, Center for Active Citizenship, 4144 Lindell Street, Room 504, St. Louis, Missouri 63108, $2, 18 pages.

Public Policy Manual for Community-based AIDS Service Programs edited by Karen B. Ringer, the AIDS Action Council, 1875 Connecticut Avenue, NW, Suite 700, Washington, DC 20009, FREE.

Allies: Organizations and Institutions

Advocacy Institute, 1730 Rhode Island Avenue, NW, Suite 600, Washington, DC 20003

Citizen Action, 1120 Nineteenth Street, NW, Suite 630, Washington, DC 20036

Common Cause, 2030 M Street, NW, Suite 300, Washington, DC 20036

Independent Sector, 1828 L Street, NW, Washington, DC 20036

League of Women Voters, 1730 M Street, NW, Washington, DC 20036

Public Citizen, 2000 Pennsylvania Avenue, NW, Washington, DC 20036

20/20 Vision, 1828 Jefferson Place, NW, Washington, DC 20036

U.S. Public Interest Research Group, 312 Pennsylvania Avenue, SE, Washington, DC 20003

FOR MORE INFORMATION ON
ACTION AT THE POLLS

Ammunition: Books and Other Resources

Be It Enacted by the People; A Citizen's Guide to Initiatives by Mike A. Males, Northern Rockies Action Group, 9 Placer Street, Helena, Montana 59601, $10, 57 pages.

A Common Cause Guide to Citizen Action: How, When and Where to Write to Your Elected Officials, Common Cause, 2030 M Street, NW, Washington, DC 20036, 1990, FREE, 9 pages.

Facts on PACs: Political Action Committees and American Campaign Finance, League of Women Voters, 1730 M Street, NW, Washington, DC 20036, Publication #297, 1984, $2, 27 pages.

Making It Work: A Guide to Training Election Workers, League of Women Voters, 1730 M Street, NW, Washington, DC 20036, Publication #271, 35¢, 15 pages.

Non Profit Organizations, Public Policy and the Political Process: A Guide to the Internal Revenue Code and Federal Election Campaign Act, Perkins Coie, 607 Fourteenth Street, NW, Suite 800, Washington, DC 20005, revised 1990, FREE, 56 pages.

The Power of the Green Vote: A Campaign Training Manual for Environmental Activists edited by Matthew C. MacWilliams, Americans for the Environment, 1400 Sixteenth Street, NW, Washington, DC 20036, 1990, $15, 60 pages.

Public Opinion Polling: A Handbook for Public Interest and Citizen Advocacy Groups by Celinda C. Lake with Pat Callbeck Harper, Island Press, Washington, DC, 1987, $19.95, 166 pages.

State Legislative and Local Campaign Manual, Republican National Committee, 310 First Street, SE, Washington, DC 20003, $5. (Books in the series: *Campaign Planning, Campaign Organization, Campaign Scheduling, Campaign Tactics, Campaign Communications, Campaigns and Computers, Campaign Finance*).

The Student Environmental Organizing Guide, Center for Environmental Citizenship, 1400 Sixteenth Street, NW, P.O. Box 24, Washington, DC 20036, 1993, FREE, 20 pages.

Take Back Your Government: A Practical Handbook for the Private Citizen Who Wants Democracy to Work by Robert A.

Heinlein, Baen Books, Riverdale, New York, 1992, $5.99, 287 pages.

Ten Steps to a Successful Registration Drive, League of Women Voters Education Fund, 1730 M Street, NW, Washington, DC 20036, Publication #948, 1992, 75¢.

Thinking of Running for Congress? A Guide for Democratic Women, Emily's List, 1112 Sixteenth Street, NW, Suite 750, Washington, DC 20036, 1991, $5, 24 pages.

Allies: Organizations and Institutions

Americans for the Environment, 1400 Sixteenth Street, NW, Washington, DC 20036

Center for Environmental Citizenship, 1400 Sixteenth Street, NW, P.O. Box 24, Washington, DC 20036

Citizens Vote, Inc., 10 East Thirty-ninth Street, Suite 601A, New York, New York 10016

Common Cause, 2030 M Street, NW, Suite 300, Washington, DC 20036

Democratic National Committee, 430 South Capitol Street, SE, Washington, DC 20003

The League of Women Voters, 1730 M Street, NW, Washington, DC 20036

Northern Rockies Action Group, 9 Placer Street, Helena, Montana 59601

Project Vote, 1611 K Street, NW, Suite 326, Washington, DC 20005

Republican National Committee, 310 First Street, SE, Washington, DC 20003

ABOUT THE AUTHOR

Diane MacEachern has been a hell-raiser for almost twenty years. She cut her activist teeth in the successful statewide referendum campaign to pass deposit legislation in the state of Michigan. She has also volunteered for candidates, lobbied for the passage of state and federal legislation, appeared on local and national television, testified before her city council and the U.S. Congress, and raised money for a variety of nonprofit causes.

In 1987, MacEachern and Maria Rodriguez founded Vanguard Communications, a full-scale communications consulting firm based in Washington, DC, that develops and implements public education campaigns on important social issues, including environmental protection, health care, energy, and economic development. Vanguard helped organize the international celebration of Earth Day in 1990; today, among many other projects, the firm is spearheading a four-year campaign to assist local communities in their efforts to reduce infant mortality.

The author of the best-selling *Save Our Planet: 750 Everyday Ways You Can Help Clean Up the Earth,* MacEachern is married to Richard Munson and has two children, Daniel and Dana. She lives in the "hell-raising" community of Takoma Park, Maryland, near Washington, DC.